THE NATURAL ANTI-GAL ANTIBODY AS FOE TURNED FRIEND IN MEDICINE

THE NATURAL ANTI-GAL ANTIBODY AS FOE TURNED FRIEND IN MEDICINE

Uri Galili, PhD
Department of Surgery, University of Massachusetts Medical School, Worcester, MA, United States (Retired)
Division of Cardiology, Department of Medicine, Rush Medical College, Chicago, IL, United States (Adjunct Professor)

Academic Press is an imprint of Elsevier
125 London Wall, London EC2Y 5AS, United Kingdom
525 B Street, Suite 1800, San Diego, CA 92101-4495, United States
50 Hampshire Street, 5th Floor, Cambridge, MA 02139, United States
The Boulevard, Langford Lane, Kidlington, Oxford OX5 1GB, United Kingdom

Library of Congress Cataloging-in-Publication Data
A catalog record for this book is available from the Library of Congress

British Library Cataloguing-in-Publication Data
A catalogue record for this book is available from the British Library

ISBN: 978-0-12-813362-0

For information on all Academic Press publications visit our website at
https://www.elsevier.com/books-and-journals

Publisher: Mica Haley
Acquisition Editor: Linda Versteeg-buschman
Editorial Project Manager: Tracy Tufaga
Senior Project Manager: Priya Kumaraguruparan
Cover designer: Matthew Limbert

Typeset by TNQ Books and Journals

To my wife Naomi and children Sharon, Shira, and Doron

Contents

14. Post Infarction Regeneration of Ischemic Myocardium by Intramyocardial Injection of α-Gal Nanoparticles

15. Regeneration of Injured Spinal Cord and Peripheral Nerves by α-Gal Nanoparticles

16. Inhalation of α-Gal/Sialic Acid Liposomes for Decreasing Influenza Virus Infection

Author Biography

Uri Galili is an immunologist who received his PhD in 1977 at the Hebrew University School of Medicine, Jerusalem, Israel. Following postdoctoral research at the Karolinska Institute, Stockholm (1977–79), he worked at Hadassah University Hospital, Jerusalem (1979–84), where he discovered anti-Gal as the most abundant natural antibody in humans. In collaboration with Bruce Macher at University of California Medical Center, San Francisco (1984–90), he identified the α-gal epitope as the mammalian antigen that binds anti-Gal, determined the unique evolution of anti-Gal and α-gal epitopes in primates, and studied the molecular basis for this evolution. In MCP-Hahnemann School of Medicine, Philadelphia (1991–99), he studied the significance of anti-Gal/α-gal epitope interaction as an immune barrier in xenotransplantation and initiated studies on harnessing anti-Gal in cancer immunotherapy and in amplifying immune response to viruses. At Rush Medical College, Chicago (1999–2004), he studied immune tolerance induction to α-gal epitopes. In the Department of Surgery at UMass Medical School, Worcester (2004–13), he developed a method for *in situ* conversion of tumors into autologous vaccines targeted to antigen presenting cells by intratumoral injection of α-gal glycolipids, performed clinical trials with this immunotherapy, and demonstrated increased immunogenicity of influenza and HIV vaccines presenting α-gal epitopes. He further developed anti-Gal-binding α-gal nanoparticles that accelerate wound and burn healing and induce tissue regeneration in internal injuries. Prof. Galili retired in 2013 and lives in Chicago. He continues his research as a volunteer Adjunct Professor at Rush Medical College on α-gal nanoparticles-induced regeneration of ischemic myocardium, postmyocardial infarction.

Email: uri.galili@rcn.com.

Preface

The natural anti-Gal antibody stands out among the multitude of natural antibodies continuously produced by the human immune system, as an immunologic "living fossil." Anti-Gal may be regarded as a relic of an evolutionary event that brought ancestral Old World primates to the brink of extinction. This book tells the evolving multidisciplinary story of anti-Gal, the carbohydrate antigen it recognizes, called the α-gal epitope, and their significance in various pathological processes. However, the main objective of this book is to introduce a new paradigm in medicine: harnessing the natural anti-Gal antibody, present in large amounts in humans, as a therapeutic agent in various clinical areas.

Anti-Gal has been known primarily as a major "foe" that prevents transplantation of porcine tissue and organ xenografts in humans. This antibody binds rapidly to α-gal epitopes on porcine cells and induces complement-mediated cytolysis, thereby causing "hyperacute" rejection of xenografts. This immune barrier to xenotransplantation has been eliminated by the elegant generation of knockout pigs for the α1,3-galactosyltransferase gene (*GGTA1*), which encodes the α1,3-galactosyltransferase enzyme synthesizing α-gal epitopes. Anti-Gal further mediates allergies to red meat and may contribute to several autoimmune phenomena. Nevertheless, because anti-Gal is present in large amounts in humans, and natural or synthetic α-gal epitopes are readily available, this antibody provides a unique opportunity for developing a variety of novel immunotherapies that could be beneficial in several areas, including cancer immunotherapy, increased vaccine immunogenicity, accelerated wound and burn healing, tissue engineering, regeneration of injured myocardium and nerves, and protection against viruses, bacteria, and protozoa. Some of the suggested immunotherapies are presented with experimental data demonstrating their efficacy. Other immunotherapies are postulated and will require experimental studies for determining feasibility, efficacy, and safety. Thus, in addition to presenting past and recent information on anti-Gal/α-gal epitopes interactions, this book discusses future directions for immunotherapy research associated with the anti-Gal antibody.

The chapters of the book are grouped in four sections: (1) Background information on anti-Gal, the α-gal epitope, their unique reciprocal evolution, anti-Gal as anti-blood group B antibody, interaction of anti-Gal with various protozoa, and induction of immune tolerance to the α-gal epitope. (2) Anti-Gal as "foe" preventing porcine xenotransplantation, causing allergic reactions to meat, and contributing to autoimmunity, (3) Experimental information on anti-Gal as "friend" amplifying the immune response to vaccines, converting *in situ* tumors into autologous vaccines in cancer immunotherapy, destroying tumor cells and pathogens via bifunctional molecules carrying α-gal epitopes, and accelerating healing of wounds and burns, and (4) Suggested hypothetical future anti-Gal-mediated therapies in tissue engineering, myocardial infarction, spinal cord and peripheral nerve injuries, and in influenza virus infection, based on the information presented in sections (1) and (3).

It is my hope that this book will stimulate researchers to further study the various therapeutic uses of anti-Gal and develop new clinical applications for this antibody, which are presently unknown.

Uri Galili
Chicago, 2017

Acknowledgment

I would like to thank research and clinical colleagues, students, and research associates who joined me in the multidisciplinary studies on anti-Gal and the α-gal epitope. Their help and input enabled me to make this exciting voyage.

SECTION 1

BACKGROUND INFORMATION ON ANTI-GAL AND THE α-GAL EPITOPE

CHAPTER

1

Anti-Gal in Humans and Its Antigen the α-Gal Epitope

OUTLINE

INTRODUCTION

The human immune system is capable of producing antibodies against an almost limitless number of antigens. These antibodies are usually produced following antigenic stimulation by an invading pathogen such as an infectious virus. After the elimination of the stimulating antigen, the concentration of these antibodies diminishes, and an immune memory is kept by memory B and T lymphocytes. Thus, at any given time, most antibodies are present at very low titers and only those against antigens stimulating the immune system at that time are found in the serum at relatively high titers. Anti-Gal differs from those elicited antibodies in that it is constantly present in large mounts, constituting ~1% of circulating immunoglobulins (Galili et al., 1984; McMorrow et al., 1997; Parker et al., 1999; Yu et al., 1999). Because anti-Gal is produced in humans without the need for any vaccination, it is referred to as a "natural

The Natural Anti-Gal Antibody as Foe Turned Friend in Medicine
http://dx.doi.org/10.1016/B978-0-12-813362-0.00001-4

antibody." It is produced as IgM, IgA, and IgG antibodies, (Hamadeh et al., 1995) and the distribution of the IgG subclasses is IgG2 > IgG1 > IgG3 > IgG4 (Galili et al., 2001). As discussed in Chapter 7, in a small proportion of the population, anti-Gal is produced also as an IgE antibody that mediates allergic reactions following exposure to meat, which contains high concentration of the carbohydrate antigen recognized by anti-Gal, the α-gal epitope. As further discussed in other chapters of this book, anti-Gal may also contribute to a variety of autoimmune and autoimmune-like phenomena in humans. Until recently, anti-Gal has formed a barrier for attempts to transplant pig organs and tissues in humans. However, because of its abundance, its ability to activate complement and thus recruit and activate macrophages and dendritic cells, and the ability of the Fc portion of immunocomplexed anti-Gal IgG to bind to Fcγ receptors on macrophages and dendritic cells, anti-Gal may be harnessed for a variety of therapeutic applications in humans.

DISCOVERY OF ANTI-GAL

Without realizing it, anti-Gal has been part of routine clinical practice for ~100 years. As discussed in Chapter 3, anti-Gal comprises >85% of anti-blood group B antibody activity. These are antibody clones that can bind to both α-gal epitopes and to blood group B antigen. In fact, the discoverer of the ABO blood groups system, Karl Landsteiner, suggested in 1925 that anti-blood group B antibody in humans can bind not only to blood group B antigen but also to a distinctly other molecule with a structure resembling the B antigen and which he found on rabbit red blood cells (RBC) and New World monkey RBC (Landsteiner and Philip-Miller, 1925). He called this molecule "B-like" antigen and in Chapter 3 it is explained why the B-like antigen is the α-gal epitope (Galili et al., 1987a,b).

The discovery of anti-Gal was the result of research on natural antibodies that bind to normal and pathologically senescent RBC and label such RBC for removal from the circulation by the reticuloendothelial system. Anti-Gal was first observed on RBC of patients with the hemoglobinopathy β-thalassemia (Galili et al., 1983). By using an assay that can detect small numbers of IgG molecules on each RBC, it was possible to demonstrate presence of several hundreds of IgG molecules bound to thalassemic RBC. These RBC display short life span in the circulation. The IgG molecules bound to thalassemic RBC could be eluted at 37°C by galactose but not by other free carbohydrates (Galili et al., 1983). That study further indicated that similar antibodies (called at that time "anti-galactosyl" antibodies), capable of binding to thalassemia RBC, could be isolated from normal sera by adsorption on rabbit RBC followed by elution with galactose. Senescent normal RBC, isolated based on their increased density (because of loss of water) also displayed the presence of IgG molecules that could be specifically eluted by galactose and even more effectively by galactose in an α-anomeric position, such as in α-methyl galactoside, whereas β-methyl galactoside displayed no such eluting capacity (Galili et al., 1984, 1986a). Normal human sera were passed on Sepharose columns presenting α-galactosyl residues on melibiose (Galα1-6Glc), to determine the source of the IgG antibodies on normal senescent and thalassemic RBC. After washes of the columns the bound antibodies were eluted with free galactose and further isolated on a protein-A Sepharose column. Approximately 1% of serum IgG (30–100 μg/mL) was found to bind to the melibiose–Sepharose column (Galili et al., 1984). Quantification studies on serum anti-Gal, based on its binding to columns of synthetic α-gal epitopes are demonstrated in Fig. 1 of Chapter 4 (Avila et al., 1989). Similar concentration of anti-Gal in the serum was measured by other investigators, as well (McMorrow et al., 1997; Parker et al., 1999; Yu et al., 1999), whereas other studies reported on lower concentrations of anti-Gal (Barreau et al., 2000; Bovin, 2013; Rispens et al., 2013). As discussed below, these differences could be the result of different affinity columns that use partial (Galα1-3Gal-R) or full (Galα1-3Galβ1-4GlcNAc-R) carbohydrate antigen binding anti-Gal (Galili and Matta, 1996; Neethling et al., 1996), use of synthetic α-gal epitopes linked to a short linker of 3 carbons versus the preferable long linker of 14 carbons, different elution conditions, linking α-gal epitopes to columns at suboptimal concentrations, as well as the use of sera-containing low affinity anti-Gal antibodies, which fail to bind while flowing through columns or which are detached from the column as a result of extensive washes of the column. Studies evaluating anti-Gal concentration by ELISA may not be suitable for this purpose because the proportion of anti-Gal binding to the solid-phase antigen coating the ELISA well is minimal, and the extent of binding is determined much more by the affinity rather than by the concentration of the antibody. As discussed in Chapter 3, anti-Gal clones in blood type A and O individuals have a wider range of antigen binding than in blood type B and AB individuals, and may display higher affinity, as well.

Studies with sickle cell anemia RBC also demonstrated anti-Gal bound *in situ* to these RBC (Galili et al., 1986b). Because of its activity with terminal galactose (Gal) linked at α anomeric position, the antibody was called "anti-α-galactosyl antibody" and later "anti-Gal antibody." Because carbohydrate chains with terminal α-galactosyl units were found to be expressed in high concentration on rabbit RBC (Eto et al., 1968), anti-Gal was studied for binding to these RBC. Indeed, >90% of the antibodies agglutinating rabbit RBC were found to bind to α-galactosyl columns such as melibiose (Galα1-6Glc) or α-methyl galactoside linked to Sepharose. In addition, the binding of this antibody to rabbit RBC was found to be completely inhibited by galactose and more effectively by α-methyl galactoside, melibiose, and stachyose (α-galactosyl tetrasaccharide) but not by other carbohydrates (Galili et al., 1984). Analysis of anti-Gal IgG binding to rabbit RBC, using rabbit anti-human IgG as secondary antibody, indicated that anti-Gal is present in the sera of most individuals tested (>95%) at titers of 1:200 to 1:1600. The antibody is further found in the cord blood in titers comparable with those in the maternal blood (Galili et al., 1984). Anti-Gal IgG reaches its lowest activity at the age of 3–6 months, and subsequently it is produced throughout life (Galili et al., 1984). In that study, anti-Gal was further found to display very low titers (5%–10% of normal titers) in severe states of immunodeficiency such a Bruton type agammaglobulinemia, multiple myeloma, and chronic lymphocytic leukemia. Later studies indicated that anti-Gal activity decreases by ~2- to 4-fold in elderly populations (>75 years) (Wang et al., 1995a). All these observations further indicated that large majority of natural antibodies in human serum, which were reported by several researchers to bind to rabbit RBC, (Landsteiner and Philip-Miller, 1925; Tönder et al., 1978) are the anti-Gal antibody.

ANTIGENIC SPECIFICITY OF ANTI-GAL

Rabbit RBC were reported to contain in their membrane two major glycolipids with terminal α-galactosyl units. These were ceramide trihexoside (CTH, with a 3-carbohydrate chain, Galα1-4Galβ1-4Glc-Cer) and ceramide pentahexoside (CPH, with a 5-carbohydrate chain,

Galα1-3Galβ1-4GlcNAcβ1-3Galβ1-4Glc-Cer) (Eto et al., 1968; Stellner et al., 1973). To determine whether anti-Gal binds to any of these glycolipids, rabbit RBC glycolipids were chromatographed on thin layer chromatography (TLC) plates and immunostained with anti-Gal. As shown in Fig. 1, anti-Gal bound to CPH and not to CTH (Galili et al., 1985). The presence in human serum of antibodies that bind to rabbit CPH was also reported by other investigators (Suzuki and Naiki, 1984). The carbohydrate trisaccharide Galα1-3Galβ1-4GlcNAc was called "Galα1-3Gal epitope" (Galili et al., 1985, 1987a,b), or "α-galactosyl epitope" (Galili et al., 1988a) and ultimately shortened to "α-gal epitope" (Galili et al., 1998).

Analysis of the various neutral glycolipids (i.e., glycolipids-lacking sialic acid) in rabbit RBC membranes further demonstrated that glycolipids with carbohydrate chains longer than the five carbohydrates of CPH, increase in size in increments of five carbohydrates and that each increase is as a new branch (also called antenna), up to eight branches, all having the α-gal epitope (Dabrowski et al., 1984; Egge et al., 1985; Hanfland et al., 1988; Honma et al., 1981). The one exception is a glycolipid with seven carbohydrates called ceramide heptahexoside that also has one α-gal epitope at its nonreducing end (Egge et al., 1985). All these glycolipids (referred to as α-gal glycolipids) readily bind the human anti-Gal antibody (Galili et al., 1987b, 2007) (see Fig. 3 in Chapter 10 for glycolipids structure and anti-Gal binding). An α-gal glycolipid with 10 carbohydrates (ceramide decahexoside) is illustrated in Fig. 2B. Anti-Gal was also found to bind to α-gal epitopes on carbohydrate chains of glycoproteins (Towbin et al., 1987; Galili, 1993; Thall and Galili, 1990), and to synthetic α-gal epitopes on glycoproteins (Stone et al., 2007a), or to synthetic α-gal epitopes linked to silica beads (Galili et al., 1985). The carbohydrate chains carrying α-gal epitopes on glycoproteins are mostly Asn (N)-linked carbohydrate chains of the complex type, as that in Fig. 2A.

Immunostaining of glycolipids on TLC plates further indicated that anti-Gal does not bind to carbohydrate chains with β-galactosyl terminal units at the nonreducing end (Galili et al.,

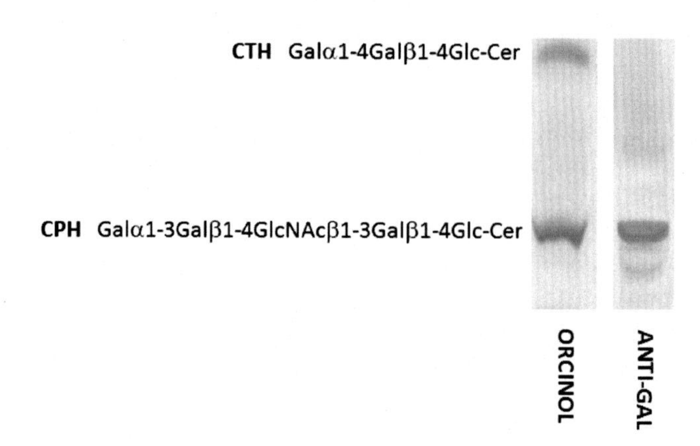

FIGURE 1 Anti-Gal specificity determined on thin layer chromatography plate by immunostaining of the two most abundant rabbit RBC glycolipids: ceramide trihexoside (CTH) with the structure Galα1-4Galβ1-4Glc-Cer and ceramide pentahexoside (CPH) with the terminal carbohydrate structure Galα1-3Galβ1-4GlcNAc-R (i.e., the α-gal epitope). Both glycolipids are stained nonspecifically by orcinol (left lane), whereas anti-Gal immunostains only CPH (right lane). *From Galili, U., 2013c. Discovery of the natural anti-gal antibody and its past and future relevance to medicine. Xenotransplantation 20, 138–147, with permission.*

1985; Teneberg et al., 1996). Analysis of anti-Gal binding to porcine cells further confirmed this antibody specificity, by demonstrating that anti-Gal interaction with α-gal epitopes on cell membrane glycolipids and glycoproteins cannot be inhibited by Galα1-2Gal oligosaccharides, by β-galactosyls, and by blood group O (Fucα1-2Galβ1-4GlcNAc), but it is effectively inhibited by Galα1-3Gal oligosaccharides of various lengths (Neethling et al., 1996). *In vivo* binding of anti-Gal to α-gal epitopes could be demonstrated in patients receiving infusion of therapeutic monoclonal antibodies. Monoclonal antibodies that carry α-gal epitopes were

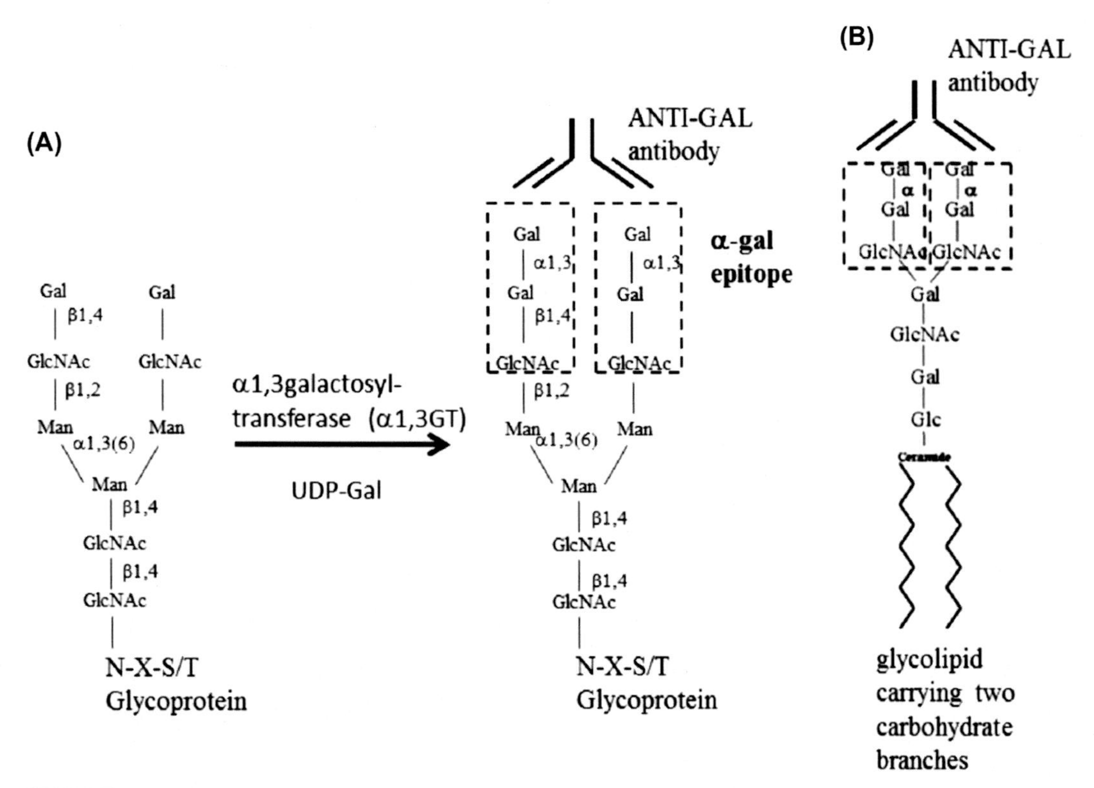

FIGURE 2 Glycoproteins (A) and glycolipids (B) with α-gal epitopes on their carbohydrate chains (*marked by dashed line rectangles*). (A) α-Gal epitopes are synthesized by α1,3galactosyltransferase (α1,3GT) within the Golgi apparatus. *N* (asparagine)-linked carbohydrate chains of glycoproteins (left structure) are synthesized when the amino acid sequence within a protein is: asparagine–any amino acid–serine or threonine (N–X–S/T). Galactose (Gal) provided by the high-energy sugar donor uridine diphosphate galactose (UDP-Gal) is linked by α1,3GT to the nascent carbohydrate chain to generate α-gal epitopes. Each carbohydrate chain may have 2–4 branches. A similar reaction results in synthesis of α-gal epitopes on glycolipids. (B) A glycolipid is comprised of a carbohydrate chain linked to ceramide that is anchored into the membrane by its fatty acid "tails." This representative glycolipid has 10 sugars in its carbohydrate chain and two branches (antennae). Each branch is capped by an α-gal epitope. Glycolipids may have 1–8 branches, some of which, or all may be capped with α-gal epitopes. α-Gal epitopes on both glycoproteins and glycolipids bind the natural anti-Gal antibody. *Gal*, galactose; *Glc, glucose*; GlcNAc, *N*-acetylglucosamine; *Man*, mannose; *N*, asparagine; *S*, serine; *T*, threonine; *UDP-Gal*, uridine diphosphate galactose; *X*, any amino acid. *Reprinted from Galili, U., 2015. Significance of the evolutionary α1,3galactosyltransferase (GGTA1) gene inactivation in preventing extinction of apes and Old World monkeys. J. Mol. Evol. 80, 1–9, with permission.*

found to display much shorter half-life in the circulation than monoclonal antibodies lacking this epitope (Borrebaeck et al., 1993). In addition, infusion of polyethylene glycol carrying multiple α-gal epitopes into monkeys was found to effectively bind anti-Gal in the circulation (Katopodis et al., 2002; Diamond et al., 2002). α-Gal epitopes on carbohydrate chains of glycolipids and glycoproteins are illustrated in Fig. 2. The α-gal epitope should not be confused with α-gal ceramide containing only the carbohydrate galactose linked to ceramide and binding to receptors on NKT cells (Barral and Brenner, 2007).

The affinity between anti-Gal and radiolabeled α-gal epitope trisaccharide was measured by equilibrium dialysis. These studies demonstrated an affinity of $\sim 10^{-6}$M, whereas the disaccharide Galα1-3Gal affinity to anti-Gal is approximately sevenfold lower than that of the trisaccharide (Galili and Matta, 1996). These differences in affinity may be the reason for detection of substantially less anti-Gal in normal human serum passed through columns presenting the disaccharide epitope rather than the trisaccharide epitope (Barreau et al., 2000; Bovin, 2013). The affinity of anti-Gal to α-gal epitopes is lower by at least 100-fold than affinity of human anti-proteins antibodies to corresponding protein antigens (e.g., anti-Rh antibody). This difference is the result of lack of electrostatically charged groups on the α-gal epitope (as on all neutral carbohydrate antigens). In the absence of these charges, there are no ionic bonds between anti-Gal and the α-gal epitope. Ionic bonds are major contributors to the interaction between anti-protein antibodies and the charged amino acids comprising the corresponding antigen. Anti-Gal interaction with the α-gal epitope is likely to be mediated primarily by hydrogen bonds, hydrophobic bonds, and van der Waals forces.

As indicated below, humans lack α-gal epitopes. However, TLC immunostaining analysis of 34 glycolipid molecules from various mammalian species demonstrated that, in addition to binding to α-gal epitopes on glycolipids, the only other glycolipid-binding anti-Gal is a glycolipid called x_2 with the structure GalNAcβ1-3Galβ1-4GlcNAcβ1-3Galβ1-4Glcβl-Cer, which displays topographical similarities to CPH (Teneberg et al., 1996) and which is found in very small amounts in human RBC membranes and in other human tissues (Kannagi et al., 1982; Thorn et al., 1992). It is possible that x_2 is a cryptic glycolipid on human RBC, which is exposed because of protease activity on the surface of macrophages of the reticuloendothelial system, when normal RBC lose flexibility after 120 days in the circulation (due to water loss). The exposure of x_2 may result in anti-Gal binding to RBC and opsonization of these cells. The opsonized RBC are phagocytosed by macrophages of the reticuloendothelial system (Galili et al., 1986a). It was further suggested that in pathologic RBC as in β-thalassemia and sickle cell anemia, x_2 is exposed already on young RBC because of the poor flexibility of these pathologic RBC and thus, prolonged exposure to reticuloendothelial macrophages. The binding of anti-Gal to the prematurely exposed x_2 glycolipids ultimately may result in phagocytosis of these anti-Gal-coated RBC within few weeks after they are released from the bone marrow (Galili et al., 1986b). As discuss in the chapter on anti-Gal and autoimmunity (Chapter 8), exposure of x_2 on other tissues may account for one of the possible mechanisms mediating anti-Gal binding to various tissues that display destruction by the immune system.

It is of interest to note that several peptides mimetic to the α-gal epitope have been identified in peptide libraries (Kooyman et al., 1996; Zhan et al., 2003; Lang et al., 2006). The observations that anti-Gal can interact with mimetic mucin peptides (Sandrin et al., 1997; Apostolopoulos et al., 1999) further raise the question of whether binding of anti-Gal to such peptides can occur *in vivo* (see Chapter 8), and whether such peptides can stimulate the immune system to produce the anti-Gal antibody.

DISTRIBUTION OF ANTI-GAL AND THE α-GAL EPITOPE IN MAMMALS

Several reports on α-gal epitopes on cells and secreted glycoproteins in various mammals were published prior to the observations on this epitope interaction with the natural anti-Gal antibody. These epitopes were found on cells, glycolipids, and secreted glycoproteins in mouse (Eckhard and Goldstein, 1983; Cummings and Kornfeld, 1984), cow and pig (Dorland et al., 1984; Chien et al., 1979), dog (Sung and Sweely, 1979), rat (Ito et al., 1984), and rabbit (Eto et al., 1968; Stellner et al., 1973; Dabrowski et al., 1984). Studies on thyroglobulin from various mammals demonstrated the presence of α-gal epitopes on thyroglobulin of calf, pig, dog, rabbit, sheep, and guinea pig but not on human thyroglobulin (Spiro and Bhoyroo, 1984). Subsequent studies quantifying carbohydrate epitopes on mouse 3T3 fibroblasts indicated that the α-gal epitope is the predominant carbohydrate epitope, presented on tetra-antennary carbohydrate chains in numbers higher even than those of sialic acid epitopes (Santer et al., 1989). Evidently, the α-gal epitope cannot be expected to be found on surface of cells in humans because the interaction of anti-Gal with it will result in an autoimmune pathology. The possibility for the occurrence of such anti-Gal-mediated autoimmunity in humans is discussed in Chapter 8. In view of the reported abundance of α-gal epitopes in the mammals mentioned above and the production of anti-Gal against it in humans, it was of interest to determine whether there is a general pattern to the distribution of α-gal epitopes and the natural anti-Gal antibody in mammals.

Expression of α-gal epitopes on RBC of various species was determined by the binding of human anti-Gal (isolated from blood type AB donors), as well as binding of the lectin *Bandeiraea (Griffonia) simplicifolia* IB4 (BS lectin), which binds specifically α-gal epitopes and blood group B antigen (Peters and Goldstein, 1979; Wood et al., 1979). As discussed in Chapter 3, blood group B antigen has the same structure as the α-gal epitope and an additional fucose linked to the penultimate galactose (Galα1-3(Fucα1-2)Galβ1-4GlcNAc-R and Galα1-3Galβ1-4GlcNAc, respectively). Presence of the natural anti-Gal antibody in the serum of the studied species was determined by agglutination of rabbit RBC using rabbit anti-human IgG as secondary antibody. Nonprimate mammals, such as rat, rabbit, cow, dog, and pig, displayed multiple α-gal epitopes on their RBC (Galili et al., 1987b). The highest number of α-gal epitopes was measured on rabbit RBC. Among primates, lemurs (prosimians that evolved in Madagascar) and New World monkeys (monkeys of South and Central America) were found to express multiple α-gal epitopes on their RBC and to lack anti-Gal in their sera. In contrast, humans, apes, and Old World monkeys (monkeys of Asia and Africa) completely lack α-gal epitopes on their RBC, but all produce anti-Gal in large amounts (Table 1), as indicated by agglutination of rabbit RBC and inhibition of the agglutination by melibiose. TLC immunostaining of glycolipids extracted from RBC membranes of squirrel monkey (a New World monkey) demonstrated the presence of the same anti-Gal binding CPH as that in rabbit RBC (Fig. 1) (Galili et al., 1987b).

The same pattern of distribution of α-gal epitopes was also observed on nucleated cells. Cells of marsupial and placental nonprimate mammals, lemurs, and New World monkeys were found to display ~10^5–10^7 α-gal epitopes/cell, whereas nucleated cells of Old World monkeys, apes, and humans completely lack α-gal epitopes (Table 1) (Galili et al., 1988a). α-Gal epitopes were also found in various numbers on carbohydrate chains of secreted glycoproteins of nonprimate mammals and New World monkeys. As many as 50 α-gal epitopes were measured on mouse laminin, 11 on bovine thyroglobulin, 6 on porcine thyroglobulin, and lower numbers on immunoglobulins and fibrinogen of these species (Arumugham et al., 1986; Spiro and Bhoyroo, 1984; Mohan and Spiro, 1986; Thall and Galili, 1990), as well as on recombinant Factor VIII produced in baby hamster kidney cells (Hironaka et al., 1992). This reciprocal distribution pattern of anti-Gal and the α-gal epitope in mammal was also confirmed by other researchers (Oriol et al., 1999; Teranishi et al., 2002). Cells from nonmammalian vertebrates, including birds, reptiles, amphibians, and fish, all were found to lack α-gal epitopes (Galili et al., 1988a). It is of interest to note that natural anti-Gal was found in chickens (McKenzie et al., 1999), an observation supporting the notion that in the absence of α-gal epitopes the immune system is likely to produce the anti-Gal antibody in response to antigenic stimulation by various bacteria of the normal flora.

Overall, these observations suggest that the α-gal epitope appeared early in mammalian evolution before the divergence of marsupial and placental mammals from a common ancestor and has been conserved in nonprimate mammals, prosimians, and New World monkeys. However, the α-gal epitope was lost in course of Old World primates (monkeys and apes) evolution, and the natural anti-Gal antibody appeared in these primates (Table 1). As discuss in Chapter 2, the synthesis of α-gal epitopes in ancestral Old World primates (but not in lemurs and New World monkeys) stopped 20–30 million years ago because of a "catastrophic evolutionary event" that led these primates to the brink of extinction (Galili and Andrews, 1995). The few primates that survived this event lacked α-gal epitopes and have ever since produced the natural anti-Gal antibody.

TABLE 1 Distribution of the natural anti-Gal antibody, α1,3galactosyltransferase, and the α-gal epitope in mammals

Group	α1,3galactosyltransferase	α-Gal epitope (Galα1-3Galβ1-4GlcNAc-R)	Natural anti-Gal antibody
1. Nonmammalian vertebrates[a]	–	–	+[b]
2. Nonprimate mammals	+	+	–
3. Prosimians (lemurs)	+	+	–
4. New World monkeys	+	+	–
5. Old World monkeys	–	–	+
6. Apes	–	–	+
7. Humans	–	–	+

[a] *Fish, amphibians, reptiles, and birds.*

[b] *Anti-Gal was found to be produced in birds (chicken). There is no information on anti-Gal production in other nonmammalian vertebrates.*

SYNTHESIS OF α-GAL EPITOPES BY α1,3GALACTOSYLTRANSFERASE

The α-gal epitope is synthesized within the Golgi apparatus of cells in nonprimate mammals, prosimians, and New World monkeys by the glycosylation enzyme UDP-Gal: 3Galβ1-4GlcNAc α1-3-galactosyltransferase, also called α1,3galactosyltransferase (α1,3GT). When the glycans (glycolipids, glycoproteins, or proteoglycans) are transported through the Golgi to the cell surface, α1,3GT in the trans-Golgi compartment utilizes UDP-Gal (nucleotide sugar donor) to link the sugar galactose to the *N*-acetyllactosamine residue (Galβ1-4GlcNAc-R) on the nascent carbohydrate chain of the glycan (acceptor substrate), as illustrated in Fig. 2A.

α1,3GT activity was originally found in rabbit bone marrow cells (Basu and Basu, 1973) and subsequently in rabbit intestinal submucosa (Betteridge and Watkins, 1983), in mouse plasmacytoma cells (Blake and Goldstein, 1981), and in bovine thymocytes (Blanken and van den Eijnden, 1985). The activity of this enzyme could be further demonstrated in New World monkeys but not in Old World monkeys or humans (Galili et al., 1988a; Thall et al., 1991). The α1,3GT is a membrane-anchored type 2 protein (i.e., the N-terminus is part of the cytoplasmic domain) with ~370 amino acids in length. α1,3GT competes in the trans-Golgi with other glycosyltransferases (e.g., sialyltransferases linking sialic acid to *N*-acetyllactosamine) for capping the nascent carbohydrate chain (Smith et al., 1990). The sequential linking of sugar units to the nascent carbohydrate chain in the Golgi is analogous to an assembly line in a car plant. The final number of α-gal epitopes on various cells in comparison to sialic acid, or other carbohydrate epitopes, varies from one type of cell to the other and is dependent on the activity of α1,3GT in the Golgi and the activities of other competing glycosyltransferases. As indicated above, α1,3GT is absent in humans and thus, the carbohydrate chains of various glycans are capped by other glycosyltransferases such as sialyltransferases, fucosyltransferase (synthesizes blood group O antigen), and A and B transferases synthesizing blood groups A and B.

The amino acid sequence of α1,3GT was determined in 1989 following cloning of the α1,3GT gene (also referred to as *GGTA1*) in mouse (Larsen et al., 1989) and bovine cells (Joziasse et al., 1989). The gene is encoded by nine exons, of which the catalytic domain at the C terminus is the largest region comprising part of exon VIII and all exon IX (the largest exon), whereas the remaining exons encode the cytoplasmic tail, the transmembrane domain, and the tether of the catalytic domain that attaches it to the cell membrane (Henion et al., 1994). Controlled truncation of the C-terminus of the New World monkey α1,3GT demonstrated that truncation of as few as three amino acids causes complete loss of this catalytic activity, whereas truncation of the regions encoded by exons I through part of exon VIII does not affect the catalytic activity of the enzyme (Henion et al., 1994). As discussed in Chapter 2, studies on the α1,3GT pseudogene in Old World monkeys, apes, and humans, indicated that very few deletion mutations led to the inactivation of α1,3GT in ancestral Old World primates, resulting in the elimination of α-gal epitopes and production of the natural anti-Gal antibody in these primates.

An additional glycosyltransferase called iGb3 synthase was found to synthesize in rats a glycolipid with the α-gal epitope structure Galα1-3Galβ1-4Glc-Cer (Keusch et al., 2000; Taylor et al., 2003). This enzymatic activity is absent in humans (Christiansen et al., 2008) and may be cryptic in mice and pigs as suggested from the ability of mice and pigs with disrupted α1,3GT gene to effectively produce the anti-Gal antibody (LaTemple and Galili, 1998; Tanemura et al., 2000; Dor et al., 2004; Fang et al., 2012; Galili, 2013a).

ANTI-GAL-PRODUCING B CELLS

As detailed in Chapter 2, once the α1,3GT gene was inactivated in ancestral Old World primates ~20–30 million years ago, monkeys and apes lacking the α-gal epitopes started producing the natural anti-Gal antibody. A present day simulation of this event can be observed in pigs in which the α1,3GT gene *GGTA1* was disrupted (i.e., knocked out) to eliminate α-gal epitopes from pig tissues and organs. As further detailed below and in Chapter 6, such knock-out pigs (referred to as GT-KO pigs) may serve as donors of xenografts lacking α-gal epitopes, thereby avoiding the anti-Gal barrier in xenotransplantation. Wild-type pigs produce α-gal epitopes on their cells, thus they are immunotolerant to this carbohydrate antigen and cannot produce the anti-Gal antibody against it. In contrast, the recently produced GT-KO pigs produce the natural anti-Gal antibody already by the age of 2 months, at titers that are even higher than those found in humans (Galili, 2013a).

Production of the natural anti-Gal antibody, as that of other natural anti-carbohydrate antibodies, such as anti-blood group A and B antibodies, is induced by continuous antigenic stimulation by normal gastrointestinal (GI) bacteria. *In vitro* analysis of purified human anti-Gal binding to various bacteria of the GI tract demonstrated binding to several bacterial strains including *Escherichia coli*, *Serratia*, *Salmonella*, and *Klebsiella* (Galili et al., 1988b). Carbohydrate analysis on the wall of some bacterial strains demonstrated the presence of cell surface carbohydrate chains that carry repeating Galα1-3Gal and Gal linked α1-3 to other carbohydrate units (Lüderitz et al., 1965; Bjorndal et al., 1971; Whitfield et al., 1991; Aucken et al., 1998). Other bacteria were reported to have an α1,3galactosyltransferase that synthesizes Galα1-3 units on the wall of *Salmonella typhimurium* (Endo and Rothfield, 1969), *Streptococcus pneumoniae* (Han et al., 2012), and *E. coli* (Chen et al., 2016). Such polysaccharide also may provide stimulatory carbohydrate antigens for anti-Gal production by B cells in lymphoid tissues along the GI tract. Accordingly, antibiotics treatment that eliminated gram-negative aerobic bacteria in baboon GI tract was found to result in marked decrease in production of the natural anti-Gal antibody (Mañez et al., 2001). Moreover, the *E. coli* O86 bacteria, which readily binds anti-Gal (Galili et al., 1988b), were found to induce anti-blood group B antibodies in humans (Springer and Horton, 1969). As discussed in Chapter 3, most of anti-blood group B antibodies in humans are anti-Gal antibodies that can also bind to blood group B antigen (Galili et al., 1987a). Accordingly, feeding of α1,3GT knockout mice (GT-KO mice), lacking the α-gal epitope, with *E. coli* O86 bacteria results in stimulation of the immune system in those mice to produce anti-Gal (Posekany et al., 2002). All these and other bacteria expressing multiple α-gal or α-gal-like epitopes are likely to continuously stimulate the human immune system to produce anti-Gal (Galili et al., 1988b).

Because the bacteria providing antigenic stimulation for anti-Gal production reside in the GI tract, it is reasonable to assume that much of the production of the natural anti-Gal antibody occurs within the lymphoid tissues lining the GI tract. Anti-Gal is found in the serum as IgG, IgM, and IgA isotypes and primarily as the IgA isotype and to a lesser extent as IgG in human body secretions including colostrum, milk, bile, and saliva (Hamadeh et al., 1995). Studies on the distribution of anti-Gal IgM-secreting cells in baboons demonstrated their presence in the spleen; however, 2 months after sensitization with α-gal epitopes on porcine xenografts, high frequencies of anti-Gal IgM and IgG secreting B cells were found in the spleen, lymph nodes, and bone marrow (Xu et al., 2006). That study further indicated that anti-Gal IgM secreting cells were CD20+, CD138−, and Ig+, whereas most anti-Gal IgG secreting plasma cells were both at early (Ig+) and late (Ig−) stages of differentiation.

The proportion of quiescent B cells capable of producing anti-Gal (anti-Gal B cells) in human blood was determined by isolation of human mononuclear cells, immortalization of the B cells by Epstein Barr virus (EBV), growth of individual B cell clones, and analysis of anti-Gal production in these clones (Galili et al., 1993). As many as 1% of immortalized B cell clones were found to produce anti-Gal *in vitro*. Analysis of immortalized B cell clones producing anti-blood group A or anti-blood group B antibodies in the individuals studied for anti-Gal B cell clones indicated that only 0.2% of the EBV immortalized B cell clones produced anti-blood group A or B antibodies (Galili et al., 1993). These studies suggested that there are approximately five times more anti-Gal B cell than B cell producing anti-blood group A or B antibodies. The heavy chains of anti-Gal in humans were found to be encoded by different IgH genes, most of which cluster within the VH3 family (Wang et al., 1995b; Yu et al., 2005).

ELICITED PRODUCTION OF ANTI-GAL

As indicated above, the natural anti-Gal antibody is likely to be continuously produced by a small proportion of anti-Gal B cells, those present in lymphoid tissues lining the GI tract. However, the many anti-Gal B cells in the blood and probably those in the lymph nodes and spleen are quiescent. The full potential for human anti-Gal B cells to produce the antibody is demonstrated in recipients of xenograft cells that present α-gal epitopes. Introduction of xenografts or implants carrying α-gal epitopes into humans results in rapid activation of the many quiescent anti-Gal B cells by the multiple α-gal epitopes on the xenograft cells (Galili et al., 2001; Stone et al., 2007a). However, the α-gal epitope itself cannot activate helper T (Th) cells that are required for providing the help for activation of anti-Gal B cells and for their isotype switch from IgM to IgG or IgA producing B cells. The inability of α-gal epitopes to activate Th cells is the result of the destruction of carbohydrate chains by glycosidases within antigen presenting cells (APC) and the protrusion of such carbohydrate chains on glycopeptides presented on class II MHC complex on APC. Protrusion of carbohydrate antigens such as the α-gal epitope from the APC prevents the subsequent interaction of accessory molecules of the T cell receptor with their ligands on the APC cell membrane, an interaction that is required for activation of the Th cell (Ishioka et al., 1993; Speir et al., 1999). Th cells, however, are effectively activated by the large number of immunogenic xenoproteins and xenopeptides of the xenograft cells (Tanemura et al., 2000; Galili, 2004). Thus, the ultimate result for exposure of patients to xenograft cells is a rapid and extensive production of elicited anti-Gal in recipients of these cells. Such extensive elicited anti-Gal antibody production was observed in ovarian carcinoma patients undergoing an experimental gene therapy treatment in which the patient received three intraperitoneal infusions of 6×10^9 mouse "packaging" cell line derived from 3T3 fibroblasts and containing replication defective virus with the transgene studied. Anti-Gal activity in the blood was determined by ELISA with synthetic α-gal epitopes linked to bovine serum albumin (α-gal BSA) as solid-phase antigen. Anti-Gal IgG titer was found to increase by ~100-fold above the natural titer of the antibody, within 2 weeks

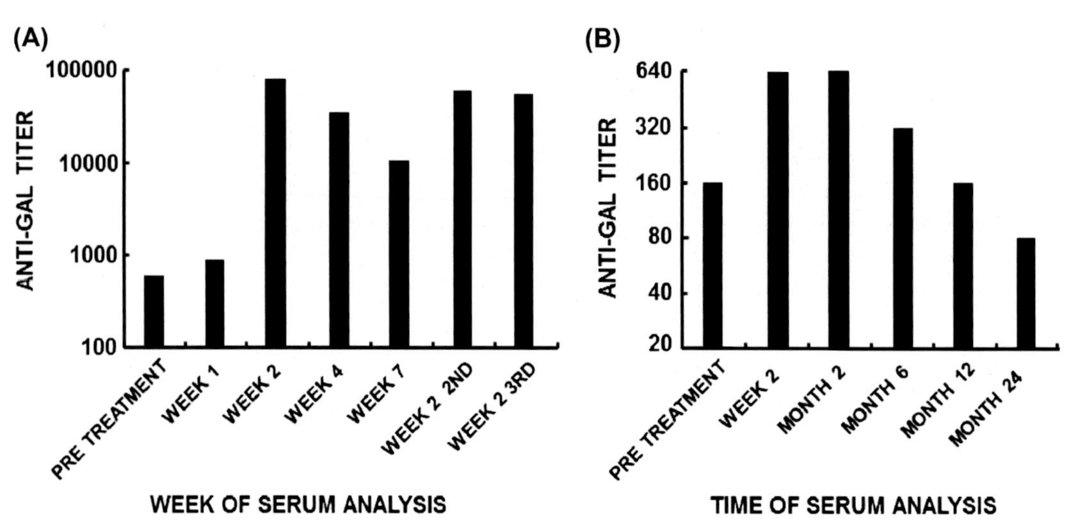

FIGURE 3 Elicited anti-Gal production following exposure of the human immune system to α-gal epitopes of mouse xenograft or porcine extracellular matrix. (A) An ovarian carcinoma patient receiving three intraperitoneal infusions of 6×10^9 3T3 derived packaging fibroblasts containing a replication defective virus, in 7-week intervals. (B) A representative patient with ruptured anterior cruciate ligament implanted with porcine processed patellar tendon enzymatically treated to remove α-gal epitopes from the soft tissue portion. However, α-gal epitopes within cavities of the bone plugs attached to the ligament are not eliminated and elicit anti-Gal response in the recipient. Anti-Gal activity was determined prior to implantation, and at various time points post implantation. The titers are presented as reciprocals of serum dilution yielding half the maximum binding in ELISA using synthetic α-gal epitopes linked to BSA (α-gal BSA) as solid-phase antigen. *(A) Modified from Galili, U., Chen, Z.C., Tanemura, M., Seregina, T., Link, C.J., 2001. Induced antibody response in xenograft recipients. Graft 4, 32–35. (B) Modified from Stone, K.R., Abdel-Motal, U.M., Walgenbach, A.W., Turek, T.J., Galili, U., 2007a. Replacement of human anterior cruciate ligaments with pig ligaments: a model for anti-non-gal antibody response in long-term xenotransplantation. Transplantation 83, 211–219.*

post intraperitoneal administration of mouse cells presenting α-gal epitopes (Fig. 3A) (Galili et al., 2001). This increase was the result of two processes. First, anti-Gal concentration in the serum increased by tenfold because of the rapid activation of many quiescent anti-Gal B cells. Second, activated anti-Gal B cells were further selected by the process of affinity maturation for the preferential expansion of B cell clones producing high affinity anti-Gal, resulting in ~10-fold increase in the affinity of this antibody (Galili et al., 2001). This resulted in an overall increase in the titer of anti-Gal by ~100-fold (Fig. 3A). Most (~90%) of the increase in anti-Gal was of the IgG2 subclass and the rest of IgG3. No significant changes in anti-Gal activity were observed in IgM or IgA isotypes following the intraperitoneal infusion of mouse 3T3 cells, suggesting that the activated anti-Gal B cells undergo rapid isotype switch into anti-Gal IgG-producing cells. As indicated above, the α-gal epitope is the most abundant carbohydrate epitope on glycoproteins of the 3T3 cell membrane (Santer et al., 1989). In contrast, injection into cancer patients of α-gal epitopes on glycolipids, in the absence of stimulatory peptides, results only in a marginal increase of anti-Gal titer by only twofold (Whalen et al., 2012).

The activation of quiescent anti-Gal B cells in humans by α-gal epitopes on xenograft cells is potent enough to overcome immune suppression by drugs that successfully prevent rejection of human allografts. This could be demonstrated in diabetic patients that were immunosuppressed and transplanted with an allogeneic kidney together with pig fetal pancreatic islet cells (Groth et al., 1994; Galili et al., 1995). The pig islet cell clusters were placed in the subcapsular space of a transplanted kidney allograft or infused into the portal vein of diabetic recipients of a kidney allograft. The immune-mediated rejection of kidney allografts in these patients was prevented by standard immunosuppression regimens used in kidney transplantation. Despite this immunosuppression that is potent enough to prevent rejection of the kidney allograft, the patients displayed an increase of 20- to 80-fold in anti-Gal titer within the period of 25–50 days post transplantation (Galili et al., 1995). As in the ovarian carcinoma patient who was infused with mouse cells (Galili et al., 2001), the increase in anti-Gal activity was mostly in the IgG isotype and to a much lesser extent of IgM and IgA isotypes. Thus, immunosuppression of T cells, which effectively prevents the rejection of kidney allografts, does not prevent activation of quiescent anti-Gal B cells by α-gal epitopes on xenografts, resulting in production of elicited anti-Gal antibody. A similar production of elicited anti-Gal despite immune suppression was observed in half of the monkeys transplanted with porcine pancreatic islet cells by infusion into the portal vein (Kang et al., 2015).

A relatively short exposure of the human immune system to α-gal epitopes on xenografts may suffice to induce the activation of anti-Gal B cells to produce the elicited anti-Gal antibodies. This could be demonstrated in patients with failed liver function, who were treated by extracorporeal perfusion of their blood through a pig liver for a period of 6–48 h (Cotterell et al., 1995; Yu et al., 1999). Subsequently, the patients received a liver allograft transplant and were immunosuppressed for preventing rejection of the allograft. A marked elevation in anti-Gal titers, measured within 10 days post the perfusion through porcine liver, indicated that activation of anti-Gal B cells occurred despite the short exposure to the α-gal epitopes in the porcine liver and despite the immunosuppression preventing the liver allograft rejection. These elicited anti-Gal antibodies also were found to increase in their avidity (Yu et al., 1999) probably because of the affinity maturation process in which anti-Gal B cell clones with mutated B cell receptors having increased affinity to α-gal epitopes, undergo preferential expansion.

The elicited anti-Gal antibody production continues as long as glycoproteins with α-gal epitopes are released from the xenograft and activate the quiescent B cells. As detailed in Chapter 13, patients with ruptured anterior cruciate ligament (ACL) were implanted with pig tendon treated for elimination of α-gal epitopes by recombinant α-galactosidase, followed by partial cross-linking with glutaraldehyde. Such tendons were used for reconstruction of the ruptured ACL (Stone et al., 2007a,b). These porcine implants also had two cancellous bone plugs that contained porcine bone marrow cells and RBC, which carried α-gal epitopes because the α-galactosidase cannot diffuse into the closed bone cavities. The patients produced for several months elevated levels of anti-Gal IgG antibody in response to glycoproteins carrying α-gal epitopes of bone marrow and blood cells released from cavities of these bone plugs (Fig. 3B). The gradual remodeling of the bone plugs into human bone resulted in slow elimination of the pig α-gal epitopes and thus decrease of anti-Gal activity, which returned to the pre-implantation level ~12 months post implantation. Therefore, in the absence of stimulation by xenogeneic α-gal epitopes, anti-Gal level returns to that of the pre-implantation level produced against bacterial of the GI flora. A temporary decrease in elicited anti-Gal activity is also observed in Fig. 3A above, reflecting elimination of the infused mouse cells, which are destroyed by anti-Gal and their cell membranes, presenting α-gal epitopes are internalized by macrophages. However, a second and third administration of the mouse cells result in rapid increase in anti-Gal titer to the maximum titer of ~80,000 (Fig. 3A).

As discussed in Chapters 6 and 13, the extensive production of elicited anti-Gal antibody is likely to be detrimental to porcine xenografts or extracellular matrix (ECM) implants that have α-gal epitopes as it exacerbates anti-Gal mediated rejection of both xenografts and implants presenting these α-gal epitopes. This detrimental immune response may be avoided using porcine xenograft or ECM implants lacking α-gal epitopes. Porcine tissues and organs lacking α-gal epitopes can be obtained from GT-KO pigs (see description below). When wild-type porcine ECM implants are used, elimination of α-gal epitopes may be achieved by enzymatic destruction of these epitopes with recombinant α-galactosidase (Stone et al., 2007a,b). Treatment with α-galactosidase is not effective in xenografts containing live cells. Although this enzyme can effectively remove α-gal epitopes, the ongoing turnover of the membrane in live cells results in reappearance of α-gal epitopes on cell membranes within 24h.

Elevation of anti-Gal activity in humans can be observed following infection with parasites such as *Trypanosoma*, *Leishmania*, and *Plasmodium* (see Chapter 4). This elevation is likely to be the result of the exposure of the immune system in infected individuals to α-gal-like epitopes presented on the cell membrane of these parasites. The reaction against these epitopes results in increase in anti-Gal titers by 5 to 30-fold or more in comparison to the natural titer of this antibody. In addition to the immune response to α-gal-like epitopes on the parasite, it is possible that damage in the intestinal wall as in megacolon in *Trypanosoma* chronic infection may increase the exposure of the immune system to bacteria of the GI flora which provide the antigenic stimulation for anti-Gal production, thereby increasing this antibody titers.

EXPERIMENTAL ANIMAL MODELS FOR STUDYING ANTI-GAL

A basic difficulty in studying anti-Gal and its harnessing for various therapeutic applications is the fact that it is naturally produced only in humans, apes, and Old World monkeys. Of these, only Old World monkeys may be considered as experimental models, which indeed have been used in xenotransplantation studies (see Chapter 6), but which are difficult and highly expensive for performing preclinical studies. Nonprimate mammals, including the usual experimental animal models such as mice, rats, rabbits, hamsters, guinea pigs, dogs, and pigs, all have active α1,3GT that synthesizes α-gal epitopes. Thus, nonprimate mammals, as well as New World monkeys, are immunotolerant to this carbohydrate antigen and cannot produce the anti-Gal antibody (Galili et al., 1987b, 1988a; Galili, 2013b). The study of anti-Gal and its manipulation for therapeutic purposes became feasible, however, with the production of α1,3GT knockout mice (GT-KO mice) (Thall et al., 1995; Tearle et al., 1996). Several years after the generation of GT-KO mice, α1,3GT knockout pigs (GT-KO pigs) were generated (Lai et al., 2002; Phelps et al., 2003; Kolber-Simonds et al., 2004; Takahagi et al., 2005). These knockout mice and pigs lack α-gal epitopes because of targeted disruption (knockout) of the α1,3GT gene, thus they do not synthesize α-gal epitopes and are not immunotolerant to it (LaTemple and Galili, 1998; Tanemura et al., 2000; Chiang et al., 2000; Dor et al., 2004; Fang et al., 2012; Galili, 2013a). GT-KO mice are usually kept in a sterile environment and are provided with sterile food. Therefore, the mice lack the GI flora which induces production of anti-Gal. Natural production of this antibody is usually low in GT-KO mice in the absence of the stimulatory GI bacteria. However, when GT-KO mice are immunized with xenogeneic cells expressing α-gal epitopes, they are induced to produce anti-Gal. Thus, 3–4 immunizations with rabbit RBC elicit anti-Gal production (LaTemple and Galili, 1998). Three immunizations with homogenates of pig tissues containing high concentrations of α-gal epitopes such as pig kidney membrane homogenate also are effective in inducing anti-Gal production in GT-KO mice (Tanemura et al., 2000).

GT-KO pigs are not kept in sterile conditions and develop shortly after birth the GI flora required for stimulation of the immune system to produce anti-Gal. Thus, natural production of anti-Gal is observed in GT-KO pigs already at the age of 1.5–2 months, and subsequently the antibody continues to be produced in titers that are at least as high as those measured in humans (Galili, 2013a). GT-KO mice have served as important experimental animal model for studying anti-Gal production and immune response to carbohydrate antigens, as well as serving as preclinical models for the variety of anti-Gal-mediated therapies discussed in this book. As described in Chapter 12, GT-KO pigs served as a large animal model validating the efficacy of α-gal nanoparticles in inducing anti-Gal-mediated acceleration of wound healing, following initial studied in GT-KO mice (Galili et al., 2010; Wigglesworth et al., 2011; Hurwitz et al., 2012). It is probable that the GT-KO pigs will serve as a suitable model for studying α-gal nanoparticles–mediated regeneration of myocardium ECM patches applied to post ischemia scars of the myocardium (discussed in Chapter 13). Thus, in addition to providing porcine organs and cells devoid of α-gal epitopes for xenotransplantation, GT-KO pigs may be useful in the future for studying various anti-Gal-mediated therapeutic applications in a large size experimental animal model.

CONCLUSIONS

Anti-Gal is a natural antibody produced in humans as ~1% of immunoglobulins and is found also in apes and Old World monkeys. The carbohydrate antigen that specifically binds to anti-Gal is the α-gal epitope with the structure Galα1-3Galβ1-4GlcNAc-R. This antigen is abundantly produced on glycoproteins, glycolipids, and proteoglycans in nonprimate mammals, prosimians, and New World monkeys, all of which lack the anti-Gal antibody and have active α1,3galactosyltranferase enzyme that synthesizes α-gal epitopes. Anti-Gal is produced in humans throughout life because of continuous immune response to α-gal-like epitopes on walls of bacteria of the natural GI flora. Anti-Gal is mostly produce by B lymphocytes (anti-Gal B cells) that reside along the GI tract. Anti-Gal B cells comprise ~1% of circulating B cells; however, most of them are quiescent. Introduction of porcine xenografts or ECM bio-implants presenting α-gal epitopes results in rapid activation of the quiescent anti-Gal B cells, thereby increasing anti-Gal titers by ~100-fold within 2 weeks. The increased production of anti-Gal continues as long as glycoproteins with α-gal epitopes are released from the xenograft or the bio-implant. Because anti-Gal is produced in large amounts in all humans that are not severely immunocompromised, it may be harnessed for a variety of therapeutic applications discussed in this book. Preclinical studies of such applications are feasible in nonprimate experimental animal models in which the α1,3galactosyltransferase gene (*GGTA1*) was disrupted (knocked out) such as knockout mice and pigs for this gene. In the absence of α-gal epitopes, these mice and pigs can produce the anti-Gal antibody.

References

Apostolopoulos, V., Sandrin, M.S., McKenzie, I.F., 1999. Carbohydrate/peptide mimics: effect on MUC1 cancer immunotherapy. J. Mol. Med. (Berl.) 77, 427–436.

Arumugham, R.G., Hsieh, T.C., Tanzer, M.L., Laine, R.A., 1986. Structures of the asparagine-linked sugar chains of laminin. Biochim. Biophys. Acta 883, 112–126.

Aucken, H.M., Wilkinson, S.G., Pitt, T.L., 1998. Re-evaluation of the serotypes of *Serratia marcescens* and separation into two schemes based on lipopolysaccharide (O) and capsular polysaccharide (K) antigens. Microbiology 144, 639–653.

Avila, J.L., Rojas, M., Galili, U., 1989. Immunogenic Galα1-3Gal carbohydrate epitopes are present on pathogenic American *Trypanosoma* and *Leishmania*. J. Immunol. 142, 2828–2834.

Barral, D.C., Brenner, M.B., 2007. CD1 antigen presentation: how it works. Nat. Rev. Immunol. 7, 929–941.

Barreau, N., Blancho, G., Boulet, C., Martineau, A., Vusio, P., Liaigre, J., et al., 2000. Natural anti-Gal antibodies constitute 0.2% of intravenous immunoglobulin and are equally retained on a synthetic disaccharide column or on an immobilized natural glycoprotein. Transpl. Proc. 32, 882–883.

Basu, M., Basu, S., 1973. Enzymatic synthesis of blood group related pentaglycosyl ceramide by an α-galactosyltransferase. J. Biol. Chem. 248, 1700–1706.

Betteridge, A., Watkins, W.M., 1983. Two α-3-D galactosyltransferases in rabbit stomach mucosa with different acceptor substrate specificities. Eur. J. Biochem. 132, 29–35.

Blake, D.D., Goldstein, I.J., 1981. An α-D-galactosyltransferase in Ehrlich ascites tumor cells: biosynthesis and characterization of a trisaccharide (α-D-galacto(1-3)-*N*-acetyllactosamine). J. Biol. Chem. 256, 5387–5393.

Bjorndal, H., Lindberg, B., Nimmich, W., 1971. Structural studies on *Klebsiella* 0 groups 1 and 6 lipopolysaccharides. Acta Chem. Scand. 25, 750.

Blanken, W.M., van den Eijnden, D.H., 1985. Biosynthesis of terminal Galα1-3Galß1-4GlcNAc-R oligosaccharide sequence on glycoconjugates: purification and acceptor specificity of a UDP-Gal: *N*-acetyllactosamine α1,3galactosyltransferase. J. Biol. Chem. 260, 12927–12934.

Borrebaeck, C.K., Malmborg, A.C., Ohlin, M., 1993. Does endogenous glycosylation prevent the use of mouse monoclonal antibodies as cancer therapeutics? Immunol. Today 14, 477–479.

Bovin, N.V., 2013. Natural antibodies to glycans. Biochem. (Mosc.) 78, 786–797.

Chen, C., Liu, B., Xu, Y., Utkina, N., Zhou, D., Danilov, L., et al., 2016. Biochemical characterization of the novel α-1,3-galactosyltransferase WclR from *Escherichia coli* O3. Carbohydr. Res. 430, 36–43.

Chiang, T.R., Fanget, L., Gregory, R., Tang, Y., Ardiet, D.L., Gao, L., et al., 2000. Anti-Gal antibodies in humans and 1,3 α-galactosyltransferase knock-out mice. Transplantation 69, 2593–2600.

Chien, J.L., Li, S.C., Li, Y.T., 1979. Isolation and characterization of a heptaglycosylceramide from bovine erythrocyte membranes. J. Lipid Res. 20, 669–673.

Christiansen, D., Milland, J., Mouhtouris, E., Vaughan, H., Pellicci, D.G., McConville, M.J., et al., 2008. Humans lack iGb3 due to the absence of functional iGb3-synthase: implications for NKT Cell development and transplantation. PLoS Biol. 6, 1527–1538.

Cotterell, A.H., Collins, B.H., Parker, W., Harland, R.C., Platt, J.L., 1995. The humoral immune response in humans following cross-perfusion of porcine organs. Transplantation 60, 861–868.

Cummings, R.D., Kornfeld, S., 1984. The distribution of repeating (Galβ1-4GlcNAcβ1-3) sequences in asparagine-linked oligosaccharides of the mouse lymphoma cell lines BW 147 and PHA 2.1. J. Biol. Chem. 259, 6253–6260.

Dabrowski, U., Hanfland, P., Egge, H., Kuhn, S., Dabrowski, J., 1984. Immunochemistry of I/i-active oligo- and polyglycosylceramides from rabbit erythrocyte membranes. Determination of branching patterns of a ceramide pentadecasaccharide by 1H nuclear magnetic resonance. J. Biol. Chem. 259, 7648–7651.

Diamond, L.E., Byrne, G.W., Schwarz, A., Davis, T.A., Adams, D.H., Logan, J.S., 2002. Analysis of the control of the anti-gal immune response in a non-human primate by galactose α1-3 galactose trisaccharide-polyethylene glycol conjugate. Transplantation 73, 1780–1787.

Dor, F.J., Tseng, Y.L., Cheng, J., Moran, K., Sanderson, T.M., Lancos, C.J., et al., 2004. α1,3-Galactosyltransferase gene-knockout miniature swine produce natural cytotoxic anti-Gal antibodies. Transplantation 78, 15–20.

Dorland, L., van Halbeek, H., Vliegenthart, J.F., 1984. The identification of terminal α(1-3)-linked galactose in *N*-acetyllactosamine type of glycopeptides by means of 500-MHz 1H-NMR spectroscopy. Biochem. Biophys. Res. Commun. 122, 859–866.

Egge, H., Kordowicz, M., Peter-Katalinic, J., Hanfland, P., 1985. Immunochemistry of I/i-active oligo- and polyglycosylceramides from rabbit erythrocyte membranes. J. Biol. Chem. 260, 4927–4935.

Eckhardt, A.E., Goldstein, I.J., 1983. Isolation and characterization of α-galactosyl containing glycopeptides from Ehrlich ascites tumor cells. Biochemistry 22, 5290–5303.

Endo, A., Rothfield, L., 1969. Studies of a phospholipid-requiring bacterial enzyme. I. Purification and properties of uridine diphosphate galactose:lipopolysaccharide α-3 galactosyltransferase. Biochemistry 8, 3500–3507.
Eto, T., Iichikawa, Y., Nishimura, K., Ando, S., Yamakawa, T., 1968. Chemistry of lipids of the posthemolytic residue or stroma of erythrocytes. XVI. Occurrence of ceramide pentasaccharide in the membrane of erythrocytes and reticulocytes in rabbit. J. Biochem. (Tokyo) 64, 205–213.
Fang, J., Walters, A., Hara, H., Long, C., Yeh, P., Ayares, D., et al., 2012. Anti-gal antibodies in α1,3-galactosyltransferase gene-knockout pigs. Xenotransplantation 19, 305–310.
Galili, U., 1993. Evolution and pathophysiology of the human natural anti-Gal antibody. Springer Semin. Immunopathol. 15, 155–171.
Galili, U., 2004. Immune response, accommodation and tolerance to transplantation carbohydrate antigens. Transplantation 78, 1093–1098.
Galili, U., 2013a. α1,3Galactosyltransferase knockout pigs produce the natural anti-Gal antibody and simulate the evolutionary appearance of this antibody in primates. Xenotransplantation 20, 267–276.
Galili, U., 2013b. Anti-Gal: an abundant human natural antibody of multiple pathogeneses and clinical benefits. Immunology 140, 1–11.
Galili, U., 2013c. Discovery of the natural anti-Gal antibody and its past and future relevance to medicine. Xenotransplantation 20, 138–147.
Galili, U., 2015. Significance of the evolutionary α1,3galactosyltransferase (*GGTA1*) gene inactivation in preventing extinction of apes and Old World monkeys. J. Mol. Evol. 80, 1–9.
Galili, U., Korkesh, A., Kahane, I., Rachmilewitz, A., 1983. Demonstration of a natural anti-galactosyl IgG antibody on thalassemic red blood cells. Blood 61, 1258–1264.
Galili, U., Rachmilewitz, E.A., Peleg, A., Flechner, I., 1984. A unique natural human IgG antibody with anti-α-galactosyl specificity. J. Exp. Med. 160, 1519–1531.
Galili, U., Macher, B.A., Buehler, J., Shohet, S.B., 1985. Human natural anti-α-galactosyl IgG. II. The specific recognition of α(1→3)-linked galactose residues. J. Exp. Med. 162, 573–582.
Galili, U., Flechner, I., Kniszinski, A., Danon, D., Rachmilewitz, E.A., 1986a. The natural anti-α-galactosyl IgG on human normal senescent red blood cells. Br. J. Haematol. 62, 317–324.
Galili, U., Clark, M.R., Shohet, S.B., 1986b. Excessive binding of the natural anti-α-galactosyl IgG to sickle red cells may contribute to extravascular cell destruction. J. Clin. Investig. 77, 27–33.
Galili, U., Buehler, J., Shohet, S.B., Macher, B.A., 1987a. The human natural anti-gal IgG. III. The subtlety of immune tolerance in man as demonstrated by crossreactivity between natural anti-Gal and anti-B antibodies. J. Exp. Med. 165, 693–704.
Galili, U., Clark, M.R., Shohet, S.B., Buehler, J., Macher, B.A., 1987b. Evolutionary relationship between the anti-Gal antibody and the Galα1-3Gal epitope in primates. Proc. Natl. Acad. Sci. U.S.A. 84, 1369–1373.
Galili, U., Shohet, S.B., Kobrin, E., Stults, C.L.M., Macher, B.A., 1988a. Man, apes, and Old World monkeys differ from other mammals in the expression of α-galactosyl epitopes on nucleated cells. J. Biol. Chem. 263, 17755–17762.
Galili, U., Mandrell, R.E., Hamadeh, R.M., Shohet, S.B., Griffiss, J.M., 1988b. Interaction between human natural anti-α-galactosyl immunoglobulin G and bacteria of the human flora. Infect. Immun. 56, 1730–1737.
Galili, U., Anaraki, F., Thall, A., Hill-Black, C., Radic, M., 1993. One percent of circulating B lymphocytes are capable of producing the natural anti-Gal antibody. Blood 82, 2485–2493.
Galili, U., Andrews, P., 1995. Suppression of α-galactosyl epitopes synthesis and production of the natural anti-Gal antibody: a major evolutionary event in ancestral Old World primates. J. Hum. Evol. 29, 433–442.
Galili, U., Tibell, A., Samuelsson, B., Rydberg, B., Groth, C.G., 1995. Increased anti-Gal activity in diabetic patients transplanted with fetal porcine islet cell clusters. Transplantation 59, 1549–1556.
Galili, U., Matta, K.L., 1996. Inhibition of anti-Gal IgG binding to porcine endothelial cells by synthetic oligosaccharides. Transplantation 62, 256–262.
Galili, U., LaTemple, D.C., Radic, M.Z., 1998. A sensitive assay for measuring α-gal epitope expression on cells by a monoclonal anti-Gal antibody. Transplantation 65, 1129–1132.
Galili, U., Chen, Z.C., Tanemura, M., Seregina, T., Link, C.J., 2001. Induced antibody response in xenograft recipients. Graft 4, 32–35.
Galili, U., Wigglesworth, K., Abdel-Motal, U.M., 2007. Intratumoral injection of α-gal glycolipids induces xenograft-like destruction and conversion of lesions into endogenous vaccines. J. Immunol. 178, 4676–4687.
Galili, U., Wigglesworth, K., Abdel-Motal, U.M., 2010. Accelerated healing of skin burns by anti-Gal/α-gal liposomes interaction. Burns 36, 239–251.
Groth, C.G., Korsgren, O., Tibell, A., Tolleman, J., Möller, E., Bolinder, J., et al., 1994. Transplantation of porcine fetal pancreas to diabetic patients. Lancet 344, 1402–1404.

Hamadeh, R.M., Galili, U., Zhou, P., Griffiss, J.M., 1995. Human secretions contain IgA, IgG and IgM anti-Gal (anti-α-galactosyl) antibodies. Clin. Diagn. Lab. Immunol. 2, 125–131.

Han, W., Cai, L., Wu, B., Li, L., Xiao, Z., Cheng, J., et al., 2012. The wciN gene encodes an α-1,3-galactosyltransferase involved in the biosynthesis of the capsule repeating unit of *Streptococcus pneumoniae* serotype 6B. Biochemistry 51, 5804–5810.

Hanfland, P., Kordowicz, M., Peter-Katalinic, J., Egge, H., Dabrowski, J., Dabrowski, U., 1988. Structure elucidation of blood group B-like and I-active ceramide eicosa- and pentacosasaccharides from rabbit erythrocyte membranes by combined gas chromatography-mass spectrometry; electron-impact and fast atom-bombardment mass spectrometry; and two-dimensional correlated, relayed-coherence transfer, and nuclear Overhauser effect 500-MHz 1H-n.m.r. spectroscopy. Carbohydr. Res. 178, 1–21.

Henion, T.R., Macher, B.A., Anaraki, F., Galili, U., 1994. Defining the minimal size of catalytically active primate α1,3galactosyltransferase: structure function studies on the recombinant truncated enzyme. Glycobiology 4, 193–201.

Hironaka, T., Furukawa, K., Esmon, P.C., Fournel, M.A., Sawada, S., Kato, M., et al., 1992. Comparative study of the sugar chains of factor VIII purified from human plasma and from the culture media of recombinant baby hamster kidney cells. J. Biol. Chem. 267, 8012–8020.

Honma, K., Manabe, H., Tomita, M., Hamada, A., 1981. Isolation and partial structural characterization of macroglycolipid from rabbit erythrocyte membranes. J. Biochem. 90, 1187–1196.

Hurwitz, Z., Ignotz, R., Lalikos, J., Galili, U., 2012. Accelerated porcine wound healing with α-gal nanoparticles. Plast. Reconstr. Surg. 129, 242–251.

Ishioka, G.Y., Lamont, A.G., Thomson, D., Bulbow, N., Gaeta, F.C., Sette, A., et al., 1993. Major histocompatibility complex class II association and induction of T cell responses by carbohydrates and glycopeptides. Springer Semin. Immunopathol. 15, 293–302.

Ito, M., Suzuki, E., Naiki, M., Sendo, F., Arai, S., 1984. Carbohydrates as antigenic determinants of tumor-associated antigens recognized by monoclonal anti-tumor antibodies produced in a syngeneic system. Int. J. Cancer 34, 689–697.

Joziasse, D.H., Shaper, J.H., Van den Eijnden, D.H., Van Tunen, A.H., Shaper, N.L., 1989. Bovine α1-3galactosyltransferase: isolation and characterization of a cDNA clone. Identification of homologous sequences in human genomic DNA. J. Biol. Chem. 264, 14290–14297.

Kang, H.J., Lee, H., Park, E.M., Kim, J.M., Shin, J.S., Kim, J.S., et al., 2015. Dissociation between anti-porcine albumin and anti-Gal antibody responses in non-human primate recipients of intraportal porcine islet transplantation. Xenotransplantation 22, 124–134.

Kannagi, R., Fukuda, M.N., Hakomori, S.-I., 1982. A new glycolipid antigen isolated from human erythrocyte membranes reacting with antibodies directed to globo-ZV-tetraosylceramide (globoside). J. Biol. Chem. 257, 4438–4442.

Katopodis, A.G., Warner, R.G., Duthaler, R.O., Streiff, M.B., Bruelisauer, A., Kretz, O., et al., 2002. Removal of anti-Galα1,3Gal xenoantibodies with an injectable polymer. J. Clin. Investig. 110, 1869–1877.

Keusch, J.J., Manzella, S.M., Nyame, K.A., Cummings, R.D., Baenziger, J.U., 2000. Expression cloning of a new member of the ABO blood group glycosyltransferases, iGb3 synthase, that directs the synthesis of isoglobo-glycosphingolipids. J. Biol. Chem. 275, 25308–25314.

Kolber-Simonds, D., Lai, L., Watt, S.R., Denaro, M., Arn, S., Augenstein, M.L., et al., 2004. Production of α1,3-galactosyltransferase null pigs by means of nuclear transfer with fibroblasts bearing loss of heterozygosity mutations. Proc. Natl. Acad. Sci. U.S.A. 101, 7335–7340.

Kooyman, D.L., McClellan, S.B., Parker, W., Avissar, P.L., Velardo, M.A., Platt, J.L., et al., 1996. Identification and characterization of a galactosyl peptide mimetic. Implications for use in removing xenoreactive anti-α-Gal antibodies. Transplantation 61, 851–855.

Lai, L., Kolber-Simonds, D., Park, K.W., Cheong, H.T., Greenstein, J.L., Im, G.S., et al., 2002. Production of α-1,3-galactosyltransferase knockout pigs by nuclear transfer cloning. Science 295, 1089–1092.

Landsteiner, K., Philip-Miller, C., 1925. Serological studies on the blood of the primates. III. Distribution of serological factors related to human isoagglutinogens in the blood of lower monkeys. J. Exp. Med. 42, 863–872.

Lang, J., Zhan, J., Xu, L., Yan, Z., 2006. Identification of peptide mimetics of xenoreactive α-Gal antigenic epitope by phage display. Biochem. Biophys. Res. Commun. 344, 214–220.

Larsen, R.D., Rajan, V.P., Ruff, M., Kukowska-Latallo, J., Cummings, R.D., Lowe, J.B., 1989. Isolation of a cDNA encoding murine UDP galactose: ßD-galactosyl-1,4-N-acetyl-D-glucosaminide α1,3-galactosyltransferase: expression cloning by gene transfer. Proc. Natl. Acad. Sci. U.S.A. 86, 8227–8231.

LaTemple, D.C., Galili, U., 1998. Adult and neonatal anti-Gal response in knock-out mice for α1,3galactosyltransferase. Xenotransplantation 5, 191–196.

Lüderitz, O., Simmons, D.A., Westphal, G., 1965. The immunochemistry of *Salmonella* chemotype VI O-antigens. The structure of oligosaccharides from *Salmonella* group U (o 43) lipopolysaccharides. Biochem. J. 97, 820–826.

Mañez, R., Blanco, F.J., Díaz, I., Centeno, A., Lopez-Pelaez, E., Hermida, M., et al., 2001. Removal of bowel aerobic gram-negative bacteria is more effective than immunosuppression with cyclophosphamide and steroids to decrease natural α-galactosyl IgG antibodies. Xenotransplantation 8, 15–23.

McKenzie, I.F., Patton, K., Smit, J.A., Mouhtouris, E., Xing, P., Myburgh, J.A., et al., 1999. Definition and characterization of chicken Galα(1,3)Gal antibodies. Transplantation 67, 864–870.

McMorrow, I.M., Comrack, C.A., Sachs, D.H., DerSimonian, H., 1997. Heterogeneity of human anti-pig natural antibodies cross-reactive with the Gal(α1,3)Galactose epitope. Transplantation 64, 501–510.

Mohan, P.S., Spiro, R.G., 1986. Macromolecular organization of basement membranes. Characterization and comparison of glomerular basement membrane and lens capsule components by immunochemical and lectin affinity procedures. J. Biol. Chem. 261, 4328–4336.

Neethling, F.A., Joziasse, D., Bovin, N., Cooper, D.K., Oriol, R., 1996. The reducing end of α-Gal oligosaccharides contributes to their efficiency in blocking natural antibodies of human and baboon sera. Transpl. Int. 9, 98–101.

Oriol, R., Candelier, J.J., Taniguchi, S., Balanzino, L., Peters, L., Niekrasz, M., et al., 1999. Major carbohydrate epitopes in tissues of domestic and African wild animals of potential interest for xenotransplantation research. Xenotransplantation 6, 79–89.

Parker, W., Lin, S.S., Yu, P.B., Sood, A., Nakamura, Y.C., Song, A., et al., 1999. Naturally occurring anti-α-galactosyl antibodies: relationship to xenoreactive anti-α-galactosyl antibodies. Glycobiology 9, 865–873.

Peters, B.P., Goldstein, I.J., 1979. The use of fluorescein-conjugated *Bandeiraea simplicifolia* B4-isolectin as a histochemical reagent for the detection of α-D-galactopyranosyl groups. Their occurrence in basement membranes. Exp. Cell Res. 120, 321–334.

Phelps, C.J., Koike, C., Vaught, T.D., Boone, J., Wells, K.D., Chen, S.H., et al., 2003. Production of α1,3-galactosyltransferase-deficient pigs. Science 299, 411–414.

Posekany, K.J., Pittman, H.K., Bradfield, J.F., Haisch, C.E., Verbanac, K.M., 2002. Induction of cytolytic anti-Gal antibodies in α-1,3-galactosyltransferase gene knockout mice by oral inoculation with *Escherichia coli* O86:B7 bacteria. Infect. Immun. 70, 6215–6222.

Rispens, T., Derksen, N.I., Commins, S.P., Platts-Mills, T.A., Aalberse, R.C., 2013. IgE production to α-gal is accompanied by elevated levels of specific IgG1 antibodies and low amounts of IgE to blood group B. PLoS One 8, e55566.

Sandrin, M.S., Vaughan, H.A., Xing, P.X., McKenzie, I.F., 1997. Natural human anti-Gal α(1,3)Gal antibodies react with human mucin peptides. Glycoconj. J. 14, 97–105.

Santer, U.V., DeSantis, R., Hård, K.J., van Kuik, J.A., Vliegenthart, J.F., Won, B., et al., 1989. N-linked oligosaccharide changes with oncogenic transformation require sialylation of multiantennae. Eur. J. Biochem. 181, 249–260.

Smith, D.F., Larsen, R.D., Mattox, S., Lowe, J.B., Cummings, R.D., 1990. Transfer and expression of a murine UDP-Gal:β-D-Gal- α1,3-galactosyltransferase gene in transfected Chinese hamster ovary cells. Competition reactions between the α1,3-galactosyltransferase and the endogenous α2,3-sialyltransferase. J. Biol. Chem. 265, 6225–6234.

Speir, J.A., Abdel-Motal, U.M., Jondal, M., Wilson, I.A., 1999. Crystal structure of an MHC class I presented glycopeptide that generates carbohydrate-specific CTL. Immunity 10, 51–61.

Spiro, R.G., Bhoyroo, V.D., 1984. Occurrence of α-D-galactosyl residues in the thyroglobulin from several species. Localization in the saccharide chains of the complex carbohydrate units. J. Biol. Chem. 259, 9858–9866.

Springer, G.F., Horton, R.E., 1969. Blood group isoantibody stimulation in man by feeding blood group-active bacteria. J. Clin. Investig. 48, 1280–1291.

Stellner, K., Saito, H., Hakomori, S., 1973. Determination of aminosugar linkage in glycolipids by methylation. Aminosugar linkage of ceramide pentasaccharides of rabbit erythrocytes and of Forssman antigen. Arch. Biochem. Biophys. 133, 464–472.

Stone, K.R., Abdel-Motal, U.M., Walgenbach, A.W., Turek, T.J., Galili, U., 2007a. Replacement of human anterior cruciate ligaments with pig ligaments: a model for anti-non-gal antibody response in long-term xenotransplantation. Transplantation 83, 211–219.

Stone, K.R., Walgenbach, A.W., Turek, T.J., Somers, D.L., Wicomb, W., Galili, U., 2007b. Anterior cruciate ligament reconstruction with a porcine xenograft: a serologic, histologic, and biomechanical study in primates. Arthroscopy 23, 411–419.

Sung, S.J., Sweely, C.C., 1979. The structure of canine intestinal trihexosylceramide. Biochim. Biophys. Acta 525, 295–298.

Suzuki, E., Naiki, M., 1984. Heterophile antibodies to rabbit erythrocytes in human sera and identification of the antigen as a glycolipid. J. Biochem. 95, 103–108.

Takahagi, Y., Fujimura, T., Miyagawa, S., Nagashima, H., Shigehisa, T., Shirakura, R., et al., 2005. Production of α1,3-galactosyltransferase gene knockout pigs expressing both human decay-accelerating factor and *N*-acetylglucosaminyltransferase III. Mol. Reprod. Dev. 71, 331–338.

Tanemura, M., Yin, D., Chong, A.S., Galili, U., 2000. Differential immune response to α-gal epitopes on xenografts and allografts: implications for accommodation in xenotransplantation. J. Clin. Investig. 105, 301–310.

Taylor, S.G., McKenzie, I.F., Sandrin, M.S., 2003. Characterization of the rat α(1,3)galactosyltransferase: evidence for two independent genes encoding glycosyltransferases that synthesize Galα(1,3)Gal by two separate glycosylation pathways. Glycobiology 13, 327–337.

Tearle, R.G., Tange, M.J., Zannettino, Z.L., Katerelos, M., Shinkel, T.A., Van Denderen, B.J., et al., 1996. The α-1,3-galactosyltransferase knockout mouse. Implications for xenotransplantation. Transplantation 61, 13–19.

Teneberg, S., Lönnroth, I., Torres Lopez, J.F., Galili, U., Olwegard Halvarsson, M., Angstrom, J., et al., 1996. Molecular mimicry in the recognition of glycosphingolipids by Galα3Galß4GlcNAcß-binding *Clostridium difficile* toxin A, human natural anti-α-galactosyl IgG and the monoclonal antibody Gal-13: characterization of a binding-active human glycosphingolipid, non-identical with the animal receptor. Glycobiology 6, 599–609.

Teranishi, K., Mañez, R., Awwad, M., Cooper, D.K., 2002. Anti-Galα1-3Gal IgM and IgG antibody levels in sera of humans and Old World non-human primates. Xenotransplantation 9, 148–154.

Thall, A., Galili, U., 1990. The differential expression of Galα1→3Galß1→4GlcNAc-R residues on mammalian secreted *N*-glycosylated glycoproteins. Biochemistry 29, 3959–3965.

Thall, A., Etienne-Decerf, J., Winand, R.J., Galili, U., 1991. The α-galactosyl epitope on mammalian thyroid cells. Acta Endocrinol. (Cph.) 124, 692–699.

Thall, A.D., Maly, P., Lowe, J.B., 1995. Oocyte Galα1,3Gal epitopes implicated in sperm adhesion to the zona pellucida glycoprotein ZP3 are not required for fertilization in the mouse. J. Biol. Chem. 270, 21437–21440.

Thorn, J.J., Levery, S.B., Salyan, M.E.K., Stroud, M.R., Cedergren, B., Nilsson, B., et al., 1992. Structural characterization of x2 glycosphingolipid, its extended form, and its sialosyl derivatives: accumulation associated with the rare blood group p phenotype. Biochemistry 31, 6509–6517.

Tönder, O., Larsen, B., Aarskog, D., Haneberg, B., 1978. Natural and immune antibodies to rabbit erythrocyte antigens. Scand. J. Immunol. 7, 245–249.

Towbin, H., Rosenfelder, G., Wieslander, J., Avila, J.L., Rojas, M., Szarfman, A., et al., 1987. Circulating antibodies to mouse laminin in Chagas disease, American cutaneous leishmaniasis, and normal individuals recognize terminal galactosyl (α1-3)-galactose epitopes. J. Exp. Med. 166, 419–432.

Whalen, G.F., Sullivan, M., Piperdi, B., Wasseff, W., Galili, U., 2012. Cancer immunotherapy by intratumoral injection of α-gal glycolipids. Anticancer Res. 32, 3861–3868.

Wang, L., Anaraki, F., Henion, T.R., Galili, U., 1995a. Variations in activity of the human natural anti-Gal antibody in young and elderly populations. J. Gerontol. Med. Sci. 50A, M227–M233.

Wang, L., Radic, M.Z., Galili, U., 1995b. Human anti-Gal heavy chain genes: preferential use of V_H3 and the presence of somatic mutations. J. Immunol. 155, 1276–1285.

Wigglesworth, K., Racki, W.J., Mishra, R., Szomolany-Tsuda, E., Greiner, D.L., Galili, U., 2011. Rapid recruitment and activation of macrophages by anti-Gal/α-gal liposome interaction accelerates wound healing. J. Immunol. 186, 4422–4432.

Whitfield, C., Richards, J.C., Perry, M.B., Clarke, B.R., MacLean, L.L., 1991. Expression of two structurally distinct D-galactan O antigens in the lipopolysaccharide of *Klebsiella pneumoniae* serotype O1. J. Bacteriol. 173, 1420–1431.

Wood, C., Kabat, E.A., Murphy, L.A., Goldstein, I.J., 1979. Immunochemical studies of the combining sites of the two isolectins, A4 and B4, isolated from *Bandeiraea simplicifolia*. Arch. Biochem. Biophys. 198, 1–11.

Xu, Y., Yang, Y.G., Ohdan, H., Ryan, D., Harper, D., Wu, C., et al., 2006. Characterization of anti-Gal antibody-producing cells of baboons and humans. Transplantation 81, 940–948.

Yu, P.B., Parker, W., Everett, M.L., Fox, I.J., Platt, J.L., 1999. Immunochemical properties of anti-Galα1-3Gal antibodies after sensitization with xenogeneic tissues. J. Clin. Immunol. 19, 116–126.

Yu, P.B., Parker, W., Nayak, J.V., Platt, J.L., 2005. Sensitization with xenogeneic tissues alters the heavy chain repertoire of human anti-Galα1-3Gal antibodies. Transplantation 80, 102–109.

Zhan, J., Xia, Z., Xu, L., Yan, Z., Wang, K., 2003. A peptide mimetic of Gal-α1,3-Gal is able to block human natural antibodies. Biochem. Biophys. Res. Commun. 308, 19–22.

CHAPTER

2

Why Do We Produce Anti-Gal: Evolutionary Appearance of Anti-Gal in Old World Primates

OUTLINE

INTRODUCTION

Anti-Gal is produced in humans throughout life in large amounts, as ~1% of immunoglobulins (Galili et al., 1984), and it binds specifically a mammalian carbohydrate antigen called the α-gal epitope with the structure Galα1-3Galβ1-4GlcNAc-R (Galili et al., 1985). Moreover, ~1% of B lymphocytes in human blood are capable of producing this antibody (Galili et al., 1993). The dedication of such a significant proportion of B cell clones to the production of one antibody raises the question whether anti-Gal has a distinct physiologic role in humans. At present, there is no clear answer to this question. The removal of senescent red blood cells (RBC) seems to be

The Natural Anti-Gal Antibody as Foe Turned Friend in Medicine
http://dx.doi.org/10.1016/B978-0-12-813362-0.00002-6

associated with binding of anti-Gal to a cryptic antigen that is exposed on circulating RBC when they reach the age of ~120 days (Galili et al., 1986a) and in some pathologic RBC, at an earlier age of the cell (Galili et al., 1983, 1986b). However, this mechanism is not applicable to New World monkeys, lemurs, and all other nonprimate mammals. The geographic distribution of anti-Gal only in Old World monkeys, apes, and humans (collectively called *catarrhines* [nostrils pointing downwards]) raises the possibility that this antibody has protected against pathogens present only in the landmass of Eurasia and Africa (the "Old World") (Galili et al., 1987a,b, 1988a). New World monkeys (called *platyrrhines* ["flat" nose with nostrils pointing sideward]) and lemurs, all synthesize α-gal epitopes and lack the anti-Gal antibody (Galili et al., 1987a, 1988a). These primates have not been reported to display higher susceptibility to infections, in comparison with Old World monkeys and apes, when kept in zoos in Asia, Africa, or Europe. This suggests that anti-Gal is not required for current protection against any particular pathogen endemic to the Old World. It has been suggested that anti-Gal may protect humans against enveloped viruses originating in nonprimate mammals and presenting α-gal epitopes, by binding to these epitopes and inducing neutralization and destruction of the viruses presenting them (Repik et al., 1994; Rother et al., 1995; Takeuchi et al., 1996, 1997). However, it has not been proven as yet that anti-Gal has a current vital protective role. Nevertheless, one may assume that the striking distribution of anti-Gal only in Old World monkeys, apes, and humans versus the synthesis of α-gal epitopes in all other mammals are associated with a major selective event in the course of Old World primate (i.e., *catarrhines*) evolution.

As detailed in Chapter 1, the glycosylation enzyme synthesizing α-gal epitopes in mammals is α1,3galactosyltransferase (α1,3GT) that was originally found in cells of rabbit (Basu and Basu, 1973; Betterige and Watkins, 1983) and subsequently in mouse, cow, and New World monkey cells (Blake and Goldstein, 1981; Blanken and van den Eijnden, 1985; Galili et al., 1988a). Studies on the expression of the biosynthetic product of α1,3GT, i.e., the α-gal epitope, in various mammals further imply that α1,3GT is active in nonprimate mammalian cells, lemurs, and New World monkeys, and it is absent in Old World monkeys, apes, and humans (Galili et al., 1987a, 1988a; Oriol et al., 1999). Furthermore, synthesis of the α-gal epitope in both marsupial and placental mammals and its absence in other vertebrates implies that the α1,3GT gene (also called *GGTA1*) and the α-gal epitope synthesized by the enzyme encoded by this gene, appeared early in mammalian evolution before marsupial and placental lineages separated from a common ancestor (Fig. 1). Since its appearance >100 million years ago, the α-gal epitope has been continuously synthesized, and it is being synthesized in nonprimate mammals. Synthesis of α-gal epitopes by α1,3GT has been conserved also in lemurs that evolved in the island of Madagascar and in New World monkeys that evolved in the South American continent, both isolated from the Old World by oceanic barriers. In contrast, α-gal epitopes are not synthesized in Old World monkeys, apes, and humans, all of which lack α1,3GT activity (Fig. 1) and all have evolved in the Old World continents of Asia, Africa, and Europe (Galili et al., 1987a, 1988a). As further discussed below, these Old World primates have the α1,3GT gene as a mutated pseudogene (Larsen et al., 1990; Galili and Swanson, 1991; Koike et al., 2002; Lantéri et al., 2002). These observations suggest that ancestral Old World primates synthesized in the distant past the α-gal epitope, similar to nonprimate mammals, lemurs, and New World monkeys. However, at a certain evolutionary period, after the geographic separation between Old World primates and New World monkeys (estimated to occur ~30 million years ago [mya]) (Dawkins, 2004; Steiper and Young, 2006; Schrago, 2007),

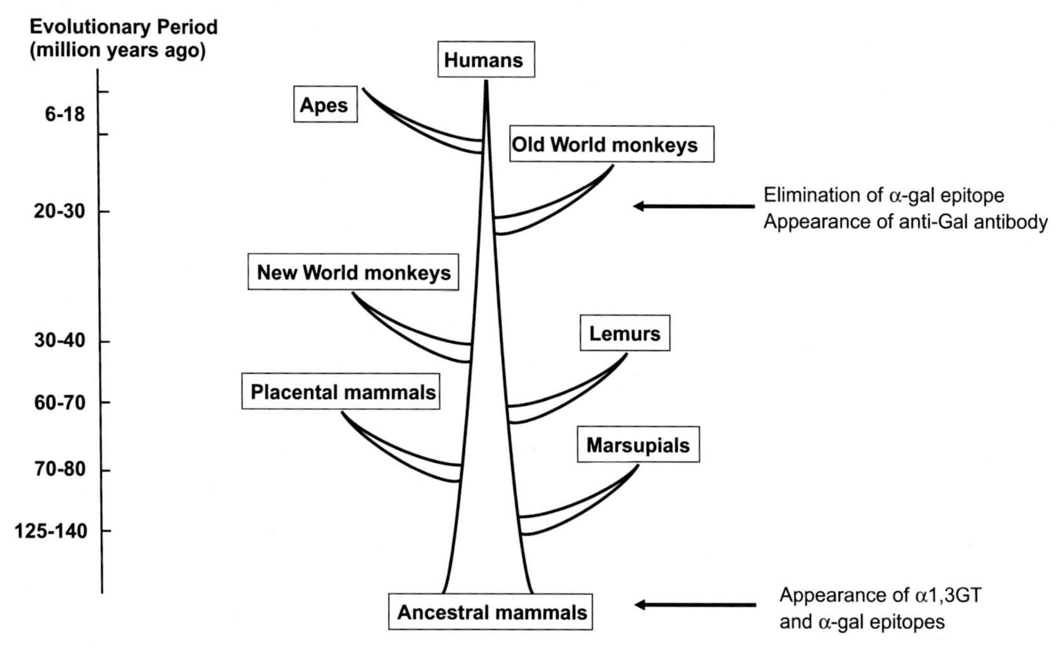

FIGURE 1 A schematic evolutionary tree describing the estimated evolutionary period in which α1,3galactosyltransferase and the α-gal epitope appeared in early mammals, and the period in which the selective pressure for elimination of primates synthesizing α-gal epitopes initiated (indicated by *arrows*). The estimated evolutionary periods for divergence events in mammals are indicated on the left. The absence of the α-gal epitope in vertebrates that are not mammals, and its synthesis in nonprimate mammals implies that α1,3GT and the α-gal epitope appeared in mammals prior to the split between marsupial and placental mammals. The absence of α1,3GT and α-gal epitopes only in Old World monkeys, apes, and humans implies that inactivation of the α1,3GT gene (*GGTA1*) and elimination of α-gal epitopes occurred after the split between New World monkeys and Old World primates. The estimates of the evolutionary periods of divergence in mammals are based on several studies (Dawkins, 2004; Schrago, 2007; Steiper and Young, 2006). *Adapted from Galili, U., 2016. Natural anti-carbohydrate antibodies contributing to evolutionary survival of primates in viral epidemics? Glycobiology 26, 1140–1150, with permission.*

an evolutionary selective process in primate populations resulted in extinction of Old World primates that synthesized α-gal epitopes. This extinction was followed by the expansion of monkey and ape populations with inactivated α1,3GT gene, which lacked α-gal epitopes, and thus could produce the natural anti-Gal antibody. This transition from α-gal epitope synthesis to elimination of primates synthesizing this carbohydrate antigen and the appearance of primates producing an antibody against the α-gal epitope is observed only in Old World primates. Such observation raises the possibility that this evolutionary event was associated with a selection process mediated by a detrimental pathogen that was endemic to the Old World. Although it is practically impossible to indentify pathogens that affected evolution of primates millions of years ago, this chapter describes several scenarios that are most likely to explain these evolutionary events in ancestral Old World primates. Understanding anti-Gal evolution requires a short discussion on production of the group of antibodies called "natural anti-carbohydrate antibodies" in response to antigenic stimulation by gastrointestinal (GI) bacteria.

NATURAL ANTI-CARBOHYDRATE ANTIBODIES TO BACTERIAL ANTIGENS

One of the major sources for constant antigenic stimulation of the human immune system is the multiple bacteria that naturally colonize the GI tract. There are at least 400 different strains of bacteria in the GI tract, and they comprise >25% of the fecal material (Stephen and Cummings, 1980; Gerritsen et al., 2011). These bacteria present a wide range of antigens that can stimulate the human immune system. The multiple different polysaccharides and oligosaccharides on these bacteria serve as a source for many carbohydrate antigens that continuously stimulate the immune system to produce a wide variety of anti-carbohydrate antibodies, without the need for active vaccination by the carbohydrate antigens, i.e., natural anti-carbohydrate antibodies (Wiener, 1951; Springer, 1971). Anti-Gal is one of these natural antibodies and it is produced in high amounts throughout life (Galili et al., 1984; Wang et al., 1995). Anti-Gal was shown to bind to several GI bacteria, as well as to their lipopolysaccharide extracts, including *Klebsiella pneumoniae*, *Serratia marcescens*, and *Escherichia coli O86* (Galili et al., 1988b). In earlier studies, feeding killed *E. coli O86* bacteria to patients with diarrhea was found to result in significant increase in the titer of anti-blood group B antibodies (Springer and Horton, 1969). As detailed in Chapter 3, >85% of anti-blood group B antibodies in humans are in fact anti-Gal antibodies that can also bind to blood group B antigen (Galili et al., 1987b). Accordingly, feeding α1,3galactosyltransferase knockout mice (GT-KO mice) with *E. coli O86* was found to induce production of the anti-Gal antibody in these mice (Posekany et al., 2002). Furthermore, production of anti-Gal in monkeys could be inhibited by administration of antibiotics that eliminate the GI bacterial flora (Mañez et al., 2001).

Although the natural anti-carbohydrate antibodies are primarily produced against bacterial carbohydrate antigens, some of these antibodies are capable of binding to mammalian carbohydrate antigens, as well (Blixt et al., 2004; Bovin et al., 2012; Bovin, 2013; Stowell et al., 2014). Accordingly, the natural anti-Gal antibody is produced against bacterial carbohydrate antigens with terminal galactosyl units linked in an alpha anomeric linkage and is capable of binding to the mammalian α-gal epitope (Galili et al., 1985; Towbin et al., 1987). Anti-Gal binds to various bacteria and bacterial lipopolysaccharides (Galili et al., 1988b); however, the exact structure of bacterial carbohydrates inducing anti-Gal production has not been identified, as yet. Galα1-3Glc and Galα1-3Gal epitopes were reported on both gram-positive and gram-negative bacteria (Han et al., 2012; Lüderitz et al., 1965). Additional examples of such antibodies in humans are anti-blood group A and B antibodies (Springer and Horton, 1969), natural antibody to *N*-glycolylneuraminic acid (called anti-Neu5Gc), which is produced in humans, and not in other Old World primates, or in nonprimate mammals (Higashi et al., 1977; Merrick et al., 1978; Zhu and Hurst, 2002; Padler-Karavani et al., 2008), and natural anti-rhamnose antibody (Chen et al., 2011; Sheridan et al., 2014; Long et al., 2014). As detailed in Chapter 3, the reason for the ability of anti-bacterial carbohydrate antibodies to bind mammalian carbohydrate antigens is that these antibodies are polyclonal, and different clones are capable of binding to various "facets" of a given carbohydrate antigen. Some of these facets are likely to be shared between mammalian and bacterial carbohydrate antigens. As discussed below, mammals produce anti-carbohydrate antibodies against many carbohydrate epitopes, provided that these epitopes are not self-antigens. Thus, if for any reason, a mammal stops synthesizing a certain carbohydrate epitope, there is high probability that it may start producing a natural antibody against that eliminated epitope, as part of the ongoing immune response against the many carbohydrate antigens on the bacteria of its natural GI flora.

Immune tolerance and anti-carbohydrate antibodies

The one factor that limits the diversity of anti-carbohydrate antibodies is the immune tolerance that prevents production of antibodies to self-carbohydrate antigens. Such production is prevented primarily by two mechanisms: (1) clonal deletion of immature B cell clones with B cell receptors, which can interact with self-antigens and (2) receptor editing in which the variable regions of immunoglobulin genes encoding antibodies to self-antigens are mutated so that the antibodies produced do not bind to self-antigens. The role of these mechanisms in prevention of anti-Gal production in animals synthesizing the α-gal epitopes as self-antigen was demonstrated in transgenic wild-type and GT-KO mice. Experimental studies in GT-KO mice indicated that the immune tolerance to the α-gal epitope is mediated by clonal deletion in which anti-Gal B cell clones, even at the stages of mature and memory B cells, are deleted following interaction of their B cell receptors with α-gal epitopes as self-antigen (Ogawa et al., 2003; Mohiuddin et al., 2003; Galili, 2004). Receptor editing mediating tolerance to α-gal epitopes was also observed in GT-KO mice in which an anti-Gal encoding gene was introduced (i.e., "knocked in") (Benatuil et al., 2008). These mice continuously produce anti-Gal without the need for their immunization. When such mice also acquired the active α1,3GT gene from wild-type mice, they ceased to produce anti-Gal because the variable regions of immunoglobulin genes encoding for anti-Gal B cell receptors were mutated at early stages of B cell development in the bone marrow. These mutations resulted in changes in the B cell receptor specificity, so it does not interact with α-gal epitopes (Benatuil et al., 2008). These immune tolerance mechanisms imply that once α-gal epitopes (and possibly other carbohydrate antigens) are eliminated because of inactivation of the gene encoding the corresponding glycosyltransferase, the immune tolerance mechanisms preventing production of antibodies against that self-antigen cease to function. Thus, the immune system is stimulated by bacteria of the GI flora to produce antibodies against the eliminated self-antigen. A present day example of a scenario in which a glycosyltransferase gene is inactivated in small human populations, and the resulting production of a natural antibody against the eliminated carbohydrate antigen is the blood group "Bombay" individuals, discussed at the end of this chapter. These rare individuals lack the blood group H (O) antigen and produce natural antibodies against this antigen.

A specific present day example for *de novo* production of the natural anti-Gal antibody following the elimination of α-gal epitopes was observed in recent years in α1,3GT knockout pigs (GT-KO pigs). As discussed in Chapter 1, these pigs were generated by disruption ("knockout") of the α1,3GT gene, with the aim of providing pig xenograft organs and tissues that lack α-gal epitopes (Lai et al., 2002; Phelps et al., 2003; Kolber-Simonds et al., 2004; Takahagi et al., 2005). Wild-type pigs present multiple α-gal epitopes as self-antigen on their cells, and thus, immune tolerance mechanisms prevent production of anti-Gal antibodies in them. However, once the α-gal epitope is eliminated by "knockout" of the α1,3GT gene in the GT-KO pig genome, these pigs naturally produce anti-Gal in high titers against GI bacteria, already at the age of 1.5–2 months (Dor et al., 2004; Fang et al., 2012; Galili, 2013). As discussed below, a similar production of anti-Gal in primates, in which the α1,3GT gene was inactivated by mutations, might have prevented the extinction of Old World primates that were exposed to highly detrimental enveloped viruses or other pathogens expressing α-gal epitopes.

POSSIBLE EVOLUTIONARY APPEARANCE OF ANTI-GAL PRODUCING PRIMATES FOLLOWING VIRAL EPIDEMICS

The selective process that eliminated α-gal epitopes from ancestral Old World monkeys and apes (Old World primates) and led to the appearance of anti-Gal producing primates occurred after Old World primates and New World monkeys diverged from a common ancestor. Studies on the mutations inactivating the α1,3GT gene suggest that this selective process initiated 20–30 mya (see below). The lack of α-gal epitopes and production of the anti-Gal antibody are uniformly observed in monkeys and apes, which evolved in all regions of the Old World (i.e., the geographic area of Eurasia-Africa). The occurrence of this selective process throughout the vast regions of Eurasia-Africa suggests that it was mediated by a highly detrimental pathogen, such as enveloped virus (Galili, 2016). Influenza virus is one current example of a virus, which potentially can become highly virulent, causing lethal infections and can effectively spread throughout human populations. Intercontinental transportation can further enable its spread over geographic barriers. Because enveloped viruses lack their own glycosylation machinery, they share the carbohydrate antigens on their envelope glycoproteins with the host cell. Many of the glycosyltransferases within the host cells reside in the Golgi apparatus. They synthesize the carbohydrate chains on cellular and viral glycoproteins in a manner similar to assembly lines in a car plant, in that there is a sequential buildup of the nascent carbohydrate chain at various compartments of the Golgi apparatus. Therefore, enveloped viruses infecting primates that synthesize α-gal epitopes are likely to have these epitopes on their envelope glycoproteins.

Hypothesis on virus-mediated selection for elimination of α-gal epitopes in primates

The proposed scenario for elimination of α-gal epitopes in Old World primates and the resulting appearance of the natural anti-Gal antibody is based on the assumption that very rare mutation event(s) occurred accidentally and randomly in one or more of ancestral Old World primate species. Such a mutation could be single base frameshift deletion resulting in a premature stop codon, which completely inactivated α1,3GT catalytic activity. Accordingly, a three amino acid deletion at the C-terminus of New World monkey α1,3GT was found to result in complete loss of catalytic activity of the enzyme (Henion et al., 1994). It is likely that offspring carrying such a mutation for several generations after it occurred were heterozygotes. They produced intact α1,3GT by the unmutated allele and synthesized α-gal epitopes (see the description of these mutations below). However, the mating of such heterozygotes resulted in homozygous offspring primates that carried two alleles of the inactivated α1,3GT gene, and therefore they lacked α-gal epitopes. As the α-gal epitope in these homozygotes became a nonself antigen, they naturally produced the anti-Gal antibody in response to the constant antigenic stimulation by carbohydrate antigens with structures similar to that of the α-gal epitope, presented on GI bacteria. These anti-Gal producing primates could evolve because the α-gal epitope turned out to be a nonessential carbohydrate epitope, similar to the observations with GT-KO pigs (Phelps et al., 2003; Kolber-Simonds et al., 2004). A scenario similar to the hypothetical one described above is presently observed in individuals of the rare blood group "Bombay" who lack the enzyme producing blood group O (H) and who naturally produce anti-H (blood group O) antibodies (Bhende et al., 1952; Watkins, 1980; Le Pendu et al., 1986; Balgir, 2005, 2007). The similarities between ancestral primate populations including small numbers of individuals lacking α-gal epitopes, prior to extinction of populations synthesizing α-gal epitopes, and present day human populations including small numbers of individuals lacking the ability to produce blood group O (i.e., blood group Bombay individuals) are further discussed at the end of this chapter.

As proposed in Fig. 2, early ancestral Old World primates synthesized α-gal epitopes similar to New World monkeys. These Old World primates could become extinct in epidemics of highly virulent enveloped viruses because they succumbed to the infections before these primates could mount a protective immune response against the infecting virus. The viruses mediating such epidemics carried α-gal epitopes on their envelope glycoproteins,

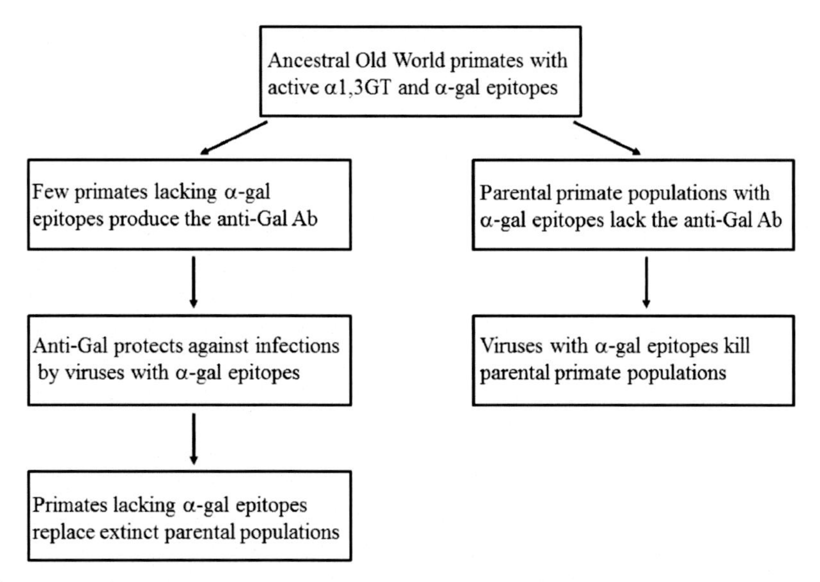

FIGURE 2 Proposed stages in the evolutionary selective process that resulted in elimination of ancestral Old World primates synthesizing α-gal epitopes and their replacement with offspring-lacking the α-gal epitope and producing the natural anti-Gal antibody. Few individuals in early Old World primate populations, who carried mutations inactivating the α1,3GT gene, produced the natural anti-Gal antibody. This antibody production is analogous to the present day rare blood type "Bombay" individuals lacking blood group O (H antigen) and producing anti-H antibodies. Epidemics by enveloped viruses presenting α-gal epitopes that were synthesized by α1,3GT of ancestral Old World primates caused the extinction of these primates, whereas offspring-lacking α-gal epitopes were protected by the natural anti-Gal antibody they produced. These offspring ultimately replaced the extinct primates that conserved active α1,3GT. Ab-antibody. *Reprinted from Galili, U., 2016. Natural anti-carbohydrate antibodies contributing to evolutionary survival of primates in viral epidemics? Glycobiology 26, 1140–1150, with permission.*

which were synthesized by α1,3GT in the infected host cells. However, the very few primates that were homozygous for the inactivated α1,3GT gene, lacked active α1,3GT enzyme, did not synthesize α-gal epitopes and produced the natural anti-Gal antibody. These few primates could have been protected by this antibody against viruses expressing α-gal epitopes. Anti-Gal protection could be mediated by several mechanisms, including: (1) neutralization and destruction of the virus by anti-Gal binding to the virus α-gal epitopes and activating the complement system, which induced complement-mediated lysis of the virus, (2) opsonization of the virus by anti-Gal could induce effective uptake and destruction of the virus by macrophages following Fc/Fcγ receptor interaction between the opsonizing anti-Gal and these cells, and (3) extensive uptake of anti-Gal opsonized viruses by macrophages and dendritic cells via Fc/Fcγ receptor interaction could result in rapid processing and presentation of immunogenic viral peptides by these antigen presenting cells (APC) that effectively transport the virus antigens to regional lymph nodes. This APC-mediated mechanism would have resulted in induction of rapid, potent humoral and cellular protective anti-virus immune responses. Thus, such an immune response could also protect against infecting viruses that "lost" their α-gal epitopes because of the initial infection of the host cells lacking α1,3GT. The ability of anti-Gal to markedly increase immunogenicity of vaccinating viruses by targeting them for effective uptake by APC is further detailed in Chapter 9 that describes the amplification of virus vaccine immunogenicity by α-gal epitopes linked to vaccinating viral glycoproteins (Abdel-Motal et al., 2006, 2007, 2010). Overall, the combined effects of the anti-Gal-mediated protective mechanisms could result in decrease in initial infecting virus burden, T cell-mediated destruction of cells infected by the virus, as well as destruction and neutralization of virus *de novo* produced in infected cells by elicited antibodies specific to virus protein antigens. The outcome of these protective mechanisms could be prevention of infective virus progression before it reaches lethal stages.

Anti-Gal IgG crosses the placenta into the fetal blood in humans (Galili et al., 1984). Anti-Gal is also present in colostrum and milk, as well as in other body secretions, primarily as the IgA isotype (class) (Hamadeh et al., 1995). Thus, it is possible that anti-Gal-mediated protection against an infectious virus that presents α-gal epitopes also occurred in newborns. In the absence of competition from parental primate populations synthesizing α-gal epitopes, the small populations of offspring-lacking α-gal epitopes and producing the natural anti-Gal antibody replaced the extinct parental Old World primate populations that conserved active α1,3GT. It should be stressed that the proposed scenario could occur with any type of enveloped virus that presented α-gal epitopes, which was endemic to the Eurasia-Africa landmass because any enveloped virus propagated in cells containing active α1,3GT is likely to present α-gal epitopes. The process of selective evolutionary elimination of α-gal epitopes, which occurred in ancestral Old World primates, may not be feasible in all mammalian species synthesizing α-gal epitopes. One example is GT-KO mice. These mice develop cataract at the age of 6–9 weeks in the absence of α-gal epitopes (Thall, 1999; Sørensen et al., 2008). Although GT-KO mice developing such cataract can survive in the protected environment of animal facilities, their survival would have been questionable in natural environments. In contrast, GT-KO pigs were not reported to develop the cataract observed in mice in the absence of α-gal epitopes.

Observations that support the hypothesis on evolution associated with viral epidemics

As indicated above, it is impossible to identify a pathogen(s) that exerted the selective pressure for evolution of primates lacking α-gal epitopes and producing the anti-Gal antibody in earlier geological periods. However, there are several observations supporting the hypothesis proposed in Fig. 2.

Anti-Gal interaction with viruses carrying α-gal epitopes

Glycoproteins are an integral part of virus envelopes. The carbohydrate chains on such glycoproteins contribute to the formation of a hydration layer that protects the virus. Carbohydrate chains of the complex type of glycoproteins (Fig. 2A in Chapter 1) are synthesized on aspargines (-*N*-) that are part of the amino acid sequence -*N*-X-S/T- in proteins. Because these carbohydrate chains are synthesized by the host cell glycosylation machinery, viruses propagated in cells containing α1,3GT usually present multiple α-gal epitopes. Thus, propagation of Eastern Equine Encephalitis virus in mouse cells resulted in production of virions carrying α-gal epitopes, whereas propagation of this virus in African Green monkey Vero cells (lacking active α1,3GT) resulted in production of virions with envelope glycoproteins lacking α-gal epitopes (Repik et al., 1994). Accordingly, influenza virus propagated in embryonated chicken eggs lacks α-gal epitopes because birds, as other nonmammalian vertebrates, lack α1,3GT. In contrast, propagation of influenza virus in bovine MDBK cells or canine MDCK cells resulted in production of virions with the envelope glycoprotein hemagglutinin carrying several α-gal epitopes per molecule (Galili et al., 1996).

α-Gal epitopes were also demonstrated on other viruses propagated in nonprimate mammalian cells, including: Friend murine leukemia virus (Geyer et al., 1984), murine Molony leukemia virus (Rother et al., 1995), porcine endogenous retrovirus (PERV) (Takeuchi et al., 1996), lymphocytic choriomeningitis virus, Newcastle disease virus, Sindbis virus, vesicular stomatitis virus (Welsh et al., 1998), and measles virus (Preece et al., 2002; Dürrbach et al., 2007). Several of these studies further showed that incubation of the viruses expressing α-gal epitopes in human serum or with purified anti-Gal antibody further resulted in anti-Gal-mediated neutralization and complement-mediated lysis of the viruses, whereas no such effects were observed in viruses lacking α-gal epitopes. As suggested in Fig. 2, it may be possible that a similar protective effect was mediated by anti-Gal in the few individuals among Old World primates that had mutations inactivating the α1,3GT gene. In contrast, populations conserving α1,3GT activity produced virions presenting α-gal epitopes and were killed by such viruses in the absence of anti-Gal. The observed anti-Gal-mediated destruction and neutralization of viruses carrying α-gal epitopes further suggested that this antibody may contribute to prevention of cross-species viral transmission from nonprimate mammals to humans (Repik et al., 1994; Rother et al., 1995; Takeuchi et al., 1996; Welsh et al., 1998; Preece et al., 2002).

Increased protective immune response by anti-Gal targeting of viruses to antigen presenting cells

As detailed in Chapter 9, anti-Gal-mediated targeting to APC of inactivated influenza virus engineered to present α-gal epitopes was found to increase anti-virus antibody response in GT-KO mice by ~100-fold, in comparison with mice immunized with inactivated influenza virus lacking α-gal epitopes (Abdel-Motal et al., 2007). Intranasal challenge of the immunized mice with a lethal dose of live influenza virus lacking α-gal epitopes resulted in death of 90% of mice immunized with virus lacking α-gal epitopes, whereas only 10% of mice immunized with virus presenting α-gal epitopes died after such challenge (Abdel-Motal et al., 2007). Similarly, anti-Gal-producing GT-KO mice immunized with gp120 of HIV carrying α-gal epitopes resulted in ~100-fold higher anti-gp120 antibody response, ~30-fold higher T cell response, and ~40-fold increase in *in vitro* HIV neutralization activity in comparison to the immune responses measured in mice immunized with gp120 lacking α-gal epitopes (Abdel-Motal et al., 2006). A similar increase in anti-virus CD8+ cytotoxic T cell response was reported in anti-Gal-producing GT-KO mice that were immunized with a mouse cell line expressing murine leukemia virus proteins and α-gal epitopes, in comparison with CD8+ T cell response in wild-type mice (i.e., mice lacking the anti-Gal antibody) and undergoing similar immunization (Benatuil et al., 2005). All these studies support the assumption that ancestral Old World primates lacking α-gal epitopes and producing the anti-Gal antibody could enhance the immune response against proteins of infecting viruses presenting α-gal epitopes, by anti-Gal-mediated targeting of the virus to APC. The enhanced immune response might have been potent enough to prevent progression of the infection to lethal stages even when the virions lacked α-gal epitopes because of growth in cells lacking active α1,3GT in anti-Gal producing hosts.

Enveloped viruses appearing after New World monkeys/Old World primates split

The hypothesis on the role of enveloped viruses in mediating the selective pressure for evolution of primates lacking α1,3GT and producing anti-Gal, includes the assumption that such viruses appeared in the Old World only after the split from New World monkeys, i.e., New World monkeys were geographically isolated in the South American continent and thus were not affected by these viruses. This assumption is supported by observations of an enveloped virus, Epstein Barr virus (EBV), which is of the Herpes virus family, and it is thought to have appeared among Old World primates after the geographic separation from New World monkeys. Therefore, this virus has influenced immune system evolution of only Old World primates. When EBV infects Old World primates, it immortalizes a proportion of their B cells. However, the immune system in Old World primates evolved to mount an extensive T cell response against EBV antigens, which in humans results in the transient infectious mononucleosis disease (Klein and Masucci, 1982; Callan, 2003). The proliferating T cells kill the majority of B cells infected by the virus and immortalized. Moreover, EBV-immortalized B cells residing in immunologic sanctuaries are destroyed upon detection by T cells if they leave such sanctuaries. As many as 90% of humans are infected by EBV. However, because of the effective T cell response against EBV infected B cells; these B cells are prevented from spreading throughout the body and from progressing into becoming lymphoma cells. In contrast, no significantly effective anti-EBV protective T cell response is observed in New World monkeys infected by EBV because the immune system in these primates was not evolutionarily exposed to infections by this virus. Therefore, many of the EBV-immortalized B cells in New World monkeys are not destroyed and progress into lethal polyconal B cell lymphomas (Epstein et al., 1973; Shope et al., 1973; Wang, 2013). Elimination of α-gal epitopes and production of the natural anti-Gal antibody may represent an analogous selective pressure mediated by a virus endemic to the Old World land mass, whereas New World monkeys evolving in South America or lemurs evolving in Madagascar have not been subjected to evolutionary selective processes by such hypothetical viruses because of geographic isolation from the Old World.

Evolutionary almost complete extinction of apes according to their fossil record

The occurrence of mutations that result in elimination of major cell surface antigens, such as the α-gal epitope and the appearance of a natural antibody against it, are very rare events in evolution. There is only one other similar event known in the evolution of Old World primates, the elimination of the sialic acid *N*-glycolylneuraminic acid (Neu5Gc) in hominins (ancestors of humans) and production of natural anti-Neu5Gc antibodies that are found only in humans (Zhu and Hurst, 2002; Padler-Karavani et al., 2008). The rest of Old World primates and nonprimate mammals synthesize Neu5Gc and lack anti-Neu5Gc antibodies (Varki, 2010). These selective processes were likely to be associated with extinction of the parental primate populations conserving the carbohydrate antigen and thus, lacking the natural antibody against it. Although it is practically impossible to associate between the fossil record from previous geological periods and biochemical/immunological changes in primate populations, there is an interesting parallelism between the suggested hypothesis on extinction of apes that conserved α-gal epitopes and the fossil record of apes. Apes were a very successful group of primates in the early Miocene (~20–23 mya) as implied from the multiple fossils of many ape species (*hominoidea*) dating to that period, which were found in Eurasia-Africa. However, the number and diversity of ape fossils from the middle Miocene (~11–16 mya) greatly declines. No fossils of apes from the late Miocene (~5–10 mya) have been found, suggesting an almost complete extinction of apes at that period (Andrews, 1992; Andrews et al., 1996; Merceron et al., 2010; Alba, 2012). These changes in ape populations have been associated with dietary adaptations because of climatic changes (Andrews and Martin, 1991; Agustí et al., 2003; Ungar and Kay, 1995). An alternative cause for this almost complete extinction of ancestral apes could be associated with the selective pressure for the evolution of apes lacking α-gal epitopes and producing the anti-Gal antibody (Galili and Andrews, 1995), possibly mediated by epidemics of viruses carrying α-gal epitopes, as suggested above. The slow decline in ape populations during the middle Miocene, toward their almost complete extinction in the late Miocene, may further suggest that the extinction of primates by viral epidemics and expansion of subpopulations lacking α-gal epitopes throughout Eurasia-Africa could have been slow processes taking millions of years. The slow pace of these changes may have been the result of the great geographical distances between various ape populations. The fossil record of Old World monkeys dating to those periods is sparse, and thus, it is difficult to determine the population changes in this group of primates during the Miocene (Miller et al., 2009).

ALTERNATIVE CAUSES FOR EVOLUTIONARY INACTIVATION OF α1,3GT IN ANCESTRAL OLD WORLD PRIMATES

The efficacy of anti-Gal in protecting against infections with viruses presenting α-gal epitopes may vary for different enveloped viruses. One example for insufficient protective activity is that of influenza virus. As indicated above, when this virus is grown is cells that have active α1,3GT (e.g., bovine MDBK cells or canine MDCK cells), the virus carries α-gal epitopes on its hemagglutinin envelope protein (Galili et al., 1996). Thus, it is likely to carry α-gal epitopes also when produced in porcine cells. Indeed PERV grown in porcine cells can be destroyed by anti-Gal in human serum (Takeuchi et al., 1996). Nevertheless, humans can be infected by influenza virus produced in pigs. Once few virions succeed in penetrating into human cells of the respiratory tract, they proliferate and carry carbohydrate chains produced by the human glycosylation machinery, i.e., chains lacking α-gal epitopes. Thus, additional scenarios for the evolutionary processes that could result in extinction of ancestral Old World primates presenting α-gal epitopes should be considered, as well. Three of these scenarios are as follows:

1. *Detrimental bacteria expressing α-gal-like epitopes*—Several bacterial strains bind the anti-Gal antibody (Galili et al., 1988b), provide antigens that elicit production of anti-Gal in humans (Almeida et al., 1991) and in GT-KO mice (Posekany et al., 2002), and display carbohydrate antigens with terminal α-galactosyls in both gram-negative and gram-positive bacteria (Lüderitz et al., 1965; Han et al., 2012). It could be hypothesized that bacterial strains that were lethal to Old World primates, which expressed antigens that elicit anti-Gal production, could generate a selective pressure for survival of primates that produced this antibody as a protective antibody, i.e., selection for individuals with inactivated α1,3GT gene.
2. *Bacterial toxins or viruses binding the α-gal epitope*—An alternative hypothesis involving bacteria could be that the lethal effects of the infecting bacteria were mediated by binding of their toxins to α-gal epitopes on host cells. A current example is enterotoxin A of *Clostridium difficile* that causes severe diarrhea. This toxin can bind to various carbohydrate receptors; however, its primary receptor on nonprimate mammalian cells is the α-gal epitope (Pothoulakis et al., 1996; Teneberg et al., 1996). It may be possible that epidemics among Old World primates by bacteria producing lethal toxin(s) that used α-gal epitopes as receptor on target cells, exerted a selective pressure for survival of individuals that lacked the α-gal epitopes and thus, were not affected by the toxin. As discussed above, once the α-gal epitope was eliminated, the immune tolerance to this antigen was lost, resulting in production of the natural anti-Gal antibody.
   A similar selective process may be envisaged if a detrimental virus "used" the α-gal epitope as a "docking receptor." Influenza virus uses sialic acid on cells as a docking receptor that enables it to attach to cell membranes and penetrate into cells. If there was a virus that used the α-gal epitope as such a receptor, it could drive the selection of primates to survival only of those lacking α-gal epitopes along a pathway similar to that described in Fig. 2, without the involvement of antibodies in the selective process. However, anti-Gal production would have been a by-product resulting from the loss of the α-gal epitope and of the immune tolerance to it. A current example for such a virus is bovine norovirus that was reported to use α-gal epitopes as a docking receptor for infecting bovine cells (Zakhour et al., 2009). In addition, Sindbis virus was found to preferentially infect cells that present α-gal epitopes and wild-type suckling mice synthesizing this epitope, in comparison with cells or suckling mice lacking α-gal epitopes (Rodriguez and Welsh, 2013).
3. *Detrimental protozoa that express α-gal or α-gal-like epitopes*—As discussed in detail in Chapter 4, several protozoa, which are parasitic in humans, were found to present cell surface carbohydrate epitopes with structures similar to the α-gal epitope. These include *Trypanosoma* (Ramasamy and Field, 2012; Milani and Travassos, 1988; Almeida et al., 1994), *Leishmania* (Avila et al., 1989; McConville et al., 1990; Ilg et al., 1992), and *Plasmodia*

(Ramasamy and Reese, 1986; Yilmaz et al., 2014). As argued above for bacteria, such protozoa pathogens could mediate the selective pressure for survival of individuals in which α-gal epitopes were eliminated and the natural anti-Gal antibody produced (Ramasamy and Rajakaruna, 1997; Yilmaz et al., 2014). These antibodies could serve as protective antibodies against infections by protozoa presenting anti-Gal-binding epitopes. Indeed, anti-Gal binding to *Trypanosoma cruzi* was shown to induce complement-mediated cytolysis of the parasite (Milani and Travassos, 1988; Almeida et al., 1991, 1994) as well as direct, complement-independent lysis (Gazzinelli et al., 1991).

Although bacteria and protozoa epidemics cannot be excluded as evolutionary causes for selection of Old World primates with inactivated α1,3GT gene, the likelihood of these scenarios may be lower than the scenario of enveloped viruses mediating such a selective pressure. The complete elimination of Old World monkeys and apes producing α-gal epitopes in all climatic regions of Eurasia-Africa suggests that the pathogen(s) had to have very high infectivity and may have not depended on secondary transmitting vectors (e.g., insects active only in certain climates). Such are characteristics of viruses that spread directly from one infected individual primate to the other, regardless the large variety of climatic environments.

MOLECULAR BASIS FOR THE EVOLUTIONARY INACTIVATION OF THE α1,3GT GENE

The elimination of the α-gal epitope in ancestral Old World primates was the result of mutations that inactivated the α1,3GT gene (*GGTA1*) in few individuals, and subsequently in small populations of ancestral primates. α1,3GT activity was not essential in ancestral primates who were homozygous for the inactivated α1,3GT gene (Galili et al., 1987a, 1988a). These observations raised the question of the mechanism that inactivated α1,3GT gene in Old World primates. This question could be addressed following the cloning of the α1,3GT gene in mouse and bovine cells (Larsen et al., 1989; Joziasse et al., 1989). The gene was found to be composed of ~1110 base pairs divided into nine exons, of which exon IX (687 bp) is the largest.

Comparison of DNA and derived protein sequences of exon IX in mouse and bovine α1,3GT and a cloned homologous human genomic sequences, indicated that in the human DNA sequence, there are two frameshift mutations caused by single base deletions, corresponding to base 822 and base 904 of the mouse α1,3GT cDNA (Fig. 3) (Larsen et al., 1989, 1990; Joziasse et al., 1989; Lantéri et al., 2002). These mutations create premature stop codons, truncating the α1,3GT enzyme by 110 and 15 amino acids at the C-terminus, respectively. Controlled truncation of a New World monkey α1,3GT cDNA indicated that elimination of as few as the last three amino acids at the C-terminus of the enzyme was sufficient to cause complete loss of catalytic activity of α1,3GT (Henion et al., 1994). This implies that α1,3GT gene in humans is a pseudogene incapable of producing an active enzyme. Sequencing of the homologous DNA region in apes revealed that orangutan and gorilla have an α1,3GT pseudogene containing only one of the two deletions, at base 904, whereas chimpanzee has both deletions, similar to humans (Fig. 3) (Galili and Swanson, 1991). The absence of any of these two deletions in Old World monkey α1,3GT pseudogenes (Rhesus, African green, and Patas monkeys in Fig. 3) suggested that the deletions appeared in apes after they and Old World monkeys diverged from a common ancestor, i.e., less than 28 mya. However, a third

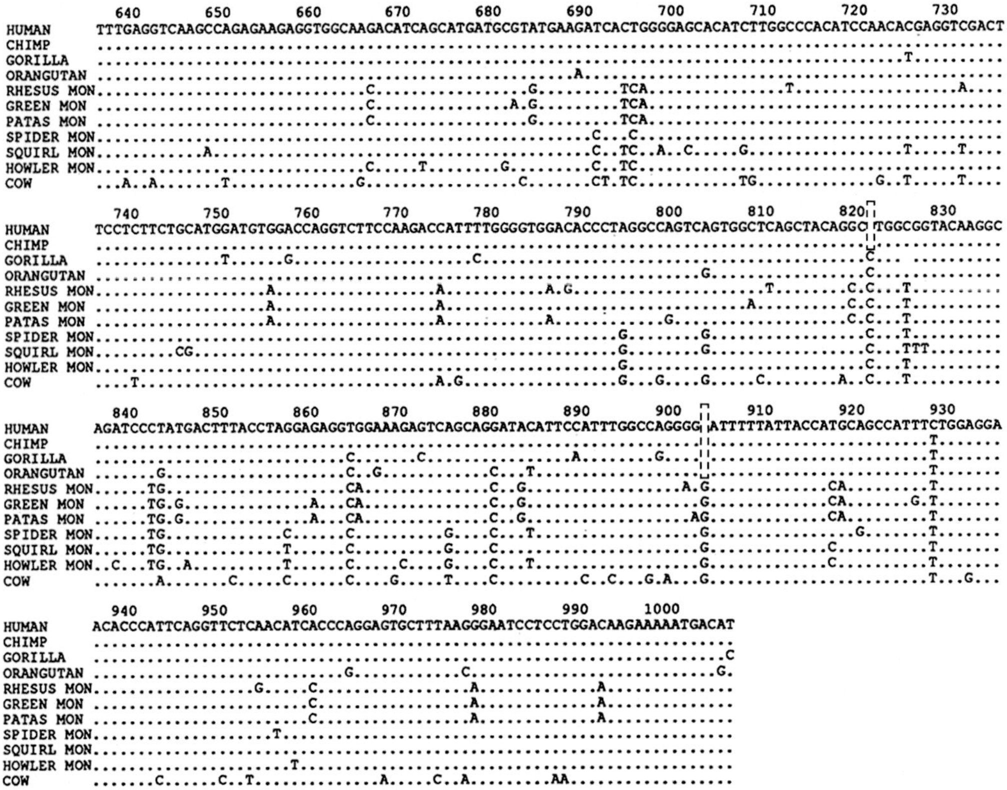

```
            640       650       660       670       680       690       700       710       720       730
HUMAN     TTTGAGGTCAAGCCAGAGAAGAGGTGGCAAGACATCAGCATGATGCGTATGAAGATCACTGGGGAGCACATCTTGGCCCACATCCAACACGAGGTCGACT
CHIMP     ....................................................................................................
GORILLA   .........................................................................................T..........
ORANGUTAN .....................................................A..............................................
RHESUS MON..............................C.................G.........TCA...............T..................A....
GREEN MON ..............................C...............A.G.........TCA.......................................
PATAS MON ..............................C.................G.........TCA.......................................
SPIDER MON.......................................................C...C........................................
SQUIRL MON............A..........................................C..TC..A..C.....G..................T.....T....
HOWLER MON..............................C....T........G..........C..TC.......................................
COW       ...A..A.......T..............G.................C.......CT.TC...........TG.............G..T.....T....

            740       750       760       770       780       790       800       810       820       830
HUMAN     TCCTCTTCTGCATGGATGTGGACCAGGTCTTCCAAGACCATTTTGGGGTGGACACCCTAGGCCAGTCAGTGGCTCAGCTACAGGC TGGCGGTACAAGGC
CHIMP     ..................................................................................... ..............
GORILLA   ..............T......G....................C..........................................C..............
ORANGUTAN ...................................................................G.................C..............
RHESUS MON...................A..................A...........A.G.....................T........C.C...T..........
GREEN MON ...................A..................A.................................A..........C.C...T..........
PATAS MON ...................A..................A...........A............G...................C.C...T..........
SPIDER MON..........................................................G........G.................C...T..........
SQUIRL MON.........CG...............................................G........G.................C...TTT........
HOWLER MON..........................................................G..........................C...T..........
COW       ....T.................................A.G.................G...G....G.....C........A..C...T..........

            840       850       860       870       880       890       900        910       920       930
HUMAN     AGATCCCTATGACTTTACCTAGGAGAGGTGGAAAGAGTCAGCAGGATACATTCCATTTGGCCAGGGG ATTTTATTACCATGCAGCCATTTCTGGAGGA
CHIMP     ................................................................... .......................T.......
GORILLA   ............................C.......C................A........G.... .......................T.......
ORANGUTAN .......G....................C..G............C...T.................. .......................T.......
RHESUS MON......TG....................CA..............C..G.................A.G..............CA........T.......
GREEN MON ......TG.G..............A...CA..............C..G...................G..............CA......G.T.......
PATAS MON ......TG.G..............A...CA..............C..G..................AG..............CA........T.......
SPIDER MON......TG.............C......C..........G....C...T..................G................G......T.......
SQUIRL MON......TG.............T......C..........G....C.......................G.............C.........T.......
HOWLER MON..C...TG..A..........T......C.....C....G....C...T..................G.............C.........T.......
COW       .......A.......C.....C......C....G.....T....C.........C..C...G.A...G.......................T...G...

            940       950       960       970       980       990      1000
HUMAN     ACACCCATTCAGGTTCTCAACATCACCCAGGAGTGCTTTAAGGGAATCCTCCTGGACAAGAAAAATGACAT
CHIMP     .......................................................................
GORILLA   ......................................................................C
ORANGUTAN ............................G............C...........................G.
RHESUS MON..................G.....C.................A.............A..............
GREEN MON ........................C.................A.............A..............
PATAS MON ........................C.................A.............A..............
SPIDER MON....................T..................................................
SQUIRL MON.......................................................................
HOWLER MON......................T................................................
COW       .......C......C..T..............A.....C..A.........AA..................
```

FIGURE 3 Aligned DNA sequences of a 370-bp region in exon IX of the αl,3GT pseudogene from humans (Larsen et al., 1990), apes including: chimpanzee, gorilla and orangutan, and Old World monkeys including: Rhesus monkey, African green monkey, and Patas monkey. These sequences are aligned with the active αl,3GT gene in New World monkeys including: Spider monkey, Squirrel monkey, and Howler monkey, and with domestic cow (described by Joziasse et al., 1989). The base numbers in this figure are according to the open reading frame of the mouse αl,3GT cDNA described by Larsen et al. (1989). The numbered base is under the second digit. Dots represent sequences identical to those of the human αl,3GT pseudogene. Note the two deletions C822 and G904 in humans and chimpanzees and G904 in humans and apes but not in other primates. *Reprinted from Galili, U., Swanson, K., 1991. Gene sequences suggest inactivation of α1-3 galactosyltransferase in catarrhines after the divergence of apes from monkeys. Proc. Natl. Acad. Sci. U.S.A. 88, 7401–7404, with permission.*

deletion was found in exon VII in rhesus monkey, in orangutan, and in humans (Koike et al., 2002). This mutation in an Old World monkey, ape, and humans suggested that the inactivation of the gene occurred before divergence of Old World monkeys and apes from a common ancestor (Koike et al., 2002). Based on these studies, it is not clear at present whether the selective process for extinction of Old World primates synthesizing the α-gal epitope and the emergence of primates lacking this epitope and producing anti-Gal, initiated before or after the split between apes and monkeys of the Old World, and thus, it is considered to initiate 20–30 mya (Fig. 1).

BLOOD GROUP BOMBAY AS A PRESENT DAY EXAMPLE FOR A RARE GLYCOSYLTRANSFERASE INACTIVATION IN HUMANS

The basic assumption in regard to the evolutionary elimination of ancestral Old World primates producing α-gal epitopes is that prior to this elimination there have been few primates that were homozygous for mutations that inactivated the α1,3GT gene (*GGTA1*) and thus produced the natural anti-Gal antibody. This assumption is supported by a similar present day example of a very rare mutation in humans, which inactivates the α1,2 fucosyltransferase gene (α1,2FT also called *FUT1*). Individuals homozygous for this mutation belong to blood group "Bombay" and are characterized by inability to synthesize the H antigen (Fucα1-2Gal-R that is blood group O) (Bhende et al., 1952; Watkins, 1980; Le Pendu et al., 1986; Balgir, 2005, 2007). These individuals are designated as Oh or h/h, in contrast to most humans who are H/H, i.e., they produce the blood group O carbohydrate antigen. In individuals who are blood group A or B, *N*-acetylgalactosamine (GalNAc) or galactose (Gal) is added α1-3 to the penultimate Gal of the H antigen, respectively. The structure of blood type "Bombay" antigen Oh is included in Fig. 1 of Chapter 3 that illustrates the α-gal epitopes and blood type ABO antigens, as well. Cloning and sequencing of the α1,2FT gene (*FUT1*) in blood group Bombay (Oh) individuals demonstrated the presence of inactivating point mutations in the coding regions of both alleles of this gene (Kelly et al., 1994; Fernandez-Mateos et al., 1998).

Blood group Bombay individuals are very rare. They are found in European populations as 1:1,000,000, whereas in India they are 1:10,000. In the absence of blood group O, blood group Bombay individuals naturally produce anti-blood group H (O) antibodies (i.e., anti-Fucα1-2Gal-R antibodies). These natural antibodies are completely absent in all other human populations. Blood group Bombay individuals also produce natural anti-A and anti-B antibodies because in the absence of blood group H, they cannot synthesize blood groups A or B antigens. Thus, blood group Bombay individuals resemble the hypothesized ancestral Old World primates who lived prior to the extinction of α-gal epitopes synthesizing primates, in the following characteristics: (1) Individuals who are homozygous for the accidently acquired mutation(s) that inactivated the α1,2FT gene, and those primates with inactivated α1,3GT gene have been very rare within their corresponding populations. (2) The homozygous individuals for the inactivated glycosyltransferase genes lack the H antigen or the α-gal epitope and produce a natural antibody against the lost carbohydrate antigen. As indicated above, anti-Gal can destroy or neutralize enveloped viruses presenting α-gal epitopes, following propagation in mammalian cells containing active α1,3GT. Similarly, enveloped viruses including severe acute respiratory syndrome coronavirus (Guillon et al., 2008), measles virus (Preece et al., 2002), and HIV (Neil et al., 2005) were found to present blood group A or B carbohydrate antigens when propagated in cells of containing α1,2FT and the corresponding A or B transferases. These viruses were further found to undergo complement-mediated inactivation in sera containing anti-A or anti-B antibodies, respectively. Thus, it would be of interest to determine whether anti-blood group H (O) antibody in the serum of blood group Bombay individuals can inactivate enveloped viruses propagated in blood group H human cells. Such an anti-viral activity of anti-blood group H (O) antibody raises a hypothetical possibility that blood group Bombay individuals may be immuno-protected better than other humans by the natural anti-blood group H (O), anti-A and anti-B antibodies against virulent enveloped viruses originating in any individual who is not blood group Bombay. This protection may be in a manner analogous to the effects of anti-Gal on enveloped viruses presenting α-gal epitopes, described in Fig. 2.

CONCLUSIONS

The natural anti-Gal antibody is one of the multiple natural anti-carbohydrate antibodies produced in humans against a wide range of carbohydrate antigens on GI bacteria. The antibody is unique to humans, apes, and Old World monkeys, and it binds specifically to a mammalian carbohydrate antigen called the α-gal epitope that is synthesized in nonprimate mammals, lemurs (prosimians) and New World monkeys by the glycosylation enzyme α1,3GT. The α1,3GT gene (*GGTA1*) appeared in mammals >100 million years ago, prior to the split between marsupial and placental mammals. This gene has been conserved in its active form, in all mammals, except for Old World monkeys, apes, and humans. Inactivation of the α1,3GT gene in ancestral Old World primates occurred 20–30 million years ago and could have been associated with epidemics of enveloped viruses in the Eurasia-Africa continent. It is suggested that prior to such epidemics, few ancestral Old World primates acquired deletion point mutations that inactivated the α1,3GT gene and eliminated α-gal epitopes. This resulted in loss of immune tolerance to the α-gal epitope and thus, in production of the anti-Gal antibody against antigens on bacteria colonizing the GI tract. This accidental inactivation of the α1,3GT gene in very small populations is analogous to the highly rare blood type "Bombay" individuals who do not synthesize blood group H (O antigen) because of inactivation of the α1,2-fucosyltransferase gene. The loss of immune tolerance to blood group H antigen has resulted in production of natural anti-blood group H antibodies in the blood group Bombay individuals. It is suggested that anti-Gal protected against infections by enveloped viruses presenting α-gal epitopes, which were lethal to the parental primate populations that conserved active α1,3GT and thus, synthesized α-gal epitopes. Alternative causes for the elimination of Old World primates synthesizing α-gal epitopes could be bacteria or protozoa parasites presenting α-gal or α-gal-like epitopes, and bacterial toxins, or detrimental viruses that used α-gal epitopes in these primates as "docking receptors." Ultimately, any of these proposed selective processes could result in extinction of Old World primates synthesizing α-gal epitopes on their cells. These ancestral primates were replaced by offspring populations lacking α-gal epitopes and producing the anti-Gal antibody, which continues to be produced by Old World monkeys, apes, and humans. New World monkeys and lemurs were protected from pathogens of the Old World by oceanic barriers, thus they continue to synthesize α-gal epitopes and lack the ability to produce the anti-Gal antibody. This scenario of few individuals in a large population having a mutation(s) that inactivates a glycosyltransferase gene thus, resulting in production of evolutionary advantageous natural antibodies against the eliminated carbohydrate antigen, may reflect one of the mechanisms inducing changes in the carbohydrate profile of various mammalian populations.

References

Abdel-Motal, U.M., Wang, S., Lu, S., Wigglesworth, K., Galili, U., 2006. Increased immunogenicity of human immunodeficiency virus gp120 engineered to express Galα1-3Galβ1-4GlcNAc-R epitopes. J. Virol. 80, 6943–6951.

Abdel-Motal, U.M., Guay, H.M., Wigglesworth, K., Welsh, R.M., Galili, U., 2007. Increased immunogenicity of influenza virus vaccine by anti-Gal mediated targeting to antigen presenting cells. J. Virol. 81, 9131–9141.

Abdel-Motal, U.M., Wang, S., Lu, S., Wigglesworth, K., Galili, U., 2010. Increased immunogenicity of HIV-1 p24 and gp120 following immunization with gp120/p24 fusion protein vaccine expressing α-gal epitopes. Vaccine 28, 1758–1765.

Agustí, J., Sanz de Siria, A., Garcés, M., 2003. Explaining the end of the hominoid experiment in Europe. J. Hum. Evol. 45, 145–153.
Alba, D.M., 2012. Fossil apes from the Vallès-Penedès basin. Evol. Anthropol. 21, 254–269.
Almeida, I.C., Milani, S.R., Gorin, P.A.J., Travassos, L.R., 1991. Complement-mediated lysis of *Trypanosoma cruzi* trypomastigotes by human anti-α-galactosyl antibodies. J. Immunol. 146, 2394–2400.
Almeida, I.C., Ferguson, M.A., Schenkman, S., Travassos, L.R., 1994. Lytic anti-α-galactosyl antibodies from patients with chronic Chagas' disease recognize novel O-linked oligosaccharides on mucin-like glycosyl-phosphatidylinositol-anchored glycoproteins of *Trypanosoma cruzi*. Biochem. J. 304, 793–802.
Andrews, P., Martin, L., 1991. Hominoid dietary evolution. Philos. Trans. R. Soc. Lond. B Biol. Sci. 334, 199–209.
Andrews, P., 1992. Evolution and environment in the Hominoidea. Nature 360, 641–646.
Andrews, P., Harrison, T., Delson, E., Bernor, R.L., Martin, L., 1996. Distribution and biochronology of European and southwest Asian Miocene catarrhines. In: Bernor, R.L., Fahlbusch, V., Mittman, H.W. (Eds.), The Evolution of Western Eurasian Neogene Mammal Faunas. Columbia University Press, New York, NY, pp. 168–207.
Avila, J.L., Rojas, M., Galili, U., 1989. Immunogenic Galα1-3Gal carbohydrate epitopes are present on pathogenic American *Trypanosoma* and *Leishmania*. J. Immunol. 142, 2828–2834.
Balgir, R.S., 2005. Detection of a rare blood group "Bombay (Oh) phenotype" among the Kutia Kondh primitive tribe of Orissa, India. Int. J. Hum. Genet. 5, 193–198.
Balgir, R.S., 2007. Identification of a rare blood group, "Bombay (Oh) phenotype," in Bhuyan tribe of Northwestern Orissa, India. Indian J. Hum. Genet. 13, 109–113.
Basu, M., Basu, S., 1973. Enzymatic synthesis of blood group related pentaglycosyl ceramide by an α-galactosyltransferase. J. Biol. Chem. 248, 1700–1706.
Benatuil, L., Kaye, J., Rich, R.F., Fishman, J.A., Green, W.R., Iacomini, J., 2005. The influence of natural antibody specificity on antigen immunogenicity. Eur. J. Immunol. 35, 2638–2647.
Benatuil, L., Kaye, J., Cretin, N., Godwin, J.G., Cariappa, A., Pillai, S., et al., 2008. Ig knock-in mice producing anti-carbohydrate antibodies: breakthrough of B cells producing low affinity anti-self antibodies. J. Immunol. 180, 3839–3848.
Betteridge, A., Watkins, W.M., 1983. Two α-3-D galactosyltransferases in rabbit stomach mucosa with different acceptor substrate specificities. Eur. J. Biochem. 132, 29–35.
Bhende, Y.M., Deshpande, C.K., Bhatia, H.M., Sanger, R., Race, R.R., Morgan, W.T., et al., 1952. A new blood group character related to the ABO system. Lancet 1, 903–904.
Blake, D.D., Goldstein, I.J., 1981. An α-D-galactosyltransferase in Ehrlich ascites tumor cells: Biosynthesis and characterization of a trisaccharide (α-D-galacto(1-3)-*N*-acetyllactosamine). J. Biol. Chem 256, 5387–5393.
Blanken, W.M., van den Eijnden, D.H., 1985. Biosynthesis of terminal Galα1-3Galß1-4GlcNAc-R oligosaccharide sequence on glycoconjugates: purification and acceptor specificity of a UDP-Gal: *N*-acetyllactosamine α1,3galactosyltransferase. J. Biol. Chem 260, 12927–12934.
Blixt, O., Head, S., Mondala, T., Scanlan, C., Huflejt, M.E., Alvarez, R., et al., 2004. Printed covalent glycan array for ligand profiling of diverse glycan binding proteins. Proc. Natl. Acad. Sci. U.S.A. 101, 17033–17038.
Bovin, N., 2013. Natural antibodies to glycans. Biochemistry (Mosc.) 78, 786–797.
Bovin, N., Obukhova, P., Shilova, N., Rapoport, E., Popova, I., Navakouski, M., et al., 2012. Repertoire of human natural anti-glycan immunoglobulins. Do we have auto-antibodies? Biochim. Biophys. Acta 1820, 1373–1382.
Callan, M.F., 2003. The evolution of antigen-specific CD8+ T cell responses after natural primary infection of humans with Epstein-Barr virus. Viral Immunol. 16, 3–16.
Chen, W., Gu, L., Zhang, W., Motari, E., Cai, L., Styslinger, T.J., et al., 2011. L-rhamnose antigen: a promising alternative to α-gal for cancer immunotherapies. ACS Chem. Biol. 6, 185–191.
Dor, F.J., Tseng, Y.L., Cheng, J., Moran, K., Sanderson, T.M., Lancos, C.J., et al., 2004. α1,3-Galactosyltransferase gene-knockout miniature swine produce natural cytotoxic anti-Gal antibodies. Transplantation 78, 15–20.
Dawkins, R., 2004. The Ancestor's Tale. Houghton Mifflin Harcourt.
Dürrbach, A., Baple, E., Preece, A.F., Charpentier, B., Gustafsson, K., 2007. Virus recognition by specific natural antibodies and complement results in MHC I cross-presentation. J. Immunol. 37, 1254–1265.
Epstein, M.A., Hunt, R.D., Rabin, H., 1973. Pilot experiments with EB virus in owl monkeys (*Aotus trivirgatus*). I. Reticuloproliferative disease in an inoculated animal. Int. J. Cancer 12, 309–318.
Fang, J., Walters, A., Hara, H., Long, C., Yeh, P., Ayares, D., et al., 2012. Anti-Gal antibodies in α1,3-galactosyltransferase gene-knockout pigs. Xenotransplantation 19, 305–310.
Fernandez-Mateos, P., Cailleau, A., Henry, S., Costache, M., Elmgren, A., Svensson, L., et al., 1998. Point mutations and deletion responsible for the Bombay H null and the Reunion H weak blood groups. Vox Sang. 75, 37–46.

Galili, U., 2004. Immune response, accommodation and tolerance to transplantation carbohydrate antigens. Transplantation 78, 1093–1098.

Galili, U., 2013. α1,3Galactosyltransferase knockout pigs produce the natural anti-Gal antibody and simulate the evolutionary appearance of this antibody in primates. Xenotransplantation 20, 267–276.

Galili, U., 2016. Natural anti-carbohydrate antibodies contributing to evolutionary survival of primates in viral epidemics? Glycobiology 26, 1140–1150.

Galili, U., Korkesh, A., Kahane, I., Rachmilewitz, A., 1983. Demonstration of a natural anti-galactosyl IgG antibody on thalassemic red blood cells. Blood 61, 1258–1264.

Galili, U., Rachmilewitz, E.A., Peleg, A., Flechner, I., 1984. A unique natural human IgG antibody with anti-α-galactosyl specificity. J. Exp. Med. 160, 1519–1531.

Galili, U., Macher, B.A., Buehler, J., Shohet, S.B., 1985. Human natural anti-α-galactosyl IgG. II. The specific recognition of α(1→3)-linked galactose residues. J. Exp. Med. 162, 573–582.

Galili, U., Flechner, I., Kniszinski, A., Danon, D., Rachmilewitz, E.A., 1986a. The natural anti-α-galactosyl IgG on human normal senescent red blood cells. Br. J. Haematol. 62, 317–324.

Galili, U., Clark, M.R., Shohet, S.B., 1986b. Excessive binding of the natural anti-α-galactosyl IgG to sickle red cells may contribute to extravascular cell destruction. J. Clin. Investig. 77, 27–33.

Galili, U., Clark, M.R., Shohet, S.B., Buehler, J., Macher, B.A., 1987a. Evolutionary relationship between the anti-Gal antibody and the Galα1-3Gal epitope in primates. Proc. Natl. Acad. Sci. U.S.A. 84, 1369–1373.

Galili, U., Buehler, J., Shohet, S.B., Macher, B.A., 1987b. The human natural anti-Gal IgG. III. The subtlety of immune tolerance in man as demonstrated by crossreactivity between natural anti-Gal and anti-B antibodies. J. Exp. Med. 165, 693–704.

Galili, U., Shohet, S.B., Kobrin, E., Stults, C.L.M., Macher, B.A., 1988a. Man, apes, and Old World monkeys differ from other mammals in the expression of α-galactosyl epitopes on nucleated cells. J. Biol. Chem. 263, 17755–17762.

Galili, U., Mandrell, R.E., Hamadeh, R.M., Shohet, S.B., Griffis, J.M., 1988b. Interaction between human natural anti-α-galactosyl immunoglobulin G and bacteria of the human flora. Infect. Immun. 56, 1730–1737.

Galili, U., Swanson, K., 1991. Gene sequences suggest inactivation of α1-3 galactosyltransferase in catarrhines after the divergence of apes from monkeys. Proc. Natl. Acad. Sci. U.S.A. 88, 7401–7404.

Galili, U., Anaraki, F., Thall, A., Hill-Black, C., Radic, M., 1993. One percent of circulating B lymphocytes are capable of producing the natural anti-Gal antibody. Blood 82, 2485–2493.

Galili, U., Andrews, P., 1995. Suppression of α-galactosyl epitopes synthesis and production of the natural anti-Gal antibody: a major evolutionary event in ancestral Old World primates. J. Hum. Evol. 29, 433–442.

Galili, U., Repik, P.K., Anaraki, F., Mozdzanowska, K., Washko, G., Gerhard, W., 1996. Enhancement of antigen presentation of influenza virus hemagglutinin by the natural anti-Gal antibody. Vaccine 14, 321–328.

Gazzinelli, R.T., Pereira, M.E., Romanha, A.J., Gazzinelli, G., Brener, Z., 1991. Direct lysis of *Trypanosoma cruzi*: a novel effector mechanism of protection mediated by human anti-Gal antibodies. Parasite Immunol. 13, 345–356.

Gerritsen, J., Smidt, H., Rijkers, G.T., de Vos, W.M., 2011. Intestinal microbiota in human health and disease: the impact of probiotics. Genes Nutr. 6, 209–240.

Geyer, R., Geyer, H., Stirm, S., Hunsmann, G., Schneider, J., Dabrowski, U., et al., 1984. Major oligosaccharides in the glycoprotein of Friend murine leukemia virus: structure elucidation by one- and two-dimensional proton nuclear magnetic resonance and methylation analysis. Biochemistry 23, 5628–5637.

Guillon, P., Clément, M., Sébille, V., Rivain, J.G., Chou, C.F., Ruvoën-Clouet, N., et al., 2008. Inhibition of the interaction between the SARS-CoV spike protein and its cellular receptor by anti-histo-blood group antibodies. Glycobiology 18, 1085–1093.

Hamadeh, R.M., Galili, U., Zhou, P., Griffis, J.M., 1995. Human secretions contain IgA, IgG and IgM anti-Gal (anti-α-galactosyl) antibodies. Clin. Diagn. Lab. Immunol. 2, 125–131.

Han, W., Cai, L., Wu, B., Li, L., Xiao, Z., Cheng, J., et al., 2012. The wciN gene encodes an α-1,3-galactosyltransferase involved in the biosynthesis of the capsule repeating unit of *Streptococcus pneumoniae* serotype 6B. Biochemistry 51, 5804–5810.

Henion, T.R., Macher, B.A., Anaraki, F., Galili, U., 1994. Defining the minimal size of catalytically active primate α1,3galactosyltransferase: Structure function studies on the recombinant truncated enzyme. Glycobiology 4, 193–201.

Higashi, H., Naiki, M., Matuo, S., Okouchi, K., 1977. Antigen of "serum sickness" type of heterophile antibodies in human sera: indentification as gangliosides with *N*-glycolylneuraminic acid. Biochem. Biophys. Res. Commun. 79, 388–395.

Ilg, T., Etges, R., Overath, P., McConville, M.J., Thomas-Oates, J., Thomas, J., et al., 1992. Structure of *Leishmania mexicana* lipophosphoglycan. J. Biol. Chem. 267, 6834–6840.
Joziasse, D.H., Shaper, J.H., Van den Eijnden, D.H., Van Tunen, A.H., Shaper, N.L., 1989. Bovine α1-3-galactosyltransferase: isolation and characterization of a cDNA clone. Identification of homologous sequences in human genomic DNA. J. Biol. Chem. 264, 14290–14297.
Kelly, R.J., Ernst, L.K., Larsen, R.D., Bryant, J.G., Robinson, J.S., Lowe, J.B., 1994. Molecular basis for H blood group deficiency in Bombay (Oh) and para-Bombay individuals. Proc. Natl. Acad. Sci. U.S.A. 91, 5843–5847.
Klein, E., Masucci, M.G., 1982. Cell-mediated immunity against Epstein-Barr virus infected B lymphocytes. Springer Semin. Immunopathol. 5, 63–73.
Koike, C., Fung, J.J., Geller, D.A., Kannagi, R., Libert, T., Luppi, P., et al., 2002. Molecular basis of evolutionary loss of the α1,3-galactosyltransferase gene in higher primates. J. Biol. Chem. 277, 10114–10120.
Kolber-Simonds, D., Lai, L., Watt, S.R., Denaro, M., Arn, S., Augenstein, M.L., et al., 2004. Production of α1,3-galactosyltransferase null pigs by means of nuclear transfer with fibroblasts bearing loss of heterozygosity mutations. Proc. Natl. Acad. Sci. U.S.A. 101, 7335–7340.
Lai, L., Kolber-Simonds, D., Park, K.W., Cheong, H.T., Greenstein, J.L., Im, G.S., et al., 2002. Production of α-1,3-galactosyltransferase knockout pigs by nuclear transfer cloning. Science 295, 1089–1092.
Lantéri, M., Giordanengo, V., Vidal, F., Gaudray, P., Lefebvre, J.-C., 2002. A complete α1,3-galactosyltransferase gene is present in the human genome and partially transcribed. Glycobiology 12, 785–792.
Larsen, R.D., Rajan, V.P., Ruff, M.M., Kukowska-Latallo, J., Cummings, D., Lowe, J.B., 1989. Isolation of a cDNA encoding a murine UDP- galactose:β-D-galactosyl-1,4-*N*-acetyl-D-glucosaminide α-1,3-galactosyltransferase: expression cloning by gene transfer. Proc. Natl. Acad. Sci. U.S.A. 86, 8227–8231.
Larsen, R.D., Rivera-Marrero, C.A., Ernst, L.K., Cummings, R.D., Lowe, J.B., 1990. Frameshift and nonsense mutations in a human genomic sequence homologous to a murine UDP-Gal:β-D-Gal(1,4)-D-GlcNAc α(1,3)-galactosyltransferase cDNA. J. Biol. Chem. 265, 7055–7061.
Le Pendu, J., Lambert, F., Gérard, G., Vitrac, D., Mollicone, R., Oriol, R., 1986. On the specificity of human anti-H antibodies. Vox Sang. 50, 223–236.
Lüderitz, O., Simmons, D.A., Westphal, G., 1965. The immunochemistry of *Salmonella* chemotype VI O-antigens. The structure of oligosaccharides from *Salmonella* group U (o 43) lipopolysaccharides. Biochem. J. 97, 820–826.
Long, D.E., Karmakar, P., Wall, K.A., Sucheck, S.J., 2014. Synthesis of α-l-rhamnosyl ceramide and evaluation of its binding with anti-rhamnose antibodies. Bioorg. Med. Chem. 22, 5279–5289.
Mañez, R., Blanco, F.J., Díaz, I., Centeno, A., Lopez-Pelaez, E., Hermida, M., et al., 2001. Removal of bowel aerobic gram-negative bacteria is more effective than immunosuppression with cyclophosphamide and steroids to decrease natural α-galactosyl IgG antibodies. Xenotransplantation 8, 15–23.
McConville, M.J., Homans, S.W., Thomas-Oates, J.E., Dell, A., Bacic, A., 1990. Structures of the glycoinositolphospholipids from *Leishmania* major. A family of novel galactofuranose-containing glycolipids. J. Biol. Chem. 265, 7385–7394.
Merceron, G., Kaiser, T.M., Kostopoulos, D.S., Schulz, E., 2010. Ruminant diets and the Miocene extinction of European great apes. Proc. Biol. Sci. 277, 3105–3112.
Merrick, J.M., Zadarlik, K., Milgrom, F., 1978. Characterization of the Hanganutziu-Deicher (serum-sickness) antigen as gangliosides containing *N*-glycolylneuraminic acid. Int. Arch. Allergy Appl. Immunol. 57, 477–480.
Milani, S.R., Travassos, L.R., 1988. Anti-α-galactosyl antibodies in chagasic patients. Possible biological significance. Braz. J. Med. Biol. Res. 21, 1275–1286.
Miller, E.R., Benefit, B.R., McCrossin, M.L., Plavcan, J.M., Leakey, M.G., El-Barkooky, A.N., et al., 2009. Systematics of early and middle Miocene Old World monkeys. J. Hum. Evol. 57, 195–211.
Mohiuddin, M.M., Ogawa, H., Yin, D.P., Galili, U., 2003. Tolerance induction to a mammalian blood group-like carbohydrate antigen by syngeneic lymphocytes expressing the antigen, II: tolerance induction on memory B cells. Blood 102, 229–236.
Neil, S.J., McKnight, A., Gustafsson, K., Weiss, R.A., 2005. HIV-1 incorporates ABO histo-blood group antigens that sensitize virions to complement-mediated inactivation. Blood 105, 4693–4699.
Ogawa, H., Yin, D.P., Shen, J., Galili, U., 2003. Tolerance induction to a mammalian blood group-like carbohydrate antigen by syngeneic lymphocytes expressing the antigen. Blood 101, 2318–2320.
Oriol, R., Candelier, J.J., Taniguchi, S., Balanzino, L., Peters, L., Niekrasz, M., et al., 1999. Major carbohydrate epitopes in tissues of domestic and African wild animals of potential interest for xenotransplantation research. Xenotransplantation 6, 79–89.

Padler-Karavani, V., Yu, H., Cao, H., Karp, F., Chokhawala, H., Varki, N., et al., 2008. Diversity in specificity, abundance, and composition of anti-Neu5Gc antibodies in normal humans: potential implications for disease. Glycobiology 18, 818–830.

Phelps, C.J., Koike, C., Vaught, T.D., Boone, J., Wells, K.D., Chen, S.H., et al., 2003. Production of α1,3-galactosyltransferase-deficient pigs. Science 299, 411–414.

Posekany, K.J., Pittman, H.K., Bradfield, J.F., Haisch, C.E., Verbanac, K.M., 2002. Induction of cytolytic anti-Gal antibodies in α-1,3-galactosyltransferase gene knockout mice by oral inoculation with *Escherichia coli* O86:B7 bacteria. Infect. Immun. 70, 6215–6222.

Pothoulakis, C., Galili, U., Castagliuolo, I., Kelly, S., Nikulasson, P.K., Brasitus, T.A., et al., 1996. A human antibody binds to α-galactose receptors and mimics the effects of Clostridium difficile toxin A in rat colon. Gastroenterology 98, 641–649.

Preece, A.F., Strahan, K.M., Devitt, J., Yamamoto, F., Gustafsson, K., 2002. Expression of ABO or related antigenic carbohydrates on viral envelopes leads to neutralization in the presence of serum containing specific natural antibodies and complement. Blood 99, 2477–2482.

Ramasamy, R., Reese, R.T., 1986. Terminal galactose residues and the antigenicity of *Plasmodium falciparum* glycoproteins. Mol. Biochem. Parasitol. 19, 91–101.

Ramasamy, R., Rajakaruna, R., 1997. Association of malaria with inactivation of α1,3-galactosyl transferase in catarrhines. Biochim. Biophys. Acta 1360, 241–246.

Ramasamy, R., Field, M.C., 2012. Terminal galactosylation of glycoconjugates in *Plasmodium falciparum* asexual blood stages and *Trypanosoma brucei* bloodstream trypomastigotes. Exp. Parasitol. 130, 314–320.

Repik, P.M., Strizki, M., Galili, U., 1994. Differential host dependent expression of α-galactosyl epitopes on viral glycoproteins: A study of Eastern equine encephalitis virus as a model. J. Gen. Virol. 75, 1177–1181.

Rodriguez, I.A., Welsh, R.M., 2013. Possible role of a cell surface carbohydrate in evolution of resistance to viral infections in Old World primates. J. Virol. 87, 8317–8326.

Rother, R.P., Fodor, W.L., Springhorn, J.P., Birks, C.W., Setter, E., Sandrin, M.S., 1995. A novel mechanism of retrovirus inactivation in human serum mediated by anti-α-galactosyl natural antibody. J. Exp. Med. 182, 1345–1355.

Schrago, C.G., 2007. On the time scale of New World primate diversification. Am. J. Phys. Anthropol. 132, 344–354.

Sheridan, R.T., Hudon, J., Hank, J.A., Sondel, P.M., Kiessling, L.L., 2014. Rhamnose glycoconjugates for the recruitment of endogenous anti-carbohydrate antibodies to tumor cells. Chembiochem 15, 1393–1398.

Shope, T., Dechairo, D., Miller, G., 1973. Malignant lymphoma in cotton-top marmosets following inoculation of Epstein-Barr virus. Proc. Natl. Acad. Sci. U.S.A. 70, 2487–2491.

Sørensen, D.B., Dahl, K., Ersbøll, A.K., Kirkeby, S., d'Apice, A.J., Hansen, A.K., 2008. Aggression in cataract-bearing α-1,3-galactosyltransferase knockout mice. Lab. Anim. 42, 34–44.

Springer, G.F., 1971. Blood-group and Forssman antigenic determinants shared between microbes and mammalian cells. Prog. Allergy 15, 9–77.

Springer, G.F., Horton, R.E., 1969. Blood group isoantibody stimulation in man by feeding blood group-active bacteria. J. Clin. Investig. 48, 1280–1291.

Steiper, M.E., Young, N.M., 2006. Primate molecular divergence dates. Mol. Phylogenet. Evol. 41, 384–394.

Stephen, A.M., Cummings, J.H., 1980. The microbial contribution to human faecal mass. J. Med. Microbiol. 13, 45–56.

Stowell, S.R., Arthur, C.M., McBride, R., Berger, O., Razi, N., Heimburg-Molinaro, J., et al., 2014. Microbial glycan microarrays define key features of host-microbial interactions. Nat. Chem. Biol. 10, 470–476.

Takahagi, Y., Fujimura, T., Miyagawa, S., Nagashima, H., Shigehisa, T., Shirakura, R., et al., 2005. Production of α1,3-galactosyltransferase gene knockout pigs expressing both human decay-accelerating factor and *N*-acetylglucosaminyltransferase III. Mol. Reprod. Dev. 71, 331–338.

Takeuchi, Y., Porter, C.D., Strahan, K.M., Preece, A.F., Gustafsson, K., Cosset, F.L., et al., 1996. Sensitization of cells and retroviruses to human serum by (α1-3) galactosyltransferase. Nature 379, 85–88.

Takeuchi, Y., Liong, S.H., Bieniasz, P.D., Jäger, U., Porter, C.D., Friedman, T., et al., 1997. Sensitization of rhabdo-, lenti-, and spumaviruses to human serum by galactosyl(α1-3)galactosylation. J. Virol. 71, 6174–6178.

Teneberg, S., Lönnroth, I., Torres Lopez, J.F., Galili, U., Halvarsson, M.O., Angstrom, J., et al., 1996. Molecular mimicry in the recognition of glycosphingolipids by Galα3Galß4GlcNAcß-binding *Clostridium difficile* toxin A, human natural anti-α-galactosyl IgG and the monoclonal antibody Gal-13: characterization of a binding-active human glycosphingolipid, non-identical with the animal receptor. Glycobiology 6, 599–609.

Thall, A.D., 1999. Generation of α1,3galactosyltransferase deficient mice. Subcell. Biochem. 32, 259–279.

Towbin, H., Rosenfelder, G., Wieslander, J., Avila, J.L., Rojas, M., Szarfman, A., et al., 1987. Circulating antibodies to mouse laminin in Chagas disease, American cutaneous leishmaniasis, and normal individuals recognize terminal galactosyl (α1-3)-galactose epitopes. J. Exp. Med. 166, 419–432.
Ungar, P.S., Kay, R.F., 1995. The dietary adaptations of European Miocene catarrhines. Proc. Natl. Acad. Sci. U.S.A. 92, 5479–5481.
Varki, A., 2010. Colloquium paper: uniquely human evolution of sialic acid genetics and biology. Proc. Natl. Acad. Sci. U.S.A. 107, 8939–8946.
Wang, F., 2013. Nonhuman primate models for Epstein-Barr virus infection. Curr. Opin. Virol. 3, 233–237.
Wang, L., Anaraki, F., Henion, T.R., Galili, U., 1995. Variations in activity of the human natural anti-Gal antibody in young and elderly populations. J Gerontol A Biol Sci Med Sci 50A, M227–M233.
Watkins, W.M., 1980. Biochemistry and genetics of the ABO, Lewis, and P blood group systems. Adv. Hum. Genet. 10, 1–136 379–385.
Welsh, R.M., O'Donnell, C.L., Reed, D.J., Rother, R.P., 1998. Evaluation of the Galα1-3Gal epitope as a host modification factor eliciting natural humoral immunity to enveloped viruses. J. Virol. 72, 4650–4656.
Wiener, A.S., 1951. Origin of naturally occurring hemagglutinins and hemolysins; a review. J. Immunol. 66, 287–295.
Yilmaz, B., Portugal, S., Tran, T.M., Gozzelino, R., Ramos, S., Gomes, J., et al., 2014. Gut microbiota elicits a protective immune response against malaria transmission. Cell 159, 1277–1289.
Zakhour, M., Ruvoën-Clouet, N., Charpilienne, A., Langpap, B., Poncet, D., Peters, T., 2009. The αGal epitope of the histo-blood group antigen family is a ligand for bovine norovirus Newbury2 expected to prevent cross-species transmission. PLoS Pathog. 5, e1000504.
Zhu, A., Hurst, R., 2002. Anti-*N*-glycolylneuraminic acid antibodies identified in healthy human serum. Xenotransplantation 9, 376–381.

CHAPTER

3

Anti-Gal Comprises Most of Anti-Blood Group B Antibodies: Landsteiner's B-Like Enigma

OUTLINE

INTRODUCTION

The presence of anti-Gal in large amounts in human serum and the absence of α-gal epitopes in humans, both imply that anti-Gal cannot bind to human cells because such interaction is likely to result in harmful autoimmune processes. Indeed, incubation of various human cells with anti-Gal was found to result in no binding of this antibody (Galili et al., 1988a). Chapter 8 proposes several mechanisms that may mediate the *in vivo* expression of α-gal, or α-gal-like epitopes on human cells, resulting in anti-Gal binding and autoimmunity. The present chapter describes the ability of many anti-Gal antibody clones in blood type A and O individuals to bind to the human blood group B antigen. The interest in the studying possible interaction between anti-Gal and blood group B antigen stems from the close similarity in the structure of these two carbohydrate antigens (α-gal epitope: Galα1-3Galβ1-4GlcNAc-R and blood group B antigen: Galα1-3[Fucα1-2]Galβ1-4GlcNAc-R). The α-gal epitope, blood groups A, B, and O (H) antigens, and blood group "Bombay" (Oh) antigen (discussed in Chapter 2)

The Natural Anti-Gal Antibody as Foe Turned Friend in Medicine
http://dx.doi.org/10.1016/B978-0-12-813362-0.00003-8

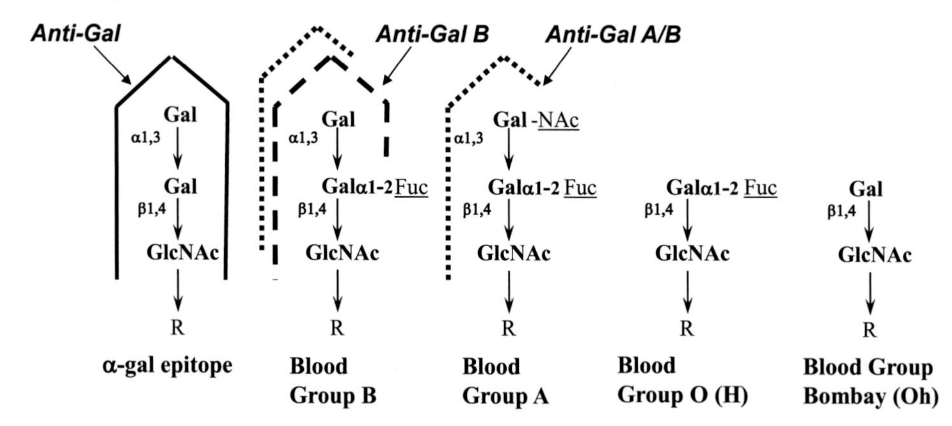

FIGURE 1 Anti-Gal antibodies and their interaction with blood group A and B antigens. The structures of blood group O (H antigen) and blood group "Bombay" (Oh also called h/h) are presented, as well. The carbohydrates common to the blood group antigens and to the α-gal epitope are presented in bold font. The additional molecules that form blood group antigens B, A and O (H) are underlined. The types of anti-Gal antibodies binding to A and B antigens are indicated as schematic lines. *Solid line*-pure anti-Gal present in all blood types and capable of binding only to the α-gal epitope; *Dashed line*-anti-Gal/B, comprises 75%–95% of natural anti-B antibodies in A and O individuals and binds to the α-gal epitope and to blood group B antigen; *Dotted line*-anti-Gal A/B, observed in O individuals and in higher titers in O recipients of blood type A kidney allograft. Anti-Gal A/B binds to α-gal epitopes and to blood groups A and B antigens. Gal-galactose; GalNAc-*N*-acetylgalactosamine; GlcNAc-*N*-acetylglucosamine; Fuc-fucose; NAc-*N*-acetyl group. *Adapted from Galili, U., 2006. Xenotransplantation and ABO incompatible transplantation: the similarities they share. Transfus. Apher. Sci. 35, 45–58, with permission.*

are illustrated in Fig. 1. As discussed below, anti-Gal comprises 75%–95% of anti-blood group B antibody activity in humans, thus, anti-Gal has been unknowingly studied since the discovery of the ABO system by Karl Landsteiner in 1901. Anti-Gal binding to blood group B antigen is feasible because it is composed of multiple antibody clones that bind to various "facets" of the α-gal epitope, some of which are also found on blood group B antigen. This diversity of anti-Gal clones also explains its ability to bind the mammalian α-gal epitope, although anti-Gal is produced in response to antigenic stimulation by bacterial carbohydrate antigens (Galili et al., 1988b; Almeida et al., 1991; Posekany et al., 2002). The studies described in this chapter further explain the enigma of the B-like blood group antigen, described 92 years ago by Landsteiner to be found on red blood cells (RBC) of New World monkeys (Landsteiner and Philip-Miller, 1925), and indicate that this B-like antigen is the α-gal epitope.

ANTI-GAL FROM BLOOD TYPE A AND O INDIVIDUALS BINDS TO BLOOD GROUP B ANTIGEN

Analysis of carbohydrate specificity of antibodies using glycolipids

Identification of anti-carbohydrate antibody specificity can be effectively achieved by immunostaining of glycolipids with known structure, which are chromatographed on thin layer chromatography (TLC) plates (Magnani et al., 1981), as demonstrated in Fig. 1 of Chapter 1. Such identification of specificity is feasible because the structure of carbohydrate chains on glycolipids can be determined by several physicochemical methods, including mass spectrometry and nuclear magnetic resonance. Binding of anti-carbohydrate antibodies to their antigens is more specific than antibody binding to proteins. Most carbohydrate epitopes (except for epitopes carrying sialic acid) are made of noncharged carbohydrates, and thus they cannot form ionic bonds with the antibody-binding sites. Therefore, interaction between epitopes, such as the α-gal epitope and blood group A and B antigens and their corresponding antibodies, requires a high degree of structural "fitting" between the carbohydrate antigen and the binding site of the antibody, to enable forces that are weaker than ionic bonds to mediate the antigen/antibody interaction. Hydrogen bonds, Van der Waals forces, and hydrophobic forces constitute the dominant forces in these interactions (Kabat, 1976), as also shown for the anti-Gal antibody binding to α-gal epitopes (Agostino et al., 2009, 2010; Plum et al., 2011). Because of the lack of ionic bonds in antibody interaction with noncharged carbohydrate antigens, the affinity of such interactions is considerably lower than that of antibody/protein interactions. For example, the affinity of human anti-Gal antibody to the α-gal epitope is ~$10^6 M^{-1}$ (Galili and Matta, 1996), whereas the affinity of anti-Rh (a human anti-protein antibody) to the Rh protein is ~$10^8 M^{-1}$ (Bloy et al., 1988). However, the extent of cross-reactivity of anti-protein antibodies with nonrelated protein antigens is much higher than that of anti-carbohydrate antibodies because of occasional similarity in charge arrays of amino acid sequences in proteins that bind specifically an antibody and proteins that are not the specific antigen for a given anti-protein antibody. One example for the high degree of specificity of anti-carbohydrate antibodies is the difference in the binding specificities of anti-blood group A and B antibodies. The difference in the structure of the corresponding two blood group antigens is the small *N*-acetyl (NAc) residue (CH_3CONH-) linked to terminal galactosyl on blood group A antigen (GalNAcα1-3[Fucα1-2]Gal-R) and absent on the terminal galactosyl of blood group B antigen (Galα1-3[Fucα1-2]Gal-R), as illustrated in Fig. 1. A mistaken transfusion of RBC with the *N*-acetyl linked to the galactosyl (i.e., blood type A) to a patient requiring RBC without that *N*-acetyl (i.e., blood type B) may have lethal consequences. Because of the high specificity of antibodies to carbohydrate antigens and the well-defined structure of glycolipids, immunostaining of glycolipids by anti-Gal and anti-blood group A or B antibodies could validate hemagglutination studies used for determining whether anti-Gal can bind to blood group B antigen.

Anti-Gal binding to blood group B antigen

As indicated above and in Fig. 1, the structure of blood group B antigen differs from that of the α-gal epitope only by a fucosyl linked α1-2 to the penultimate galactose to form the structure Galα1-3(Fucα1-2)Galβ1-4GlcNAc-R of blood group B antigen. Based on this structural similarity, the α-gal epitope is occasionally referred to as "linear B" because of the absence of branching fucose of blood group B in the α-gal epitope. The term "linear B" may not be suitable in describing the α-gal epitope for the following reasons: (1) "Linear B" has been the term used for almost 100 years to describe the writing system used by the Mycenaean civilization during the Late Bronze Age, between c.1450 and c.1100 BCE. (2) Linear B suggests a biosynthetic relationship between the α-gal epitope and blood group B antigen. However, the glycosyltransferases that synthesize these two antigens and the acceptors for their synthesis are different. Blood group B transferase can link the terminal α1,3galactosyl to the carbohydrate chains with α1-2fucosyl (i.e., Fucα1-2Galβ1-4 GlcNAc-R), which is blood group O (H antigen)

(Hakomori, 1981), whereas α1,3GT can link the α1,3galactosyl to the carbohydrate chains only in the absence of the fucosyl (i.e., to Galβ1-4GlcNAc-R) (see Chapters 1 and 2).

The similarity in structure between the α-gal epitope and blood group B antigen raised the question whether anti-Gal can also bind to blood group B antigen. Evidently, no such antibody can exist in individuals of blood type B or AB because such anti-Gal antibodies would bind to RBC of the individual producing the antibody, resulting in an autoimmune anemia. However, it was of interest to determine whether blood type A or O individuals can produce anti-Gal antibodies that also bind to blood group B antigen (Galili et al., 1987a). For this purpose, anti-Gal was purified from heat inactivated normal human plasma of various blood type donors by affinity chromatography on columns consisting of silica beads presenting synthetic α-gal epitopes (Synsorb beads). After extensive washes with PBS, the antibodies retained on the column were eluted by incubation with 0.5 M melibiose (Galα1-6Glc) at 37°C. The free melibiose was removed by repeated dialysis against large volumes of PBS and the purified anti-Gal brought to a concentration of 100 μg/ml in PBS, which is the approximate concentration in human serum (Galili et al., 1984, 1985, 1987a). Isoelectric focusing analysis of the purified anti-Gal antibody indicated that it is highly polyclonal with clones having pI values that range from 4.0 to 8.5 (Galili et al., 1984; Tinguely et al., 2002).

For the isolation of anti-blood group A or B antibodies, plasma from blood type B or A individuals were adsorbed on equal volumes of washed packed A type or B type RBC, respectively. The RBC were washed repeatedly with PBS, lysed by hypotonic shock and after additional washes, the antibodies bound to the RBC membranes were eluted with glycine-HCI buffer (pH 2.6) and brought immediately to pH 7.4. Activity of the isolated antibodies was confirmed by agglutination of RBC of the corresponding blood type. Activities of the various anti-Gal and anti-blood group antibody preparations were also determined by immunostaining of the α-gal glycolipid ceramide pentahexoside (CPH from rabbit RBC membranes) with the structure Galα1-3Galβ1-4GlcNAcβ1-3Galβ1-4Glc-Cer (as in Fig. 1 of Chapter 1) and the corresponding blood group B antigen with the structure Galα1-3(Fucα1-2)Galβ1-4GlcNAcβ1-3Galβ1-4Glc-Cer. The results were further validated in hemagglutination studies with the various RBC including rabbit RBC that lack blood group B antigen, as indicated by the ability of rabbits to produce anti-blood group B antibodies (Landsteiner and Philip-Miller, 1925).

Anti-Gal purified from blood type B or AB individuals bound only to CPH and to rabbit RBC (as shown in Fig. 1 of Chapter 1) and did not immunostain blood group B antigen (Table 1). This antibody bound to blood group B glycolipid only after the glycolipid was treated with the

TABLE 1 "Pure" anti-Gal, anti-Gal/B and anti-Gal A/B antibody activities in sera of various blood type individuals

Blood type	Pure anti-Gal	Anti-Gal/B	Anti-Gal A/B
A	+	+	–
B	+	–	–
AB	+	–	–
O	+	+	[a]±

[a] *Anti-Gal A/B antibody activity is relatively weak in healthy individuals, but it markedly increases in blood type O recipients of kidney allografts from blood type A donors (Galili et al., 2003).*

enzyme α-fucosidase that cleaves the fucose from blood group B antigen, thus converting it into the α-gal epitope (Galili et al., 1987a). Anti-Gal purified from plasma of blood type A and O individuals also bound to CPH and to rabbit RBC. However, in addition to these interactions, anti-Gal from A individuals bound to blood group B glycolipid and agglutinated blood type B RBC. Anti-Gal from blood type O individuals also bound to blood group B glycolipid and agglutinated B type RBC, as well as displayed weak agglutinating activity of A type RBC (Table 1; Fig. 1) (Galili et al., 1987a). In agglutination studies with anti-blood group A and B antibodies, anti-blood group A agglutinated only blood type A RBC, whereas anti-blood group B agglutinated blood type B RBC and rabbit RBC. The ability of anti-Gal in blood type A and O individuals to bind to blood group B antigen was also reported by McMorrow et al. (1997).

The difference in specificity range of anti-Gal antibody from individuals with various blood types, regarding the ability to bind both to the α-gal epitope and to blood group B antigen implies that there are multiple anti-Gal antibody clones that bind to different "facets" of the 3-dimensional structure of the α-gal epitope. Based on studies of the carbohydrate-binding region of various anti-carbohydrate antibodies (Kabat, 1956; Ramsland et al., 2004; Evans et al., 2011), as well as studies on anti-Gal (Agostino et al., 2009, 2010; Plum et al., 2011), the antibody clones are likely to have groove, cavity, or pocket-like binding sites of various shapes formed by the complementarity determined regions (CDRs) of the light and heavy chains of each immunoglobulin clone. According to the patterns of interaction with α-gal epitopes on rabbit RBC and with blood group A and B antigens (referred to as "A antigen" and "B antigen," respectively), the anti-Gal antibody clones may be divided into three mains subsets: (1) "Pure" anti-Gal, (2) Anti-Gal/B, and (3) Anti-Gal A/B (Fig. 1 and Table 1).

"Pure" anti-Gal antibody clones

Many of the anti-Gal clones are represented by the schematic solid line "pocket" of anti-Gal combining site in Fig. 1. These clones may be regarded as "pure" anti-Gal antibody clones capable of binding only to the α-gal epitope and not to B antigen. Pure anti-Gal antibody clones are found in humans of all blood types and are the only anti-Gal clones produced in blood type B and AB individuals (Table 1). Pure anti-Gal is not able to bind to B antigen because the fucosyl linked α1-2 to the penultimate galactose in B antigen prevents this epitope from penetrating the antibody pocket or cavity and perform the antigen/antibody interaction. Cleaving of this fucosyl on B antigen glycolipid by α-fucosidase enabled the subsequent pure anti-Gal binding to this glycolipid on TLC plates (Galili et al., 1987a). The pocket or cavity-binding site of pure anti-Gal may be of various sizes; however, in most clones it may be of a size that can contain the trisaccharide free α-gal epitope Galα1-3Galβ1-4GlcNAc. This is suggested by equilibrium dialysis affinity studies with humans' anti-Gal. The affinity of Galα1-3Galβ1-4 Glc (i.e., trisaccharide with Glc instead of GlcNAc at the reducing end) decreased by 30%–40% and that of the disaccharide Galα1-3Gal displayed approximately seven-fold lower affinity to anti-Gal than the affinity of the trisaccharide α-gal epitope (Galili and Matta, 1996). The binding site of pure anti-Gal may even reach the size of a tetrasaccharide. This is suggested from the characteristics of a monoclonal anti-Gal antibody called Gal-13 (Galili et al., 1987b). This antibody was produced by hybridoma cells that were the fusion product of wild-type mouse spleen cells (i.e., mouse synthesizing α-gal epitopes) from a mouse that was extensively immunized with rabbit RBC and SP2/0 myeloma cells, also synthesizing α-gal epitopes. Gal-13 monoclonal antibody binds to α-gal epitopes on glycolipids but not to α-gal epitopes on glycoproteins (Galili et al., 1987b). This specificity implies that a tetrasaccharide Galα1-3Galβ1-4GlcNAcβ1-3Gal-R of a glycolipid but not a tetrasaccharide Galα1-3Galβ1-4GlcNAcβ1-2Man-R of a glycoprotein can penetrate the pocket or cavity of Gal-13 and form an antigen/antibody interaction (see comparison in structure of α-gal epitopes on carbohydrate chains on glycolipids and glycoproteins in Fig. 2 of Chapter 1). In contrast, the monoclonal antibody M86 (Galili et al., 1998) binds to α-gal epitopes both on glycolipids and glycoproteins, suggesting that pocket or cavity of this antibody can contain only the trisaccharide Galα1-3Galβ1-4GlcNAc, and therefore, it is not affected by the fourth carbohydrate whether it is mannose or galactose. All these observations suggest that there is a variety of shapes and sizes of the binding site in various anti-Gal antibody clones. This variety may explain the ability of many anti-Gal antibody clones to bind also to B antigen and few clones of this antibody can bind even to A antigen.

Anti-Gal/B antibody clones

The ability of a subset of anti-Gal antibody clones to bind both to the α-gal epitope and to B antigen implies that the fucosyl linked to the penultimate galactose of B antigen does not prevent the binding site in some of anti-Gal clones from interacting with "facets" of B antigen, which do not include the fucosyl branch. Because of the ability of these antibody clones to bind to α-gal epitope and to B antigen, these clones are referred to as "anti-Gal/B" clones (schematically illustrated in Fig. 1 as the dashed line capable of binding both to α-gal epitope and to B antigen).

Adsorption studies for removal of anti-Gal and of anti-blood group B antibodies from plasma of various donors could provide an approximation of the proportion of pure anti-Gal and of anti-Gal/B clones out of the total anti-Gal and anti-B antibody clones in humans. Adsorption of plasma from A or O individuals on blood type B RBC results in decrease of anti-Gal activity by ~50% (Galili et al., 1987a). This suggests that approximately half of anti-Gal antibody clones in A and O individuals can bind both to α-gal epitope and to B antigen, i.e., they have the anti-Gal/B specificity. For evaluating the proportion of anti-blood group B antibody clones that are in fact anti-Gal clones, the changes in anti-blood group B activity (i.e., agglutination of blood type B RBC) were determined in A and O plasma following their adsorption on natural α-gal epitopes on rabbit RBC or on synthetic α-gal epitopes linked to silica beads. As shown in Fig. 2, such adsorptions resulted in a decrease of 75%–95% in ability of the remaining antibodies to agglutinate blood type B RBC. These findings imply that most of the natural anti-blood group B antibody clones in blood type A and O individuals are in fact anti-Gal antibody clones that bind to the B antigen at a facet that does not include the fucosyl linked to the penultimate galactose. Therefore, these anti-Gal antibody clones bind to α-gal epitopes on rabbit RBC and on silica beads, as well as to the α-gal epitope core structure in the B antigen, i.e., they are anti-Gal/B antibodies (Fig. 1 and Table 1). The remaining anti-blood group B antibody clones, which do not bind to α-gal epitopes, may be regarded as "pure anti-blood group B antibodies" that require the presence of the branching fucosyl to bind the antigen. As discussed below, immunization of rabbits with human blood type B RBC results in production of only "pure anti-blood group B antibodies" because the α-gal epitope as a self-antigen in rabbits prevents production of pure anti-Gal or anti-Gal/B antibodies. Anti-Gal/B antibody clones are absent in blood type B and AB individuals

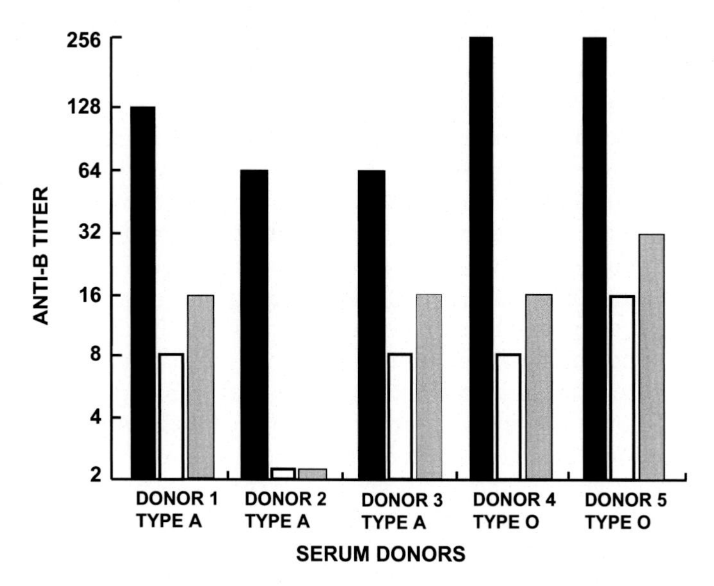

FIGURE 2 Demonstration of anti-Gal/B antibodies in sera of blood types A and O healthy individuals. The activity of these antibodies is indicated by the decrease in anti-blood group B antibody activity (determined by hemagglutination of human blood type B RBC) following adsorption of sera on equal volume of packed rabbit RBC, which present multiple α-gal epitopes (*open columns*), or on synthetic α-gal epitopes linked to silica beads (*gray columns*). Titers are presented as the reciprocal of the highest serum dilution yielding agglutination of blood type B RBC. Anti-blood group B antibody activity in the original serum is presented as *closed columns*. Note that removal of anti-Gal antibodies by these adsorptions results in elimination of 75%–95% of anti-blood group B antibody activity. *Based on data in Galili, U., Buehler, J., Shohet, S.B., Macher, B.A., 1987a. The human natural anti-Gal IgG. III. The subtlety of immune tolerance in man as demonstrated by cross reactivity between natural anti-Gal and anti-B antibodies. J. Exp. Med. 165, 693–704.*

(Table 1) because the immune tolerance mechanisms prevent appearance of such autoreactive B cell clones in these individuals (see Chapter 5). This absence of anti-Gal/B clones in blood type B and AB individuals may be associated with the reports on lower anti-Gal titers in the serum of a proportion of these individuals in comparison with anti-Gal titers in blood type A and O individuals (Buonomano et al., 1999; McMorrow et al., 1997; Bernth-Jensen et al., 2011).

Anti-Gal A/B antibody clones

Purification of anti-Gal from plasma of blood type O individuals by affinity chromatography on columns with synthetic α-gal epitopes resulted in isolation of anti-Gal antibodies that also displayed a weak agglutinating activity with blood type A RBC (Galili et al., 1987a). This suggested that the spectrum of anti-Gal specificities in individuals lacking both A and B antigens also includes antibody clones that interact with Galα1-3Galβ1-4GlcNAc core portion of blood group A and B, at a facet that does not include the *N*-acetyl group (CH_3CONH-) linked to the terminal α-galactosyl, as well as the fucosyl branching from the penultimate galactose. This anti-Gal specificity is referred to as anti-Gal A/B. The activity of anti-Gal A/B in blood type O individuals is at least 10-fold lower than that of anti-Gal/B antibody clones. The binding site pocket or cavity in anti-Gal A/B antibody clones is schematically illustrated as the dotted line in Fig. 1. It is of interest to note that the titers of pure anti-Gal, anti-Gal/B and anti-Gal

A/B antibodies were found to increase in blood type O recipients of ABO incompatible kidney allograft from blood type A or B donors (Galili et al., 2003). These observations suggest that the B cell clones producing these three types of anti-Gal antibodies are activated to expand and secrete their antibodies following transplantation of ABO incompatible allografts, even under immunosuppression regimens that prevent rejection because of anti-HLA immune response.

The observations on anti-Gal/B and anti-Gal A/B antibodies demonstrate the spectrum of anti-Gal specificities. Because these are polyclonal antibodies produced by B cells with different immunoglobulin genes, and thus, with a variety of amino acid sequences in their CDRs (Wang et al., 1995; Kearns-Jonker et al., 1999), their binding sites can interact with various facets of the α-gal epitope, some of which are found also on A and B antigens. In addition, these studies exemplify the potential of the immune tolerance mechanisms in fine differentiation between self- and nonself carbohydrate antigens. Whereas in blood type O individuals, the spectrum of anti-Gal clones also includes anti-Gal/B and anti-Gal A/B clones, in blood type A individuals, the immune tolerance prevents the production of anti-Gal A/B clones and in blood type B and AB individuals, the immune tolerance prevents production of both anti-Gal A/B and anti-Gal/B antibody clones, thus restricting the production only to pure anti-Gal antibody clones.

LANDSTEINER'S B-LIKE ANTIGEN IS THE α-GAL EPITOPE

The finding that 75%–95% of anti-blood group B antibodies in blood type A and O individuals are in fact anti-Gal antibody clones capable of binding also to B antigen (Fig. 2) further explains the enigma of the B-like antigen described 92 years ago, by Karl Landsteiner, the discoverer of the ABO blood group antigens (Landsteiner and Philip-Miller, 1925). Landsteiner's study focused on blood groups in New World monkeys and was the third in a series of three studies on blood groups in apes and monkeys, published back to back in Journal of Experimental Medicine in 1925. Landsteiner and Philip-Miller (1925), and later Owen (1954) made two observations, which appeared contradictory to each other: (1) They found rabbit RBC to adsorb a large proportion, but not all, of the anti-blood group B reactivity in sera of blood type A individuals, and (2) They further found that immunization of rabbits with human blood type B RBC resulted in production of anti-blood group B antibodies that could agglutinate human B type RBC, even after adsorption of these antibodies on blood type O RBC. Thus, the antigen on rabbit RBC interacting with human anti-blood group B antibody could not be the same blood group B antigen on human RBC. Therefore, this antigen was designated "B-like" antigen. The finding that anti-Gal/B antibodies comprised most anti-blood group B antibody activity in humans implies that the B-like antigen on the rabbit RBC is the α-gal epitope, which is abundant on these RBC (~2×10^6 epitopes/cell) and it readily binds anti-Gal/B antibodies (Galili et al., 1987a,c). The anti-blood group B antibodies produced in the immunized rabbits are antibodies with specificity like that of the minority human "pure" anti-blood group B antibodies, i.e., antibodies that require the presence of the branching fucosyl of the B antigen to bind to human blood type B RBC. Rabbits can produce these pure anti-blood group B antibodies because they lack the B antigen (i.e., fucosylated α-gal epitopes).

The primary objective of the study of Landsteiner and Philip-Miller (1925) was to determine the blood types of New World monkeys. They found that the B-like antigen is expressed on RBC of New World monkeys, but not of Old World monkeys or apes, i.e., they found that

New World monkey RBC bound anti-blood group B antibodies produced in humans, but not anti-blood group B antibodies produced in rabbits. These observations of Landsteiner were confirmed by other researchers studying blood group antigens in primates (Gengozian, 1964; Froehlich et al., 1977; Socha and Ruffie, 1983). Anti-Gal from blood type AB individuals (i.e., pure anti-Gal) was found to bind to RBC of all New World monkey species tested and TLC immunostaining demonstrated the binding of human anti-Gal to the same CPH in New World monkey RBC as that in rabbit RBC (Galili et al., 1987c) (as in Fig. 1 of Chapter 1). Accordingly, α1,3GT catalytic activity was demonstrated in New World monkey nucleated cells as in other nonprimate mammals (Galili et al., 1988a). These studies imply the B-like antigen observed by Landsteiner and Phillip-Miller on rabbit and New World monkey RBC is the α-gal epitope, which binds pure anti-Gal as well as anti-Gal/B antibodies that composed most of anti-blood group B antibody activity in humans. The reasons for the appearance of these anti-Gal antibodies after ancestral Old World primates split from New World monkeys are discussed in Chapter 2.

CONCLUSIONS

Anti-Gal is an abundant natural polyclonal antibody in humans constituting ~1% of circulating immunoglobulins. Although anti-Gal binds specifically to the α-gal epitope with the structure Galα1-3Galβ1-4GlcNAc-R, different clones of this antibody "recognize" various facets of the α-gal epitope. All anti-Gal antibody clones produced in blood type B and AB individuals bind only to the α-gal epitope. However, ~50% of anti-Gal clones in blood type A and O individuals bind both to the α-gal epitope and to blood group B antigen (Galα1-3[Fucα1-2]Galβ1-4GlcNAc-R), which has the same structure as the α-gal epitope, but includes an additional fucosyl, linked α1-2 to the penultimate galactose. This binding is performed by anti-Gal antibody clones that interact with facets of blood group B antigen that do not include the fucosyl and thus have the same spatial structure as the α-gal epitope. These anti-Gal antibodies are called anti-Gal/B antibodies and they comprise 75%–95% of human antibodies binding to blood group B antigen. The remaining 5%–25% anti-blood group B antibodies "require" the presence of the fucosyl linked to the penultimate galactose for binding to blood group B antigen. Anti-Gal/B antibodies bind to rabbit RBC and to New World monkey RBC, all lacking blood group B antigen but presenting multiple α-gal epitopes. The binding of anti-Gal/B antibodies to these RBC implies that the carbohydrate antigen on rabbit and New World monkey RBC, called by Karl Landsteiner in 1925, the "B-like" blood group antigen, is the α-gal epitope. This epitope was evolutionary conserved in nonprimate mammals, prosimians, and New World monkeys but is absent in Old World monkeys, apes, and humans.

References

Agostino, M., Sandrin, M.S., Thompson, P.E., Yuriev, E., Ramsland, P.A., 2009. In silico analysis of antibody-carbohydrate interactions and its application to xenoreactive antibodies. Mol. Immunol. 47, 233–246.

Agostino, M., Sandrin, M.S., Thompson, P.E., Yuriev, E., Ramsland, P.A., 2010. Identification of preferred carbohydrate binding modes in xenoreactive antibodies by combining conformational filters and binding site maps. Glycobiology 20, 724–735.

Almeida, I.C., Milani, S.R., Groin, P.A.J., Travassos, L.R., 1991. Complement-mediated lysis of *Trypanosoma cruzi* trypomastigotes by human anti-α-galactosyl antibodies. J. Immunol. 146, 2394–2400.

Bernth-Jensen, J.M., Møller, B.K., Jensenius, J.C., Thiel, S., 2011. Biological variation of anti-αGal-antibodies studied by a novel time-resolved immuno-fluorometric assay. J. Immunol. Methods 373, 26–35.

Bloy, C., Blanchard, D., Lambin, P., Goossens, D., Rouger, P., Salmon, C., et al., 1988. Characterization of the D, C, E and G antigens of the Rh blood group system with human monoclonal antibodies. Mol. Immunol. 25, 925–930.

Buonomano, R., Tinguely, C., Rieben, R., Mohacsi, P.J., Nydegger, U.E., 1999. Quantitation and characterization of anti-Galα1-3Gal antibodies in sera of 200 healthy persons. Xenotransplantation 6, 173–180.

Evans, D.W., Müller-Loennies, S., Brooks, C.L., Brade, L., Kosma, P., Brade, H., et al., 2011. Structural insights into parallel strategies for germline antibody recognition of lipopolysaccharide from Chlamydia. Glycobiology 21, 1049–1059.

Froehlich, J.W., Socha, W.W., Wiener, A.S., Moor-Jankowski, J., Thorginton, R.W., 1977. Blood groups of mantled howler monkey. J. Med. Primatol. 6, 219–231.

Galili, U., 2006. Xenotransplantation and ABO incompatible transplantation: the similarities they share. Transfus. Apher. Sci. 35, 45–58.

Galili, U., Rachmilewitz, E.A., Peleg, A., Flechner, I., 1984. A unique natural human IgG antibody with anti-α-galactosyl specificity. J. Exp. Med. 160, 1519–1531.

Galili, U., Macher, B.A., Buehler, J., Shohet, S.B., 1985. Human natural anti-α-galactosyl IgG. II. The specific recognition of α(1-3)-linked galactose residues. J. Exp. Med. 162, 573–582.

Galili, U., Buehler, J., Shohet, S.B., Macher, B.A., 1987a. The human natural anti-Gal IgG. III. The subtlety of immune tolerance in man as demonstrated by cross reactivity between natural anti-Gal and anti-B antibodies. J. Exp. Med. 165, 693–704.

Galili, U., Basbaum, C., Shohet, S.B., Buehler, J., Macher, B.A., 1987b. Identification of erythrocyte Galα1-3Gal glycosphingolipids with a mouse monoclonal antibody Gal-13. J. Biol. Chem. 262, 4683–4688.

Galili, U., Clark, M.R., Shohet, S.B., Buehler, J., Macher, B.A., 1987c. Evolutionary relationship between the anti-Gal antibody and the Galα1-3Gal epitope in primates. Proc. Natl. Acad. Sci. U.S.A. 84, 1369–1373.

Galili, U., Shohet, S.B., Kobrin, E., Stults, C.L.M., Macher, B.A., 1988a. Man, apes, and Old World monkeys differ from other mammals in the expression of α-galactosyl epitopes on nucleated cells. J. Biol. Chem. 263, 17755–17762.

Galili, U., Mandrell, R.E., Hamadeh, R.M., Shohet, S.B., Griffiss, J.M., 1988b. Interaction between human natural anti-α-galactosyl immunoglobulin G and bacteria of the human flora. Infect. Immun. 56, 1730–1737.

Galili, U., Matta, K.L., 1996. Inhibition of anti-Gal IgG binding to porcine endothelial cells by synthetic oligosaccharides. Transplantation 62, 256–262.

Galili, U., LaTemple, D.C., Radic, M.Z., 1998. A sensitive assay for measuring α-gal epitope expression on cells by a monoclonal anti-Gal antibody. Transplantation 65, 1129–1132.

Galili, U., Ishida, H., Toma, K., Tanabe, H., 2003. Anti-Gal A/B, a novel anti-blood group antibody identified in recipients of ABO incompatible kidney allografts. Transplantation 74, 1574–1580.

Gengozian, N., 1964. Human A-like and B-like antigens on red cells of marmosets. Proc. Soc. Exp. Biol. Med. 117, 858–861.

Hakomori, S.-I., 1981. Blood group ABH and Ii antigens of human erythrocytes: chemistry, polymorphism, and their developmental change. Semin. Hematol. 18, 39–62.

Kabat, E.A., 1956. Heterogeneity in the extent of the combining regions of human anti-dextran. J. Immunol. 77, 377–385.

Kabat, E.A., 1976. Structural Concepts in Immunology and Immunochemistry, second ed. Holt, Rinehart and Winston General Book, New York, p. 547.

Kearns-Jonker, M., Swensson, J., Ghiuzeli, C., Chu, W., Osame, Y., Starnes, V., et al., 1999. The human antibody response to porcine xenoantigens is encoded by IGHV3–11 and IGHV3–74 IgVH germline progenitors. J. Immunol. 163, 4399–4412.

Landsteiner, K., Philip-Miller, C., 1925. Serological studies on the blood of the primates. III. Distribution of serological factors related to human isoagglutinogens in the blood of lower monkeys. J. Exp. Med. 42, 863–872.

Magnani, J.L., Brockhaus, M., Smith, D.F., Ginsburg, V., Blaszczyk, M., Mitchell, K.F., et al., 1981. A monosialoganglioside is a monoclonal antibody-defined antigen of colon carcinoma. Science 212, 55–56.

McMorrow, I.M., Comrack, C.A., Nazarey, P.P., Sachs, D.H., DerSimonian, H., 1997. Relationship between ABO blood group and levels of Galαl-3Galactose-reactive human immunoglobulin G. Transplantation 64, 546–549.

Owen, R.D., 1954. Heterogeneity of antibodies to human blood groups in normal and immune sera. J. Immunol. 73, 29–39.

Plum, M., Michel, Y., Wallach, K., Raiber, T., Blank, S., Bantleon, F.I., et al., 2011. Close-up of the immunogenic α1,3-galactose epitope as defined by a monoclonal chimeric immunoglobulin E and human serum using saturation transfer difference (STD) NMR. J. Biol. Chem. 286, 43103–43111.

Posekany, K.J., Pittman, H.K., Bradfield, J.F., Haisch, C.E., Verbanac, K.M., 2002. Induction of cytolytic anti-Gal antibodies in α-1,3-galactosyltransferase gene knockout mice by oral inoculation with *Escherichia coli* O86:B7 bacteria. Infect. Immun. 70, 6215–6222.

Ramsland, P.A., Farrugia, W., Bradford, T.M., Mark Hogarth, P., Scott, A.M., 2004. Structural convergence of antibody binding of carbohydrate determinants in Lewis Y tumor antigens. J. Mol. Biol. 340, 809–818.

Socha, W.W., Ruffie, J., 1983. Blood Groups of Primates: Theory, Practice, Evolutionary Meaning. Alan R. Liss, Inc., New York, pp. 39–51.

Tinguely, C., Schaller, M., Carrel, T., Nydegger, U.E., 2002. Spectrotype analysis and clonal characteristics of human anti-Galα1-3Gal antibodies. Xenotransplantation 9, 252–259.

Wang, L., Radic, M.Z., Galili, U., 1995. Human anti-Gal heavy chain genes: preferential use of V_H3 and the presence of somatic mutations. J. Immunol. 155, 1276–1285.

CHAPTER

4

Anti-Gal Interaction With *Trypanosoma*, *Leishmania*, and *Plasmodium* Parasites

OUTLINE

INTRODUCTION

One area in which the natural anti-Gal antibody may be considered to have protective functions in humans is infection by protozoan parasites including *Trypanosoma*, *Leishmania*, and *Plasmodium*. Studies have been performed in the last 30 years on anti-Gal and these parasites, including analysis of this antibody activity in infected patients, *in vitro* interaction of anti-Gal with the parasites, and identification of α-gal-like epitopes, which may bind anti-Gal to these parasites. The α-gal-like epitopes are carbohydrate epitopes that may have a structure resembling α-gal epitopes, with terminal α-galactosyl linked to penultimate carbohydrates in an α1-2, α1-3, or α1-6 linkage, and which can bind some of the anti-Gal antibody clones in human serum. These studies have suggested that anti-Gal binds to the parasites in infected individuals and that such interaction may result in killing of these protozoa and their neutralization. This chapter reviews studies supporting this assumption and suggests a method that may improve the protective effects of anti-Gal against these protozoan infections.

The Natural Anti-Gal Antibody as Foe Turned Friend in Medicine
http://dx.doi.org/10.1016/B978-0-12-813362-0.00004-X

ANTI-GAL INTERACTION WITH *Trypanosoma cruzi* IN CHAGAS' DISEASE

Chagas' disease is caused by *Trypanosoma cruzi*

Chagas' disease (American trypanosomiasis) is endemic in South America where ~11 million patients are infected with *Trypanosoma cruzi*, the parasite causing this disease, and about 100 million are at risk (Coura and Vinas, 2010). The parasite epimastigotes proliferate in the gut of the triatomine bug of the Reduviidae family and is introduced as metacyclic trypomastigotes into the human blood circulation following scratching into the skin feces of this bug. In the acute phase of the disease, trypomastigotes invade red blood cells (RBC) in which the parasite differentiates into proliferating amastigotes. Subsequently, the amastigotes filling infected RBC differentiate into nondividing trypomastigotes that rupture the RBC membrane and invade into additional RBC to repeat the expansion of *T. cruzi* in the blood. *T. cruzi* evades the protective immune response mounted by the host against the parasite by invading nucleated cells of the liver, colon, esophagus, heart, and peripheral nerve system as amastigotes. Following this invasion, *T. cruzi* amastigotes induce the chronic phase of the disease in which the parasite sheds its antigens, which elicit extensive chronic inflammations. These autoimmune-like inflammations cause hepatomegaly, megacolon, and cardiomyopathy that gradually destroy the infected tissues, resulting in a chronic disease that ends with heart failure in many of the patients.

Anti-Gal production in patients with Chagas' disease

The interest in measuring anti-Gal activity in patients with Chagas' disease emerged from a study demonstrating the extensive elevation of antibodies to mouse laminin (Szarfman et al., 1982). Laminin is a glycoprotein found in large amounts, primarily in basement membranes and extracellular matrix of many tissues (Timpl et al., 1978). Identification of antibodies against laminin in Chagasic patients was of interest because it could help in understanding the autoimmune-like phenomena observed in Chagas' disease. The findings that mouse and bovine laminins carry many α-gal epitopes (Rao et al., 1983; Mohan and Spiro, 1986; Arumugham et al., 1986) prompted researchers to determine whether the elevated anti-laminin antibodies bind to the multiple α-gal epitopes on laminin. Studies on antibodies in the blood of rhesus monkeys infected with *T. cruzi*, or with *Trypanosoma rhodesiense* and in blood of patients with American cutaneous leishmaniasis indicated that the antibody binding to laminin is the anti-Gal antibody, which is produced in much higher titers than natural anti-Gal in healthy human sera (Towbin et al., 1987). This elevated anti-Gal activity was subsequently confirmed in several independent studies, both in acute and in chronic phases of Chagas' disease (Milani and Travassos, 1988; Gazzinelli et al., 1988; Avila et al., 1989; González et al., 1996; Antas et al., 1999; Brito et al., 2016; Schocker et al., 2016). These studies were performed with various methods measuring anti-Gal activity including agglutination of rabbit RBC and ELISA with various solid-phase antigens presenting multiple α-gal epitopes, including mouse laminin, virus-like particles expressing these epitopes, and synthetic α-gal epitopes linked to bovine serum albumin (α-gal-BSA).

An example of anti-Gal analysis in Chagasic patients is included in Fig. 1. The titers of anti-Gal IgG were determined in patients with chronic Chagas' disease and in patients with active

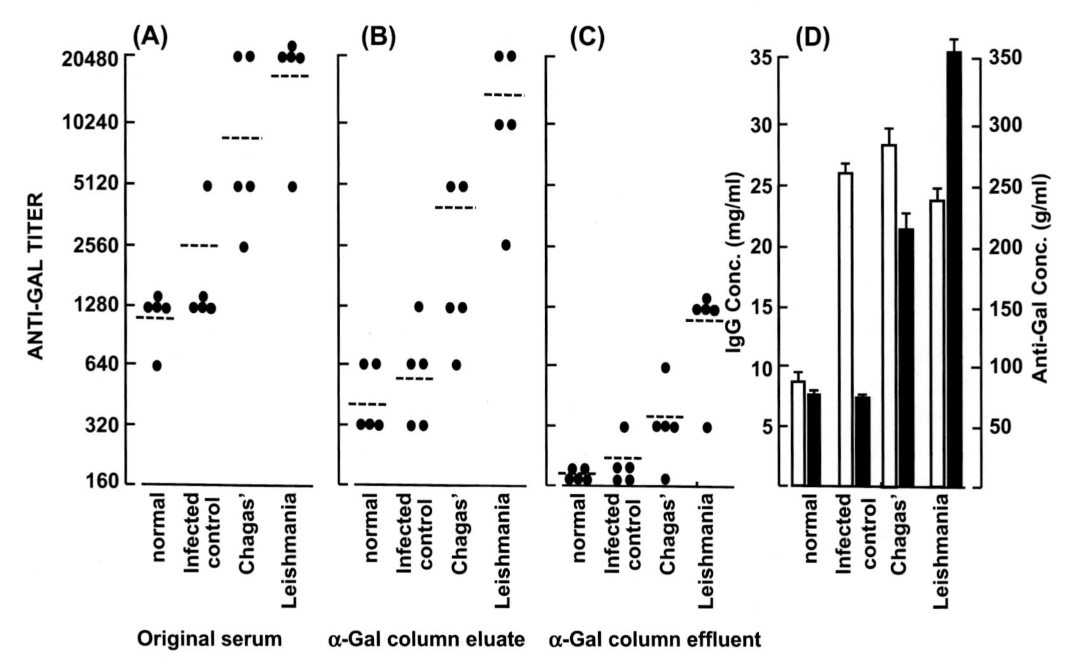

FIGURE 1 Anti-Gal IgG titers in 5 representative sera out of 9 healthy individuals, 23 patients with acute bacterial infections of β-hemolytic group A *Streptococcus*, *Haemophilus influenza* or *Staphylococcus aureus*, 31 patients with chronic Chagas' disease and 31 patients with active nontreated American cutaneous leishmaniasis. Anti-Gal titers were determined by indirect hemagglutination assay with rabbit RBC, using rabbit anti-human IgG as secondary antibody. (A) Anti-Gal titers in the original serum sample. (B) Anti-Gal titers in eluates from serum past through columns with synthetic α-gal epitopes linked to Synsorb silica beads. (C) Anti-Gal titers in serum effluents from these columns. (D) Serum total IgG concentration (*open columns*) and anti-Gal IgG concentration (*closed columns*) in all sera studied, presented as mean ± S.E. IgG concentration was determined by Mancini radial immunodiffusion assay with rabbit anti-humans IgG antibody. *Dashed lines* in (A)–(C) represent the mean titer in each group. *Adapted from Avila, J.L., Rojas, M., Galili, U., 1989. Immunogenic Galα1-3Gal carbohydrate epitopes are present on pathogenic American* Trypanosoma *and* Leishmania. *J. Immunol. 142, 2828–2834, with permission.*

nontreated American cutaneous leishmaniasis by indirect agglutination of rabbit RBC using rabbit anti-human IgG antibody as secondary antibody (Avila et al., 1989). These RBC express an abundance of α-gal epitopes (Eto et al., 1968; Stellner et al., 1973; Egge et al., 1985; Galili et al., 1987a; Ogawa and Galili, 2006) and thus are convenient for performing hemagglutination assays for measuring anti-Gal. As shown in Fig. 1A, the titer of anti-Gal IgG in Chagasic patients was ~10-fold higher and in patients with leishmaniasis ~20-fold higher than that in healthy individuals (*Leishmania* interaction with anti-Gal is discussed below in this chapter). Passing of the sera through columns of Synsorb silica beads with linked synthetic α-gal epitopes resulted in adsorption of ~90% of the hemagglutinating activity, indicating that this assay indeed measures anti-Gal activity (Fig. 1B). Activity of the isolated Chagasic anti-Gal, which was eluted from α-gal columns (α-gal column eluate), was ~10-fold higher than that of anti-Gal isolated from a similar volume of normal human serum. The binding of anti-Gal to the α-gal column further correlated with the decrease in this antibody activity in the serum effluent after passing through the α-gal column (α-gal column effluent in Fig. 1C).

As discussed in Chapter 1, the increase in anti-Gal activity in Chagasic sera seems to be the result of a combined increase in concentration of the antibody (the result of activation of quiescent anti-Gal B cells) and increase in affinity of anti-Gal (because of affinity maturation of anti-Gal B cells). The combined increase in concentration and affinity of anti-Gal is suggested from measures of the concentration of anti-Gal IgG versus concentration of total IgG in the various sera (Fig. 1D). The concentration of both anti-Gal and of total IgG in Chagasic patients increased by an average of approximately threefold (Fig. 1D), whereas the titer of the antibody increased by 10-fold (Fig. 1A). In patients with bacterial infections, including group A *Streptococcus*, *Haemophilus influenza*, and *Staphylococcus aureus*, anti-Gal activities and anti-Gal concentrations did not differ significantly from those observed in healthy individuals, despite the increase in total IgG concentration by approximately threefold (Fig. 1) (Avila et al., 1989).

A potentially useful experimental model for studying anti-Gal response to *T. cruzi* is the α1,3galactosyltransferase knockout (GT-KO) mouse, which lacks α-gal epitopes (Thall et al., 1995; Tearle et al., 1996). Anti-Gal production could be experimentally induced by inoculation of GT-KO mice with *T. cruzi* trypomastigotes (Avila, 1999). Similar studies in wild-type mice demonstrated no anti-Gal production because the α-gal epitope is a self-antigen in these mice. As discussed below, this model may serve for identification of probiotic bacteria that can induce continuous elevation of anti-Gal activity to increase anti-Gal-mediated protection against infections by *T. cruzi* in endemic areas.

Trypanosoma cruzi molecules binding anti-Gal

The extensive increase in anti-Gal activity in acute and chronic phases of Chagas' disease suggested that the infecting *T. cruzi* parasite may present immunizing α-gal-like epitopes on this protozoan cell surface. Studies with Chagasic anti-Gal isolated by affinity chromatography, as in Fig. 1B, indicated that this antibody readily binds to *T. cruzi* trypomastigotes (Travassos et al., 1988; Milani and Travassos, 1988; Travassos and Almeida, 1993; Almeida et al., 1991, 1993, 1994a, 1994b). A monoclonal anti-Gal antibody, called Gal-13 specific for α-gal epitopes on glycolipids (Galili et al., 1987b), was also found to readily bind to *T. cruzi* in ELISA (Avila et al., 1989). Immunostaining of *T. cruzi in* electron microscopy, using gold particles, demonstrated extensive Chagasic anti-Gal extracellular binding to trypomastigotes and amastigotes grown in Vero cells (Old World monkey cells lacking endogenous α-gal epitopes), whereas binding of purified anti-Gal from healthy individuals to the parasite was poor (Souto-Padron et al., 1994). Electron microscopy studies further indicated that anti-Gal binds to mucin-like molecules on the parasite cell membrane (Pereira-Chioccola et al., 2000). Poor binding of anti-Gal from normal human serum to the trypomastigotes versus effective binding of Chagasic anti-Gal was also observed in other binding assays such as immunostaining of blotted proteins (Almeida et al., 1994a, 1994b). It is possible that the difference between the two preparations of anti-Gal is the higher concentration and higher affinity of the Chagasic anti-Gal, which may be the result affinity maturation. Alternatively, there could be preferential expansion of anti-Gal B cell clones producing anti-Gal that "fits" specifically the α-gal-like epitopes on *T. cruzi* (see Chapter 3 for discussion on various anti-Gal antibody clones).

Several studies aimed to determine the identity of the *Trypanosoma* carbohydrate epitopes binding the purified Chagasic anti-Gal demonstrated the presence of Galα1-3Galβ1-4GalNAc on glycoinositol phospholipid (GPI)-anchored proteins (Almeida et al., 1994a, 1994b; Soares et al., 2012). Additional terminal α-galactosyls in α-gal-like sequences, including Galα1-2Gal and Galα1-6Gal linked to variant surface glycoproteins and GPI, expressed on *T. Brucei*, were also found to bind Chagasic anti-Gal (Almeida et al., 1994b). An independent support for the expression of α-gal-like epitopes on both *T. cruzi* and *T. Brucei* has been provided by the observation on binding of the lectin *Bandeiraea* (*Griffonia*) *simplicifolia IB4* to these parasites (Avila et al., 1989; Ramasamy and Field, 2012). This lectin binds specifically to the α-gal epitope on mammalian cells and to blood group B antigen in which there is a fucose linked to the penultimate galactose of the α-gal epitope (Wood et al., 1979).

Protective effects of anti-Gal in the acute stage of Chagas' disease

Anti-Gal may have a protective effect in the acute stage of Chagas' disease. This is suggested by *in vitro* studies in which incubation of *T. cruzi* trypomastigotes with Chagasic anti-Gal (purified by affinity chromatography) and with complement resulted in complement-mediated cytolysis of the parasite (Milani and Travassos, 1988; Almeida et al., 1991). Similar cytolysis by anti-Gal from rhesus monkey infected with *T. cruzi* could be inhibited in these studies by mouse laminin. These findings and the observations on the anti-Gal identity of elevated anti-mouse laminin antibodies confirmed the assumption that the cytolytic anti-mouse laminin antibodies are indeed anti-Gal antibodies binding to *T. cruzi* and inducing its complement-mediated cytolysis (Travassos et al., 1988; Milani and Travassos, 1988; Almeida et al., 1991). Interestingly, *T. cruzi* can coat itself with sialic acid originating from host cells by the enzyme *trans*-sialidase (Schenkman et al., 1991). The negative charge provided by the sialic acid reduces the extent of anti-Gal binding and complement-mediated cytolysis by this antibody (Pereira-Chioccola et al., 2000). It is possible that the protective negatively charged layer linked to mucin-like molecules on the surface of *T. cruzi* "repels" anti-Gal, which also carries negative charges of sialic acid on the carbohydrate chains linked to the Fc portion of the antibody.

T. cruzi was also found to be killed by Chagasic anti-Gal in the absence of complement, by forming clusters of glycoconjugates carrying α-gal or α-gal-like epitopes (Gazzinelli et al., 1991). Such clusters are formed primarily by anti-Gal IgA and IgM isotypes that have 4 and 10 binding sites, respectively. These clusters are likely to cause perturbation and destruction of the membrane structure leading to direct cytolysis. A similar type of direct cytolysis was observed in mouse cell lines binding the α-gal epitope–specific lectin *B.* (*Griffonia*) *simplicifolia IB4* (Kim et al., 1993), as well as with *T. cruzi* binding this lectin (Gazzinelli et al., 1991). The complement-mediated and the direct cytolysis mechanisms may contribute to protection against infection by *T. cruzi* and may attenuate the acute stage of the disease.

In view of the protective effects of anti-Gal, it is possible that the higher incidence of Chagas' disease in infants and in elderly individuals is associated with lower variety of anti-Gal clones and lower titers of this antibody. This lower activity and diversity of anti-Gal antibody clones may provide less protection against the infecting parasite than in adult populations. Experimental support to the assumption that anti-Gal may contribute to immune protection against infections by *T. cruzi* is provided by studies in GT-KO mice.

As mentioned above, intraperitoneal inoculation of these mice with *T. cruzi* induced production of anti-Gal, not seen in wild-type mice (Avila, 1999). Accordingly, the extent of parasitemia in the GT-KO mice was found to be lower by ~70% than that in wild-type mice (Avila, 1999). In addition, the decrease in the number of trypomastigotes in wild-type mice evaluated 30 days post infection was only ~30% of the peak value, whereas in anti-Gal producing GT-KO mice, the decrease was of 65% in the number of parasites (Avila, 1999). It is probable that anti-Gal is only one of several mechanisms protecting against *T. cruzi* infection. This can be inferred from observations on the immune protection against the parasite in wild-type mice (i.e., mice incapable of producing anti-Gal) that are immunized against *T. cruzi* (Avila, 1999).

Possible anti-Gal contribution to autoimmune-like damage in chronic Chagas' disease

The marked increase in anti-Gal titers in *T. cruzi* infected individuals, the possible expansion of anti-Gal clones interacting with α-gal-like epitopes on *T. cruzi* and the production of antibodies against proteins of the parasite, all are likely to result in effective destruction of the parasite in the circulation. However, *T. cruzi* evades this protective immune response by invading into nucleated cells of various tissues in which the amastigotes of the parasite are shielded from such protective antibodies. This transition results in the conversion of the protective immune response into a detrimental one, which mediates extensive autoimmune-like inflammatory reactions that cause hepatomegaly, megacolon, and cardiomyopathy. These chronic inflammatory reactions cause gradual impairment of the infected organs, which may become lethal, mostly because of heart failure. Studies of such cardiomyopathies have indicated that there are multiple mechanisms that may contribute to the inflammatory processes, which result in fibrosis of the myocardium in Chagas' disease patients (cf. Bonney and Engman, 2015). Moreover, the absence of parasites in a significant proportion of myocardial sections in these patients led investigators already in the 1950s to suggest that part the immune response resulting in cardiomyopathies in Chagas' disease patients may be mediated by autoimmune reactivity against self-antigens of the myocardium (Laranja et al., 1956). Such autoimmune mechanisms may be elicited by the chronic inflammatory reactivity in the heart of patients in the chronic disease stage. However, the 10–20-fold higher anti-Gal activity observed in chronic Chagasic patients, in comparison with sera from healthy individuals (see Fig. 1 and Towbin et al., 1987; Travassos et al., 1988; Milani and Travassos, 1988; Avila et al., 1989; Gazzinelli et al., 1991; González et al., 1996; Antas et al., 1999; Schocker et al., 2016; Brito et al., 2016), suggests the presence of *T. cruzi* parasites producing α-gal-like epitopes in infected tissues. These epitopes may stimulate anti-Gal B cells to produce anti-Gal at elevated titers, like the elicited production of anti-Gal in patients exposed to xenoglycoproteins (Galili et al., 1995, 2001; Stone et al., 2007a). Although the amastigotes are difficult to detect in histological sections of infected tissues, it is possible that the interaction between the elicited anti-Gal antibody and α-gal-like epitopes shed from the parasite contributes to the extensive inflammatory reactions observed in the infected tissues. This assumption is supported by observations on Chagasic anti-Gal binding to cells lacking α-gal epitopes after such cells are incubated with *T. cruzi* antigens (Yokoyama-Yasunaka et al., 1998). This study suggests that anti-Gal binds to epitopes shed from *T. cruzi*, which adhere to normal cells of the host and induce anti-Gal-mediated destruction of such cells. The binding of anti-Gal to *T. cruzi* antigens attached to normal cells may result in destruction of the cells either by complement-dependent cytolysis or by antibody-dependent cell cytolysis of anti-Gal coated cells, similar to the cytolytic effects of anti-Gal in recipients of xenograft presenting α-gal epitopes (see Chapter 6). Moreover, Vero cells (Old World monkey cells lacking α-gal epitopes) infected with *T. cruzi* were found by scanning electron microscopy with gold particles to bind *in vitro* Chagasic anti-Gal, as well as *B. simplicifolia IB4* lectin, which binds to α-gal epitopes (Souto-Padron et al., 1994). These findings suggested that the *T. cruzi* α-gal-like epitopes appear *in vivo* on the cell membrane of the *T. cruzi* infected cells bind anti-Gal and thus provide the basis for autoimmune-mediated destruction of such cells.

An additional mechanism for anti-Gal-mediated inflammatory reactions in Chagas' disease may be associated with the extracellular vesicles formed by the parasite. *T. cruzi*, trypomastigotes were reported to shed high numbers extracellular vesicles (Goncalves et al., 1991; Trocoli Torrecilhas et al., 2009, Torrecilhas et al., 2012), with the size of ~150nm (Nogueira et al., 2015). These vesicles are enriched with various glycoconjugates that carry multiple α-gal-like epitopes, such as mucins (Trocoli Torrecilhas et al., 2009; Nogueira et al., 2015). The extensive ability of extracellular vesicles presenting multiple α-gal-like epitopes to mediate inflammatory reactions can be inferred from similar abilities of α-gal nanoparticles expressing multiple α-gal epitopes (described in Chapters 12 and 13).

α-Gal nanoparticles are prepared from rabbit RBC glycolipids, phospholipids, and cholesterol and have a size range like that of extracellular vesicles (see Fig. 5 in Chapter 12). These nanoparticles present ~10^{15} α-gal epitopes/mg (Wigglesworth et al., 2011). Administration of small numbers of α-gal nanoparticles is sufficient to activate the complement system by anti-Gal/α-gal epitope interaction and induce production of complement cleavage chemotactic factors such as C5a and C3a. These chemotactic factors induce rapid recruitment of macrophages, which reach the α-gal nanoparticles within 24h (Fig. 3 in Chapter 12). If intracellular amastigotes infecting heart or other tissues in the chronic stage of Chagas' disease continuously produce such extracellular vesicles, the interaction of anti-Gal with these vesicles may induce similar chronic processes of complement activation, recruitment of macrophages, and gradual damage to the infected tissue. Adhesion of such vesicles to cell membranes of healthy cells may further result in subsequent anti-Gal binding and destruction of the cells in a manner similar to anti-Gal binding to bacterial fragments that attach to human cells (Galili et al., 1988). Such chronic inflammation may further result in activation of CD4+ and CD8+ T cells that may join the macrophages and react against normal autologous antigens on cells of the infected tissue, resulting in a "real" autoimmune disease that contributes to the destruction of the infected organ.

A suggested approach for increasing long-term anti-Gal protection against *Trypanosoma cruzi*

The complement-dependent and independent cytolytic activities of Chagasic anti-Gal on *T. cruzi* trypomastigotes, described above, raise the possibility that elevated titers of anti-Gal in individuals living in *T. cruzi* endemic regions may provide improved protection against invasion of the parasite by killing it upon its penetration into the circulation. Anti-Gal from normal human serum was found to bind well to several bacterial strains, including *Serratia marcescens*, *Klebsiella pneumonia* and *Escherichia coli O86* (Galili et al., 1988). Based on these observations, a human volunteer was immunized with a vaccine prepared of *S. marcescens*. This immunization elicited anti-Gal antibody response at a level that was cytolytic to trypomastigotes (Almeida et al., 1991). That study further reported that antibodies lytic to *T. cruzi* could be removed from serum of Chagasic patients by adsorption on *S. marcescens*. This suggests that immunization with *S. marcescens* vaccine may result in protection against *T. cruzi* infection. However, the elevated activity of anti-Gal following administration of vaccines carrying α-gal or α-gal-like epitopes is likely to be temporary. Previous studies in humans demonstrated increase in anti-Gal activity following administration of xenograft cells expressing α-gal epitopes, but the level of these antibodies decreases within several weeks after destruction of the immunizing cells (see Fig. 3A in Chapter 1) (Galili et al., 2001). As long as there is a source of immunizing α-gal glycoconjugates, there is increased production of anti-Gal antibodies. Accordingly, elimination in baboons of gram negative aerobic bacteria from the gastrointestinal (GI) tract (i.e., many of the bacteria stimulating for anti-Gal production) by antibiotics resulted in natural anti-Gal IgG activity decrease within several weeks (Mañez et al., 2001). Thus, standard immunization with α-gal or α-gal-like containing vaccine may elevate anti-Gal temporarily but will not suffice for maintaining continuous protective elevated titers of anti-Gal for prevention of infection by *T. cruzi* in endemic regions.

Bacteria, such as *E. coli O86*, were found to induce anti-Gal production in GT-KO mice, when introduced into the GI tract (Posekany et al., 2002; Yilmaz et al., 2014). Similarly, feeding of blood type A and O individuals with *E. coli O86* was reported to induce elevation in production of anti-blood group B antibodies in patients with diarrhea (Springer and Horton, 1969). As discussed in Chapter 3, 75%–95% of the antibody clones comprising anti-blood group B antibodies are anti-Gal antibodies capable of binding also to blood group B antigens (anti-Gal/B) (Galili et al., 1987c). All these studies suggest that it may be possible to achieve elevated titers of the natural anti-Gal antibodies, at levels that may be protective against *T. cruzi* infection. Continuous production of anti-Gal, at titers high enough to cause lysis of *T. cruzi* invading the circulation, may be achieved by introduction of bacteria presenting α-gal-like antigens, which can become an integral part of the normal microbiome in the GI tract and induce extensive production of anti-Gal. Production of anti-Gal at high titers for prolonged periods was found to have no detrimental effects in humans (Stone et al., 2007a) or in monkeys (Stone et al., 2007b). Early studies in humans indicated that *E. coli O86* may not be suitable for eliciting long-term-elevated anti-Gal response because it was reported to be a pathogen causing hospital-associated epidemics of enteropathogenic *E. coli* gastroenteritis (Pal et al., 1969), thus precluding its use for therapeutic purpose.

Identification of bacteria that are not harmful and which induce elevated anti-Gal production may be feasible by screening of various bacterial strains. Initial screening may include identification of nonpathogenic bacteria that bind anti-Gal, feeding of GT-KO mice or of Old World monkeys with various killed bacteria and determining both the extent of anti-Gal production and the efficacy of that anti-Gal in mediating cytolysis of *T. cruzi*. Subsequent analyses, including determination of toxicity to host, measuring *in vivo* protection against infection by *T. cruzi*, and integration into the microbiome, might help in narrowing the choice for the appropriate bacterial strain. Ultimately, such bacterial strain(s) stimulating anti-Gal production may be administered to the GI tract by methods currently used for delivery of probiotic bacteria, such as food supplements. These considerations may also be applicable to inducing production of anti-Gal, which might be protective against infections by additional parasites that bind anti-Gal, such as *Leishmania* and *Plasmodium*, as described below.

ANTI-GAL INTERACTION WITH *Leishmania* PARASITES

Anti-Gal production in patients infected with *Leishmania*

Expression of α-gal-like epitopes on cell membranes is not limited only to *Trypanosoma* but is found also in several *Leishmania* strains. Presence of such epitopes on *Leishmania* was first suggested in studies demonstrating marked elevation of anti-Gal in patients Chagas' disease and those with American cutaneous leishmaniasis caused by *Leishmania* strains such as *Leishmania mexicana* and *Leishmania braziliensis* (Towbin et al., 1987). The skin lesions in cutaneous leishmaniasis begin at the site of parasite entrance as a small papule, which develops into a nodule that ulcerates and eventually expands into a lesion with elevated borders and a sharp crater. An example of the extensive production of anti-Gal in infected patients is included in Fig. 1 (Avila et al., 1989). A similar extensive increase in anti-Gal titers was observed in Old World Cutaneous Leishmaniasis patients infected by *Leishmania major* and *Leishmania tropica* parasites (Al-Salem et al., 2014). The ability of the invading parasite to induce extensive anti-Gal production was further confirmed in the experimental animal model of GT-KO mice immunized with killed *L. major* parasites (Pearse et al., 1998) or inoculated with live *L. Mexicana* parasites (Avila, 1999). In both studies, GT-KO mice displayed marked increase in anti-Gal titers following exposure to *Leishmania* antigens. The increase in agglutination of rabbit RBC and in anti-Gal binding to mouse laminin in serum of patients and of mice infected with *Leishmania*, as well as rapid rejection of wild-type mouse heart (i.e., mouse expressing α-gal epitopes) in the *Leishmania*-immunized GT-KO mice (Pearse et al., 1998), all imply that anti-Gal clones capable of binding to the mammalian α-gal epitope comprise a large proportion of the antibodies produced as part of the immune response against the parasite.

α-Gal-like epitopes on *Leishmania*

The observations on the elicited anti-Gal production in patients infected with various *Leishmania* strains motivated researchers to determine whether α-gal-like epitopes can be identified on cell membranes of *Leishmania* parasites. Initial studies evaluated the binding of purified human natural anti-Gal and of monoclonal anti-Gal antibody Gal-13 (Galili et al., 1987b) to *L. mexicana* epimastigotes and promastigotes coating ELISA wells as solid-phase antigen. Both antibodies displayed specific binding to the parasite as well as to lipid fractions extracted from the parasites (Avila et al., 1989). Analysis of carbohydrate epitopes extracted from *L. mexicana* and *L. braziliensis* demonstrated the presence of relatively large amounts of an α-gal-like epitope with the structure Galα1-3Man (Avila and Rojas, 1990). Similarly, analysis of the glycolipids in *Leishmania* demonstrated that some of the glycoinositol phospholipids from *L. major, L. mexicana,* and *L. braziliensis* have Galα1-3Gal, Galα1-6Gal, and Galα1-3Man terminating epitopes (Schneider et al., 1993; McConville et al., 1993; Almeida et al., 1994b; Farias et al., 2013).

It is possible that these carbohydrate epitopes are contributing to the antigenic stimulation by *Leishmania* that induces the immune system to increase the titers of anti-Gal, as observed in patients with Old World and New World leishmaniasis. Similar to the observations in Chagasic patients, the elevation in anti-Gal titers in patients infected with *L. mexicana* exceeds by several fold the elevation in concentrations of this antibody in the serum (Fig. 1). This suggests that such increased anti-Gal activity is the result of higher concentration, as well as higher affinity of the antibody in comparison with sera from healthy individuals. It is not clear, however, whether the interaction of anti-Gal with these carbohydrate epitopes on parasites in infected patients has a protective effect against the *Leishmania* parasite.

ANTI-GAL INTERACTION WITH *Plasmodium* IN MALARIA

Researchers studying malaria causing *Plasmodia* reported the presence of α-gal-like epitopes on these parasites. Studies with the α-gal-specific lectin *B. simplicifolia* demonstrated binding of this lectin to the membrane of asexual blood stages of *Plasmodium falciparum* (Ramasamy and Reese, 1987). Incubation of the parasite with α-galactosidase (an enzyme that removes terminal α-galactosyl units) eliminated this binding, thus, demonstrating the specificity of the lectin binding to α-gal carrying carbohydrate chains. Yilmaz et al. (2014) confirmed these observations and further showed that treatment of the parasite with phospholipase C partially cleaved the carbohydrate chains carrying the lectin-binding epitopes, thus decreasing lectin binding. Accordingly, terminal α-galactosyl units were found on four glycoproteins of *P. falciparum*. These glycoproteins were found to bind antibodies from malaria-immune sera and treatment of the glycoproteins with α-galactosidase eliminated binding of the antibodies (Jakobsen et al., 1987). These binding patterns suggested that antibodies binding to *P. falciparum* may also include some anti-Gal clones. Indeed, studies measuring anti-Gal activity in sera reported that anti-Gal activity was significantly elevated in many subjects living in malaria endemic areas and in patients with acute *P. falciparum* malaria in comparison to subjects living in areas where the incidence of *P. falciparum* malaria was scarce (Ravindran et al., 1988; Ramasamy, 1988; Satapathy and Ravindran, 1996). IgG subclass typing in these studies by cell-ELISA with rabbit RBC as solid-phase antigen revealed IgG3 to be the predominant type with anti-Gal activity in *P. falciparum*-infected patients' serum, whereas IgG2 was found to be dominant in the nonendemic control serum (Satapathy and Ravindran, 1996). Observations of marked elevation in anti-Gal titers in malaria patients were reported by Yilmaz et al. (2014), as well.

Anti-Gal isolated from malaria-immune human sera and from nonimmune human sera was found to inhibit the reinvasion and growth of *P. falciparum* in cultures when anti-Gal was added to AB serum depleted of anti-Gal used in these cultures (Ramasamy and Rajakaruna, 1997). In this study, inactivation of complement components by heating the anti-Gal deficient sera used for culturing the parasites abolished growth inhibition mediated by the added anti-Gal antibody. These observations were further supported by studies demonstrating anti-Gal-mediated protection of GT-KO mice from transmission of *Plasmodium* parasites by mosquito (Yilmaz et al., 2014). In the study of Yilmaz et al. (2014), anti-Gal production in GT-KO mice was elicited by GI inoculation with *E. coli O86* bacteria or by immunization with rabbit RBC membranes. Anti-Gal producing GT-KO mice were found to be protected from infection by the *Anopheles* mosquito or from infection by intradermal injection of the parasite. This protection was dependent on complement activation, which was shown *in vitro* to cause cytolysis of the parasite. In view of these anti-Gal protective effects against *Plasmodium* infection, Yilmaz et al. (2014) suggested that the reported low infection rates for human experimental malaria induced by *P. falciparum*-infected mosquitoes (Verhage et al., 2005) may be associated with anti-Gal-mediated cytolysis of the parasite. These protective effects further prompted researchers to suggest that *Plasmodium* spp. could be considered as a possible cause for the evolutionary selection of Old World primates that lack α-gal epitopes and produce the natural anti-Gal antibody (Ramasamy, 1994; Ramasamy and Rajakaruna, 1997; Yilmaz et al., 2014).

The *in vitro* cytolytic effects of anti-Gal on *P. falciparum* (Ramasamy and Rajakaruna, 1997) and the *in vivo* protective effects of this antibody in GT-KO mice (Yilmaz et al., 2014) raise the possibility that increased titers of anti-Gal in populations living in malaria endemic areas may contribute to natural protection against *Plasmodium* spp. infections in endemic areas. As argued in the discussion above regarding a similar immunotherapeutic approach in *T. cruzi* endemic areas, immunization with vaccines containing α-gal-like epitopes may effectively elevate anti-Gal activity above the natural level only for several weeks, until the vaccinating material is eliminated. Thus, it would be of interest to identify nonpathogenic bacteria that may be incorporated into the natural microbiome and continuously stimulate the immune system to produce anti-Gal at levels that are high enough to destroy infecting parasites and thus lower the incidence of infection by *Plasmodium* spp.

CONCLUSIONS

The natural anti-Gal antibody binds to several protozoan parasites, including *Trypanosoma*, *Leishmania*, and *Plasmodium*. The binding is to α-gal-like epitopes presented by these parasites mostly on glycoinositol phospholipids and glycoproteins. These epitopes further stimulate the immune system of infected individuals to produce anti-Gal at elevated titers. *In vitro* studies with sera from infected patients and *in vivo* studies in GT-KO mice producing anti-Gal demonstrated complement-dependent and independent lysis and neutralization of *Trypanosoma* and *Plasmodium* parasites. These observations suggest that anti-Gal may contribute to natural protection against infections by these parasites. The high rates of infections suggest, however, that in a substantial proportion of populations in endemic regions, protective activity of the natural anti-Gal antibody is insufficient. Thus, significant numbers of infecting parasites succeed in evading the antibody by invading RBC or nucleated cells where they are protected from anti-Gal. Moreover, shedding of *Trypanosoma* cell membranes in the chronic phase of Chagas' disease may cause anti-Gal-mediated autoimmune-like phenomena that could be lethal, in particular, because of heart failure. In view of these considerations, it may be of interest to determine whether constant elevation of anti-Gal activity in individuals in endemic regions may contribute to protection against such infections. It is not clear whether immunization with vaccines inducing increased production of anti-Gal may be helpful because anti-Gal activity usually returns to the preimmunization level, within few months post immunization. It may be possible that providing nonpathogenic bacteria (e.g., as food supplement) that elevate anti-Gal production, and which become an integral part of the microbiome, will result in continuous increase in anti-Gal titers to levels that enable effective destruction of invading parasites, thus increasing protection against such infections.

References

Almeida, I.C., Milani, S.R., Gorin, P.A.J., Travassos, L.R., 1991. Complement-mediated lysis of *Trypanosoma cruzi* trypomastigotes by human anti-α-galactosyl antibodies. J. Immunol. 146, 2394–2400.

Almeida, I.C., Krautz, G.M., Krettli, A.U., Travassos, L.R., 1993. Glycoconjugates of *Trypanosoma cruzi*: a 74 kD antigen of trypomastigotes specifically reacts with lytic anti-α-galactosyl antibodies from patients with chronic Chagas disease. J. Clin. Lab. Anal. 7, 307–316.

Almeida, I.C., Ferguson, M.A., Schenkman, S., Travassos, L.R., 1994a. GPI-anchored glycoconjugates from *Trypanosoma cruzi* trypomastigotes are recognized by lytic anti-α-galactosyl antibodies isolated from patients with chronic Chagas' disease. Braz. J. Med. Biol. Res. 27, 443–447.

Almeida, I.C., Ferguson, M.A., Schenkman, S., Travassos, L.R., 1994b. Lytic anti-α-galactosyl antibodies from patients with chronic Chagas' disease recognize novel O-linked oligosaccharides on mucin-like glycosyl-phosphatidylinositol-anchored glycoproteins of *Trypanosoma cruzi*. Biochem. J. 304, 793–802.

Al-Salem, W.S., Ferreira, D.M., Dyer, N.A., Alyamani, E.J., Balghonaim, S.M., Al-Mehna, A.Y., et al., 2014. Detection of high levels of anti-α-galactosyl antibodies in sera of patients with Old World cutaneous leishmaniasis: a possible tool for diagnosis and biomarker for cure in an elimination setting. Parasitology 141, 1898–1903.

Antas, P.R., Medrano-Mercado, N., Torrico, F., Ugarte-Fernandez, R., Gómez, F., Correa Oliveira, R., et al., 1999. Early, intermediate, and late acute stages in Chagas' disease: a study combining anti-galactose IgG, specific serodiagnosis, and polymerase chain reaction analysis. Am. J. Trop. Med. Hyg. 61, 308–314.

Arumugham, R.G., Hsieh, T.C.-Y., Tanzer, M.L., Laine, R.A., 1986. Structure of the asparagine-linked sugar chains of laminin. Biochim. Biophys. Acta 883, 112–126.

Avila, J.L., 1999. α-Galactosyl-bearing epitopes as potent immunogens in Chagas' disease and leishmaniasis. Subcell. Biochem. 32, 173–213.

Avila, J.L., Rojas, M., Galili, U., 1989. Immunogenic Galα1-3Gal carbohydrate epitopes are present on pathogenic American *Trypanosoma* and *Leishmania*. J. Immunol. 142, 2828–2834.

Avila, J.L., Rojas, M., 1990. A galactosyl(α1-3)mannose epitope on phospholipids of *Leishmania mexicana* and *L. braziliensis* is recognized by trypanosomatid-infected human sera. J. Clin. Microbiol. 28, 1530–1537.

Bonney, K.M., Engman, D.M., 2015. Autoimmune pathogenesis of Chagas' heart disease: looking back, looking ahead. Am. J. Pathol. 185, 1537–1547.

Brito, C.R., McKay, C.S., Azevedo, M.A., Santos, L.C., Venuto, A.P., Nunes, D.F., et al., 2016. Virus-like particle display of the α-gal epitope for the diagnostic assessment of Chagas' disease. 2, 917–922.

Coura, J.R., Vinas, P.A., 2010. Chagas' disease: a new worldwide challenge. Nature 465, S6–S7.

Egge, H., Kordowicz, M., Peter-Katalinic, J., Hanfland, P., 1985. Immunochemistry of I/i-active oligo- and polyglycosylceramides from rabbit erythrocyte membranes. J. Biol. Chem. 260, 4927–4935.

Eto, T., Iichikawa, Y., Nishimura, K., Ando, S., Yamakawa, T., 1968. Chemistry of lipids of the post hemolytic residue or stroma of erythrocytes. XVI. Occurrence of ceramide pentasaccharide in the membrane of erythrocytes and reticulocytes in rabbit. J. Biochem. (Tokyo) 64, 205–213.

Farias, L.H., Rodrigues, A.P., Silviers, F.T., Seabra, S.H., DaMatta, R.A., Saraiva, E.M., et al., 2013. Phosphatidylserine exposure and surface sugars in two *Leishmania (Viannia) braziliensis* strains involved in cutaneous and mucocutaneous leishmaniasis. J. Infect. Dis. 207, 537–543.

Galili, U., Clark, M.R., Shohet, S.B., Buehler, J., Macher, B.A., 1987a. Evolutionary relationship between the anti-Gal antibody and the Galα1-3Gal epitope in primates. Proc. Natl. Acad. Sci. U.S.A. 84, 1369–1373.

Galili, U., Basbaum, C., Shohet, S.B., Buehler, J., Macher, B.A., 1987b. Identification of erythrocyte Galα1-3Gal glycosphingolipids with a mouse monoclonal antibody Gal-13. J. Biol. Chem. 262, 4683–4688.

Galili, U., Buehler, J., Shohet, S.B., Macher, B.A., 1987c. The human natural anti-Gal IgG. III. The subtlety of immune tolerance in man as demonstrated by crossreactivity between natural anti-Gal and anti-B antibodies. J. Exp. Med. 165, 693–704.

Galili, U., Mandrell, R.E., Hamadeh, R.M., Shohet, S.B., Griffiss, J.M., 1988. Interaction between human natural anti-α-galactosyl immunoglobulin G and bacteria of the human flora. Infect. Immun. 56, 1730–1737.

Galili, U., Tibell, A., Samuelsson, B., Rydberg, B., Groth, C.G., 1995. Increased anti-Gal activity in diabetic patients transplanted with fetal porcine islet cell clusters. Transplantation 59, 1549–1556.

Galili, U., Chen, Z.C., Tanemura, M., Seregina, T., Link, C.J., 2001. Understanding the induced antibody response. Graft 4, 32–35.

Gazzinelli, R.T., Galvao, L.C.M., Dias, J.C.P., Gazzinelli, G., Brener, Z., 1988. Anti-laminin and specific antibodies in acute Chagas' disease. Trans. R. Soc. Trop. Med. Hyg. 82, 574–576.

Gazzinelli, R.T., Pereira, M.E., Romanha, A.J., Gazzinelli, G., Brener, Z., 1991. Direct lysis of *Trypanosoma cruzi*: a novel effector mechanism of protection mediated by human anti-Gal antibodies. Parasite Immunol. 13, 345–356.

Goncalves, M.F., Umezawa, E.S., Katzin, A.M., de Souza, W., Alves, M.J., Zingales, B., et al., 1991. *Trypanosoma cruzi*: shedding of surface antigens as membrane vesicles. Exp. Parasitol. 72, 43–53.

González, J., Neira, I., Gutiérrez, B., Anacona, D., Manque, P., Silva, X., et al., 1996. Serum antibodies to *Trypanosoma cruzi* antigens in Atacameños patients from highland of northern Chile. Acta Trop. 60, 225–236.

Jakobsen, P.H., Theander, T.G., Jensen, J.B., Mølbak, K., Jepsen, S., 1987. Soluble *Plasmodium falciparum* antigens contain carbohydrate moieties important for immune reactivity. J. Clin. Microbiol. 25, 2075–2079.

Kim, M., Rao, M.V., Tweardy, D.J., Prakash, M., Galili, U., Gorelik, E., 1993. Lectin-induced apoptosis of tumor cells. Glycobiology 3, 447–453.

Laranja, F.S., Dias, E., Nobrego, G., Marinda, A., 1956. Chagas' disease: a clinical, epidemiologic and pathologic study. Circulation 14, 1035–1060.

Mañez, R., Blanco, F.J., Díaz, I., Centeno, A., Lopez-Pelaez, E., Hermida, M., et al., 2001. Removal of bowel aerobic gram-negative bacteria is more effective than immunosuppression with cyclophosphamide and steroids to decrease natural α-galactosyl IgG antibodies. Xenotransplantation 8, 15–23.

McConville, M.J., Collidge, T.A., Ferguson, M.A., Schneider, P., 1993. The glycoinositol phospholipids of *Leishmania mexicana* promastigotes. Evidence for the presence of three distinct pathways of glycolipid biosynthesis. J. Biol. Chem. 268, 15595–15604.

Milani, S.R., Travassos, L.R., 1988. Anti-α-galactosyl antibodies in chagasic patients. Possible biological significance. Braz. J. Med. Biol. Res. 21, 1275–1286.

Mohan, P.S., Spiro, R.G., 1986. Macromolecular organization of basement membranes. Characterization and comparison of glomerular basement membrane and lens capsule components by immunochemical and lectin affinity procedures. J. Biol. Chem. 261, 4328–4336.

Nogueira, P.M., Ribeiro, K., Silveira, A.C., Campos, J.H., Martins-Filho, O.A., Bela, S.R., et al., 2015. Vesicles from different *Trypanosoma cruzi* strains trigger differential innate and chronic immune responses. J. Extracell. Vesicles 4, 28734.

Ogawa, H., Galili, U., 2006. Profiling terminal *N*-acetyllactoamines of glycans on mammalian cells by an immunoenzymatic assay. Glycoconj. J. 23, 663–674.

Pal, S.C., Rao, C.K., Kereselidze, T., Krishnaswami, A.K., Murty, D.K., Pandit, C.G., et al., 1969. An extensive community outbreak of enteropathogenic *Escherichia coli* O86: B7 gastroenteritis. Bull. World Health Organ. 41, 851–858.

Pearse, M.J., Witort, E., Mottram, P., Han, W., Murray-Segal, L., Romanella, M., et al., 1998. Anti-Gal antibody-mediated allograft rejection in α1,3-galactosyltransferase gene knockout mice: a model of delayed xenograft rejection. Transplantation 66, 748–754.

Pereira-Chioccola, V.L., Acosta-Serrano, A., Correia de Almeida, I., Ferguson, M.A., Souto-Padron, T., Rodrigues, M.M., et al., 2000. Mucin-like molecules form a negatively charged coat that protects *Trypanosoma cruzi* trypomastigotes from killing by human anti-α-galactosyl antibodies. J. Cell Sci. 113, 1299–1307.

Posekany, K.J., Pittman, H.K., Bradfield, J.F., Haisch, C.E., Verbanac, K.M., 2002. Induction of cytolytic anti-Gal antibodies in α-1,3-galactosyltransferase gene knockout mice by oral inoculation with *Escherichia coli O86:B7* bacteria. Infect. Immun. 70, 6215–6222.

Rao, C.N., Goldstein, I.J., Liotta, L.A., 1983. Lectin-binding domains on laminin. Arch. Biochem. Biophys. 227, 118–124.

Ramasamy, R., 1988. Binding of normal human immunoglobulins to *Plasmodium falciparum*. Indian J. Med. Res. 87, 584–593.

Ramasamy, R., 1994. Is malaria linked to the absence of α-galactosyl epitopes in Old World primates? Immunol. Today 15, 140.

Ramasamy, R., Reese, R.T., 1987. Terminal galactose residues and the antigenicity of *Plasmodium falciparum* glycoproteins. J. Clin. Microbiol. 25, 2075–2079.

Ramasamy, R., Rajakaruna, R., 1997. Association of malaria with inactivation of α1,3-galactosyl transferase in catarrhines. Biochim. Biophys. Acta 1360, 241–246.

Ramasamy, R., Field, M.C., 2012. Terminal galactosylation of glycoconjugates in *Plasmodium falciparum* asexual blood stages and *Trypanosoma brucei* bloodstream trypomastigotes. Exp. Parasitol. 130, 314–320.

Ravindran, B., Satapathy, A.K., Das, M.K., 1988. Naturally-occurring anti-α-galactosyl antibodies in human *Plasmodium falciparum* infections: a possible role for autoantibodies in malaria. Immunol. Lett. 19, 137–141.

Satapathy, A.K., Ravindran, B., 1996. A quantitative cell-ELISA for α-galactose specific antibodies in human malaria. J. Immunoass. 17, 245–256.

Schenkman, S., Jiang, M.S., Hart, G.W., Nussenzweig, V., 1991. A novel cell surface trans-sialidase of *Trypanosoma cruzi* generates a stage-specific epitope required for invasion of mammalian cells. Cell 65, 1117–1125.

Schneider, P., Rosat, J.P., Ransijn, A., Ferguson, M.A., McConville, M.J., 1993. Characterization of glycoinositol phospholipids in the amastigote stage of the protozoan parasite *Leishmania major*. Biochem. J. 295, 555–564.

Schocker, N.S., Portillo, S., Brito, C.R., Marques, A.F., Almeida, I.C., Michael, K., 2016. Synthesis of Galα(1,3)Galβ(1,4) GlcNAcα-, Galβ(1,4)GlcNAcα- and GlcNAc-containing neoglycoproteins and their immunological evaluation in the context of Chagas' disease. Glycobiology 26, 39–50.

Soares, R.P., Torrecilhas, A.C., Assis, R.R., Rocha, M.N., Moura e Castro, F.A., Freitas, G.F., et al., 2012. Intraspecies variation in *Trypanosoma cruzi* GPI mucins: biological activities and differential expression of α-galactosyl residues. Am. J. Trop. Med. Hyg. 87, 87–96.

Souto-Padron, T., Almeida, I.C., de Souza, W., Travassos, L.R., 1994. Distribution of α-galactosyl-containing epitopes on *Trypanosoma cruzi* trypomastigote and amastigote forms from infected Vero cells detected by Chagasic antibodies. J. Eukaryot. Microbiol. 41, 47–54.

Springer, G.F., Horton, R.E., 1969. Blood group isoantibody stimulation in man by feeding blood group-active bacteria. J. Clin. Investig. 48, 1280–1291.

Stellner, K., Saito, H., Hakomori, S.-I., 1973. Determination of aminosugar linkage in glycolipids by methylation. Aminosugar linkage of ceramide pentasaccharides of rabbit erythrocytes and of Forssman antigen. Arch. Biochem. Biophys. 133, 464–472.

Stone, K.R., Abdel-Motal, U.M., Walgenbach, A.W., Turek, T.J., Galili, U., 2007a. Replacement of human anterior cruciate ligaments with pig ligaments: a model for anti-non-gal antibody response in long-term xenotransplantation. Transplantation 83, 211–219.

Stone, K.R., Walgenbach, A.W., Turek, T.J., Somers, D.L., Wicomb, W., Galili, U., 2007b. Anterior cruciate ligament reconstruction with a porcine xenograft: a serologic, histologic, and biomechanical study in primates. Arthroscopy 23, 411–419.

Szarfman, A., Terranova, V.P., Rennard, S.I., Foidart, J.M., de Fatima Lima, M., Scheinman, J.I., et al., 1982. Antibodies to laminin in Chagas' disease. J. Exp. Med. 155, 1161–1171.

Tearle, R.G., Tange, M.J., Zannettino, Z.L., Katerelos, M., Shinkel, T.A., Van Denderen, B.J., et al., 1996. The α-1,3-galactosyltransferase knockout mouse. Implications for xenotransplantation. Transplantation 61, 13–19.

Thall, A.D., Maly, P., Lowe, J.B., 1995. Oocyte Galα1,3Gal epitopes implicated in sperm adhesion to the zona pellucida glycoprotein ZP3 are not required for fertilization in the mouse. J. Biol. Chem. 270, 21437–21440.

Timpl, R., Martin, G.R., Bruckner, P., Wick, G., Wiedemann, H., 1978. Nature of the collagenous protein in a tumor basement membrane. Eur. J. Biochem. 84, 43–52.

Torrecilhas, A.C., Schumacher, R.I., Alves, M.J., Colli, W., 2012. Vesicles as carriers of virulence factors in parasitic protozoan diseases. Microbes Infect. 14, 1465–1474.

Towbin, H., Rosenfelder, G., Wieslander, J., Avila, J.L., Rojas, M., Szarfman, A., et al., 1987. Circulating antibodies to mouse laminin in Chagas' disease, American cutaneous leishmaniasis, and normal individuals recognize terminal galactosyl (α1-3)-galactose epitopes. J. Exp. Med. 166, 419–432.

Travassos, L.R., Milani, S.R., Oliveira, T.G., Takaoka, D., Gorin, P.A., 1988. Immunobiological responses to short carbohydrate epitopes in *Trypanosoma cruzi*. Mem. Inst. Oswaldo Cruz 83 (Suppl. 1), 427–430.

Travassos, L.R., Almeida, I.C., 1993. Carbohydrate immunity in American trypanosomiasis. Springer Sem. Immunopathol. 15, 183–204.

Trocoli Torrecilhas, A.C., Tonelli, R.R., Pavanelli, W.R., da Silva, J.S., Schumacher, R.I., de Souza, W., et al., 2009. Trypanosoma cruzi: parasite shed vesicles increase heart parasitism and generate an intense inflammatory response. Microbes Infect. 11, 29–39.

Verhage, D.F., Telgt, D.S., Bousema, J.T., Hermsen, C.C., van Gemert, G.J., van der Meer, J.W., et al., 2005. Clinical outcome of experimental human malaria induced by *Plasmodium falciparum*-infected mosquitoes. Neth. J. Med. 63, 52–58.

Wigglesworth, K., Racki, W.J., Mishra, R., Szomolany-Tsuda, E., Greiner, D.L., Galili, U., 2011. Rapid recruitment and activation of macrophages by anti-Gal/α-gal liposome interaction accelerates wound healing. J. Immunol. 186, 4422–4432.

Wood, C., Kabat, E.A., Murphy, L.A., Goldstein, I.J., 1979. Immunochemical studies of the combining sites of the two isolectins, A4 and B4, isolated from *Bandeiraea simplicifolia*. Arch. Biochem. Biophys. 198, 1–11.

Yilmaz, B., Portugal, S., Tran, T.M., Gozzelino, R., Ramos, S., Gomes, J., et al., 2014. Gut microbiota elicits a protective immune response against malaria transmission. Cell 159, 1277–1289.

Yokoyama-Yasunaka, J.K., Piazza, R.M., Umezawa, E.S., Stolf, A.M., 1998. Reactivity of chagasic anti-gal antibodies with noninfected cells treated with *Trypanosoma cruzi* secreted/excreted antigens. J. Clin. Lab. Anal. 12, 108–114.

CHAPTER

5

Anti-Gal B Cells Are Tolerized by α-Gal Epitopes in the Absence of T Cell Help

OUTLINE

INTRODUCTION

The interest in developing methods for induction of immune tolerance that prevents the production of the natural anti-Gal antibody emerged primarily from two areas in the field of transplantation: (1) the clinical experience with ABO incompatible allograft transplantation and (2) xenotransplantation studies in Old World monkey recipients of porcine xenografts such as heart or kidneys, in which hyperacute rejection was determined to be

The Natural Anti-Gal Antibody as Foe Turned Friend in Medicine
http://dx.doi.org/10.1016/B978-0-12-813362-0.00005-1

mediated by anti-Gal/α-gal epitope interaction (detailed in Chapter 6). The α-gal epitope and blood groups A and B antigens have similar structures (Fig. 1 in Chapter 3). Therefore, understanding of the immune response to the α-gal epitope and its prevention by immune tolerance, benefited from the clinical experience gained in transplantation of ABO incompatible kidney allografts, which initiated several years before xenotransplantation became an area of wide interest. The research on induction of tolerance to the α-gal epitope could progress with the generation of α1,3galactosyltransferase knockout (GT-KO) mice lacking this carbohydrate antigen and producing the anti-Gal antibody that interacts with this epitope. Whereas the significance of anti-Gal as a barrier in xenotransplantation is discussed in detail in Chapter 6, the present chapter discusses the mechanisms participating in production of anti-Gal, induction of tolerance in B cells producing this antibody, and the possibility of applying this research to induction of tolerance to A and B blood group antigens in allograft transplantation.

LESSONS FROM ABO INCOMPATIBLE ALLOGRAFT TRANSPLANTATION IN HUMANS

From the early days of kidney allograft transplantation in humans, it has been well established that kidney grafts presenting an incompatible blood group antigen (e.g., kidney from blood type B donor transplanted in blood type A or O recipient) are rapidly rejected. This rejection is mediated by the corresponding natural anti-blood group A or B (A/B) antibodies that bind to the incompatible blood group antigen on the endothelial cells of the graft. This interaction activates complement, damages the endothelial cells of the graft, and induces platelet aggregation, resulting in collapse of the vascular bed and "hyperacute" rejection of the graft (Starzl, 1964; Wilbrandt et al., 1969; Chopek et al., 1987; Cooper, 1990). Transplantation of kidney from a living donor across the ABO barrier became feasible with the development of methods for removal of natural anti-blood group (called here anti-AB) antibodies by plasmapheresis (Alexandre et al., 1987; Bannett et al., 1987; Slapak et al., 1990), removal of these antibodies by adsorption on columns expressing the corresponding synthetic blood group antigen (Mendez et al., 1992; Yamashita et al., 1993; Tanabe et al., 1998; Kobayashi et al., 2000; Takahashi, 2001; Shishido et al., 2001) and destruction of B cells by rituximab (Nydegger et al., 2005; Tydén et al., 2005). Recipients of ABO incompatible kidney allograft, in whom the anti-A/B antibodies were removed prior to transplantation and who underwent splenectomy, displayed three types of immune response to the incompatible blood group antigens on the graft: (1) rejection, (2) accommodation, and (3) tolerance (Latinne et al., 1989; Ishida et al., 2000; Park et al., 2003; Holgersson et al., 2014; Zschiedrich et al., 2015). Accommodation and tolerance were observed also in infant recipients of ABO incompatible heart grafts (West et al., 2001; Urschel et al., 2013).

Rejection—In a proportion of recipients of ABO incompatible kidneys, there is a rapid production of anti-blood group antibodies against the incompatible antigen on the kidney allograft. In these patients, binding of the elicited anti-blood group antibodies to the corresponding ABO incompatible antigens on graft endothelial cells activates complement, causes destruction of endothelial cells, and induces the formation of microthrombi that occlude the blood vessels, resulting in ischemia and rejection of the graft (Takahashi, 2001; Rydberg, 2001; Takahashi et al., 2010).

Accommodation—The second group of recipients also produces anti-A/B antibodies against the incompatible blood group antigen of the kidney allograft. However, the antibodies produced do not induce destruction of the graft as that observed in the rejection group (Platt, 1994; Park et al., 2003). This phenomenon is referred to as "accommodation" (Platt and Bach, 1991). Studies of accommodation in experimental animal models of rats transplanted with guinea pig or hamster heart proposed that the accommodation observed in these models is the result of activation of genes encoding for protective anti-apoptotic and cytoprotective proteins within the graft cells (Bach et al., 1991, 1997; Soares et al., 1999). Induction of accommodation was reported also to be associated with inhibition of complement activity in the recipient (Yuzawa et al., 1995; Suhr et al., 2000; West et al., 2001; Ding et al., 2008), decrease in expression of the incompatible carbohydrate antigen on the graft (Ulfvin et al., 1993; Yuzawa et al., 1995; Rydberg, 2001), and production of antibodies that may protect the graft from rejection (Hasan et al., 1992). Additional research is required to determine whether all these mechanisms contribute to induction of accommodation in every donor/recipient combination and for better understanding of each of these mechanisms.

Tolerance—In the third group, recipients of ABO incompatible kidney allografts display diminished or no production of antibodies to the incompatible blood group antigen of the allograft (Tanabe et al., 1998; West et al., 2001; Takahashi et al., 2010; Fehr and Stussi, 2012). Thus, these recipients seem to acquire immune tolerance to the incompatible blood group antigen.

ANTI-GAL IgG PRODUCTION IS T CELL-DEPENDENT BUT α-GAL EPITOPES DO NOT ACTIVATE T CELLS

Experimental studies on anti-Gal production became feasible with the generation of GT-KO mice, which lack the α-gal epitope (Thall et al., 1995). These mice require immunization with immunogenic antigens that carry α-gal epitopes (e.g., rabbit red blood cell (RBC) membranes or pig kidney homogenate) for production of the anti-Gal antibody (LaTemple and Galili, 1998; Tanemura et al., 2000a). Production of anti-Gal by immunization with α-gal epitopes differs from antibody production in response to immunogenic proteins in that the immunogenic α-gal epitope cannot activate helper T (Th) cells (Tanemura et al., 2000a). Nevertheless, T cell help is required for anti-Gal IgG production (Tanemura et al., 2000a). Activation of Th cells can be achieved using immunogenic proteins. When a protein vaccine is administered into any mammal, it is internalized by antigen presenting cells (APC) that process the protein and generate peptides that are presented on class II MHC molecules. The presented peptides engage the corresponding T cell receptor (TCR) for activation of Th cells. These activated Th cells provide the help to B cells that engage the vaccinating protein

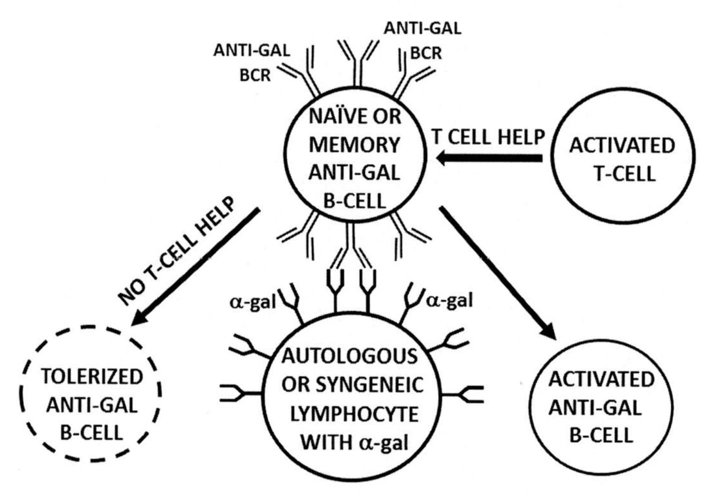

FIGURE 1 Illustration of the significance of Th cells in anti-Gal B cells activation or tolerization following interaction with α-gal epitopes. When B cell receptors (BCRs) on anti-Gal B cells engage α-gal epitopes on autologous or syngeneic cells, and activated Th cells provide help to B cells, anti-Gal B cells are activated to produce the anti-Gal antibody. In the absence of T cell help, anti-Gal B cells engaging α-gal epitopes are tolerized. Anti-Gal B cells may produce anti-Gal IgM antibodies if they engage bacteria or particles presenting high concentrations of α-gal or α-gal-like epitopes. *Reprinted from Mohiuddin, M., Ogawa, H., Yin, D.P., Galili, U., 2003a. Tolerance induction to a mammalian blood group like carbohydrate antigen by syngeneic lymphocytes expressing the antigen: II. Tolerance induction on memory B cells. Blood 102, 229–236, with permission.*

via their B cell receptor (BCR) and are activated to produce antibodies. The α-gal epitope is a carbohydrate antigen that cannot activate Th cells but can activate B cells. This could be demonstrated in GT-KO mice. Immunization with a syngeneic wild-type (WT) mouse kidney homogenate does not elicit an anti-Gal antibody response (Tanemura et al., 2000a), although these kidneys contain high concentration of α-gal epitopes (Tanemura et al., 2000b). In contrast, immunization of GT-KO mice with pig kidney membranes (PKM) homogenate or with rabbit RBC membranes elicits an effective production of anti-Gal IgG antibodies (LaTemple and Galili, 1998; Tanemura et al., 2000a) because the xenoproteins in pig kidney or in rabbit RBC provide immunogenic peptides that activate the Th cells. The help provided by activated Th cells is required for the activation of B cells capable of producing the anti-Gal antibody (anti-Gal B cells) (Galili, 2004) (Fig. 1).

To determine whether there are any T cells-recognizing α-gal epitopes in mice that were repeatedly immunized with PKM and induced to produce anti-Gal, the spleen lymphocytes were coincubated with syngeneic tumor mouse cells engineered to express α-gal epitopes. No T cell proliferation was observed in such cell cultures, whereas T cells extensively proliferated when incubated with pig cells and cell membranes expressing or lacking α-gal epitopes (Tanemura et al., 2000a). These findings indicated that no T cells-recognizing the α-gal epitopes can be found even after repeated immunization of GT-KO mice with homogenates of PKM. In accord with the observations in GT-KO mice, immunization of humans with α-gal glycolipids from rabbit RBC resulted in only marginal (approximately two-fold) increase in elicited anti-Gal response (Whalen et al., 2012). In contrast, administration of pig cells (porcine fetal islet cell clusters) into humans resulted in 30–80-fold increase in anti-Gal titer within several weeks post transplantation, despite the treatment of these recipients with immunosuppressive drugs (Galili et al., 1995). Similarly, immunization of GT-KO mice with α-gal glycolipids anchored in the wall of liposomes made of phospholipids and cholesterol results in no significant elicited anti-Gal antibody response because of the lack of immunogenic peptides that can activate Th cells (Tanemura et al., 2000a). However, if this immunization is performed together with keyhole limpet cyanine (KLH), which is a potent immunogenic protein, anti-Gal IgG is produced. This production is enabled because KLH peptides activate Th cells that provide the help required by anti-Gal B cells that engage the α-gal epitope via their BCR and are activated to produce the anti-Gal IgG antibody (Fig. 1). Accordingly, blockade of costimulatory signals of Th cells by anti-CD40L (also called anti-CD154) was found to prevent activation of anti-Gal B cells to produce anti-Gal IgG antibody, whereas anti-Gal IgM antibody production was not prevented (Tanemura et al., 2000a; Cretin et al., 2002). In view of the similarity in the structure of the α-gal epitope and blood group A and B antigens, it is likely that production of anti-blood group A and B IgG antibodies in humans also requires Th cell activation, but these carbohydrate antigens themselves cannot activate Th cells for this purpose.

The inability of the α-gal epitope and of blood group A and B antigens to activate T cells is probably the result of their degradation within APC, as well as the hydrophilic "protrusion outward" of carbohydrate antigens from the cell surface. Glycolipids and glycoproteins are digested in the endocytic compartment of APC by glycosidases that cleave carbohydrate chains into nonimmunogenic monosaccharides. If such glycosidases are not produced, (as in inherited metabolic storage diseases) accumulation of the undigested carbohydrate chains impairs the function of cells in the reticuloendothelial system. Thus, many of the α-gal epitopes or of blood group A and B antigens on glycoconjugates internalized into APC are likely to be destroyed before peptides carrying carbohydrate chains reach the cell surface in association with MHC molecules. Moreover, because complex carbohydrate chains such as the α-gal epitope are highly hydrophilic, they protrude away from the cell surface. Crystallography studies have shown that because of this protrusion, the interaction between such carbohydrate epitopes on the cell surface of APC and TCR on T cells occurs at a distance that prevents interaction of TCR accessory molecules with the corresponding ligands on APC (Speir et al., 1999). For this reason, carbohydrate antigens comprised of a chain of three or more carbohydrate units are considered incapable of activating Th cells (Ishioka et al., 1993; Speir et al., 1999).

In contrast to anti-Gal IgG response, anti-Gal IgM antibody production does not require T cell help; however, it is much weaker than the IgG response because expansion of anti-Gal B cell clones is limited in absence of T cell help. In addition, anti-Gal IgM production requires a high concentration of α-gal epitopes on the immunizing bacteria or particle. Thus, immunization with liposomes presenting α-gal glycolipids or with mouse WT kidney membranes result in production of low titers of anti-Gal IgM (Tanemura et al., 2000a). Similarly, immunizations of GT-KO mice lacking T cells with a soluble antigen carrying multiple α-gal epitopes resulted in moderate anti-Gal IgM production (Cretin et al., 2002). Production of anti-Gal IgM antibodies was also achieved by feeding GT-KO mice with *Escherichia coli O86* bacteria that present multiple α-gal or α-gal-like epitopes on their lipopolysaccharides (Posekany et al., 2002). However, anti-Gal IgG titers in these mice were much lower than those of IgM titers. Thus, the multiple epitopes on *E. coli O86* can activate anti-Gal B cells along the gastrointestinal tract to produce anti-Gal IgM antibodies; however, the Th cell activation in these mice is insufficient for inducing expansion of the anti-Gal B cell clones and their class switch into IgG-producing cells.

Overall, the observations in GT-KO mice suggest that the α-gal epitope and possibly other carbohydrate antigens on long carbohydrate chains such as *N*-(asparagine) linked carbohydrate chains of the "complex type" on glycoproteins, or carbohydrate chains on glycolipids cannot activate T cells. Mounting an effective IgG antibody response to such antigens requires activation of Th cells by immunogenic peptides that may be provided by xenoglycoproteins carrying α-gal epitopes or any other immunogenic protein molecule (Tanemura et al., 2000a; Galili, 2004). However, as described below, prolonged exposure of anti-Gal B cells to α-gal epitopes on various cells in the absence of T cell help has significant consequences for these B cells that may explain the prevention of anti-blood group antibody response in recipients of ABO mismatched allografts and the production of accommodating antibodies.

TOLERANCE INDUCTION IN ANTI-GAL B CELLS BY SYNGENEIC LYMPHOCYTES PRESENTING α-GAL EPITOPES

Anti-Gal tolerance in α1,3galactosyltransferase knockout mice by bone marrow chimerism

Studies on induction of T cell tolerance indicated that mixed bone marrow chimerism in mice results in tolerance to allo-antigens (Mayumi and Good, 1989; Sharabi and Sachs, 1989). Moreover, mixed bone marrow chimerism between mouse and rat resulted in tolerance induction to xenoantigens between the two species (Lee et al., 1994). These observations prompted researchers to determine whether chimerism with WT mouse bone marrow (i.e., bone marrow expressing α-gal epitopes) in the GT-KO mouse model can induce immune tolerance that prevents anti-Gal antibody production. Indeed, a lasting state of tolerance was achieved by administration into lethally irradiated GT-KO mice a mixture of bone marrow cells from GT-KO mice and from WT mice (Yang et al., 1998; Ohdan et al., 1999; Sykes et al., 2005). The GT-KO mice with the mixed bone marrow failed to produce anti-Gal in response to immunization with pig cells presenting α-gal epitopes and displayed lack of anti-Gal B cells in their spleen. Tolerance induction to α-gal epitopes was also achieved in GT-KO mouse recipients of syngeneic bone marrow cells transduced with the α1,3galactosyltransferase (α1,3GT) gene (Bracy et al., 1998; Bracy and Iacomini, 2000). These bone marrow cells induced tolerance that prevented production of the anti-Gal antibody, therefore the mice could be grafted with WT hearts expressing α-gal epitopes with no rejection of the graft. These studies raised the question whether induction of tolerance to the α-gal epitope is unique to bone marrow cells or can it be induced also by other mouse cells presenting this epitope? This question is addressed in the sections below.

Tolerance induction in naïve anti-Gal B cells by wild-type lymphocytes presenting α-gal epitopes

Induction of tolerance was primarily determined by the ability of GT-KO mice to produce anti-Gal following 2–4 intraperitoneal immunizations with 50 mg PKM homogenate. This homogenate contains multiple α-gal epitopes, many of which are carried by laminin of the glomeruli (a basement membrane protein with 50–70 α-gal epitopes), and others are present on the brush border epithelium of the proximal tubules of the nephrons, as well as on other cell surfaces and extracellular matrix proteins (Tanemura et al., 2000b). Four PKM immunizations resulted in production of anti-Gal in control GT-KO mice at titers (serum dilution yielding 50% of maximum absorbance in ELISA) of ~1:400 with synthetic α-gal epitopes linked to bovine serum albumin (α-gal BSA) as solid-phase antigen. Inability of treated mice to produce anti-Gal following such immunization indicated that the mice were tolerized. The state of tolerance was further validated by the heterologous transplantation of a WT heart into the abdominal cavity of the treated mice and monitoring the heart function by daily palpation to confirm lack of rejection (Pearse et al., 1998; Ogawa et al., 2003). Mice producing anti-Gal following PKM immunizations, rejected the transplanted WT heart within 30 min to few hours because of anti-Gal binding to the multiple α-gal epitopes on the WT heart endothelial cells and cardiomyocytes, complement-mediated destruction of these cells, platelets aggregation, and occlusion of the blood vessels (Fig. 2). Transplanted WT heart functioning for weeks or months and subsequent lack of immunoglobulin deposits in the explanted heart subjected to immunohistology (Fig. 2) reflected a state of tolerance to the α-gal epitope.

To determine whether cells, other than bone marrow cells, can induce tolerance to the α-gal epitope and prevent anti-Gal production, lymphocytes from syngeneic WT mice were introduced intravascular into GT-KO mice. Two weeks after administration of WT lymphocytes, the mice received four weekly immunizations, each of 50 mg PKM. GT-KO mice receiving 20×10^6 or 2×10^6 WT lymphocytes did not produce anti-Gal following the PKM immunizations, whereas mice that did not receive the WT lymphocytes readily produced anti-Gal (Ogawa et al., 2003). As expected, WT mouse hearts from C57BL/6 mice transplanted into the tolerized mice were not rejected even after 2 months. Immunostaining of the hearts removed after 2 months displayed no immunoglobulin deposits, whereas WT hearts rejected within 30 min in anti-Gal-producing mice displayed IgM deposits on blood vessel walls and occluded blood vessels (Fig. 2). Similar deposits were observed in rejected WT hearts immunostained for IgG, C3, and C5 with the corresponding peroxidase-coupled secondary antibodies. In addition, enzyme-linked immunospot (ELISPOT) analysis of anti-Gal B cells that actively secrete the antibody demonstrated multiple such cells in control mice that were immunized four times with PKM, whereas no or very few such cells were found in mice tolerized with WT lymphocytes and receiving similar PKM immunizations (Ogawa et al., 2003). Overall, these findings indicated that the presentation of α-gal epitopes on syngeneic cells (i.e., WT C57BL/6 lymphocytes) for a period of 2 weeks, in the absence of any T cell help, resulted in induction of tolerance on naïve anti-Gal B cells as illustrated in Fig. 1.

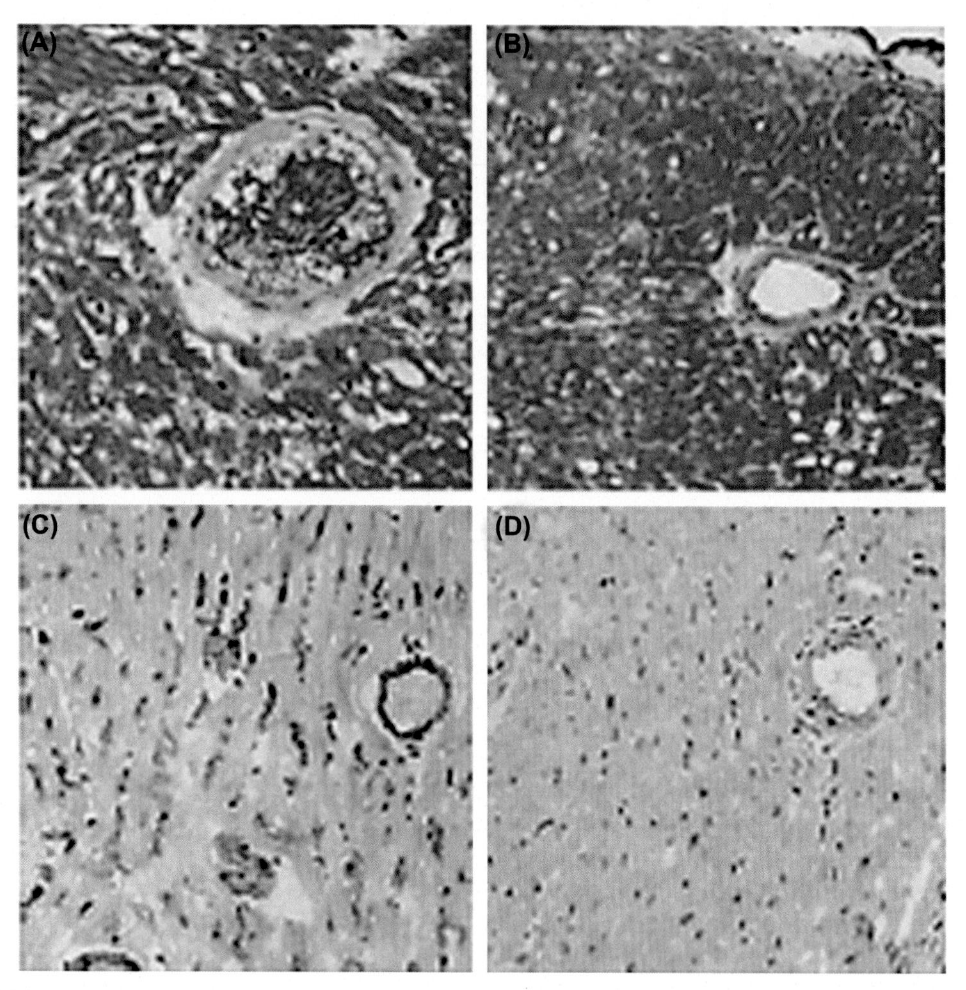

FIGURE 2 Rejection and tolerance in α1,3galactosyltransferase knockout mice transplanted with wild-type (WT) heart. Hearts transplanted heterotopically in mice producing anti-Gal are rejected within 30 min (A and C). Hearts transplanted into mice tolerized by WT lymphocytes presenting α-gal epitopes were not rejected, despite repeated pig kidney membranes immunizations. Hearts in the tolerized mice were removed for histologic inspection 2 months posttolerance induction (B and D). The section was stained either by hematoxylin–eosin (H&E) (A and B) or immunostained with peroxidase-coupled anti-mouse IgM antibodies (C and D). The WT heart that was hyperacute rejected (A and C) displays edema of the myocardium, occlusion of the blood vessels, and binding of anti-Gal IgM to the endothelial cells, as well as to cells in the extravascular areas. The hearts removed after 2 months in the tolerized mice (B and D) display normal histology of the myocardium and no IgM binding to any of the cells (original magnification ×200). *Reprinted from Ogawa, H., Yin, D.P., Shen, J., Galili, U., 2003. Tolerance induction to a mammalian blood group-like carbohydrate antigen by syngeneic lymphocytes expressing the antigen. Blood 101, 2318–2320, with permission.*

Tolerance induction in memory anti-Gal B cells by wild-type lymphocytes presenting α-gal epitopes

The basic mechanisms of B cell tolerance to self-antigen include clonal deletion, receptor editing in the differentiation stage of immature B cells (i.e., B cells in the bone marrow), and anergy of B cells (Nossal, 1994; Radic and Zouali, 1996; Goodnow, 1996; Nemazee, 2000). However, after the initial activation of anti-blood group A and B B cells or of anti-Gal B cells, which usually occurs 3–6 months after birth (Galili et al., 1984), the lymphocytes producing these natural anti-carbohydrate antibodies are a mixture of naïve and memory B cells. Thus, it was of interest to determine whether memory anti-Gal B cells, which are at an advanced stage of maturation, can also be tolerized by α-gal epitopes on WT lymphocytes, i.e., in the absence of T cell help.

Tolerance induction in memory anti-Gal B cells requires the use of B cells from GT-KO mice that were immunized repeatedly with PKM, i.e., B cells that were induced to undergo isotype (class) switch into IgG or IgA-producing cells. However, such mice also produce anti-Gal that rapidly destroys lymphocytes presenting α-gal epitopes, like anti-Gal-mediated hyperacute rejection of porcine xenografts (Chapter 6). To overcome the obstacle of anti-Gal antibody-mediated destruction of cells presenting α-gal epitopes, 20×10^6 lymphocytes from GT-KO mice receiving four PKM immunizations (i.e., lymphocytes including memory anti-Gal B cells) were administered into lethally irradiated naïve GT-KO mice together with 20×10^6 GT-KO bone marrow cells (for production of various blood cell lineages) and 20×10^6 lymphocytes from WT mice. The recipients of this mixture of cells could not produce anti-Gal following two PKM immunizations, starting 14 days after this adoptive transfer (Mohiuddin et al., 2003a). In contrast, irradiated recipients of only 20×10^6 lymphocytes from the PKM immunized mice and 20×10^6 GT-KO bone marrow cells produced anti-Gal at high titers when immunized with PKM 2 weeks post adoptive transfer. This implied that in the absence of tolerizing WT lymphocytes, the transferred memory anti-Gal B cells were readily activated by α-gal epitopes of PKM and produced anti-Gal. ELISPOT studies with α-gal BSA as solid-phase antigen demonstrated the presence of ~100 anti-Gal secreting B cells per 10^6 lymphocytes in spleens of PKM immunized GT-KO mice, whereas in the tolerized mice only a background of 10 cells per 10^6 lymphocytes was detected (Mohiuddin et al., 2003a). Tolerance induction was also observed when 2×10^6 WT lymphocytes were administered into the irradiated GT-KO recipients; however, 0.2×10^6 WT lymphocytes did not induce tolerance. Induction of tolerance to α-gal epitopes also was observed in nonirradiated naïve GT-KO mice that received a mixture of 20×10^6 lymphocytes from PKM immunized GT-KO mice, and 20×10^6 WT lymphocytes, then immunized with 50 mg PKM on days 14 and 21 post adoptive transfer of the lymphocytes (Mohiuddin et al., 2003a).

Tolerance induction in anti-Gal B cells was found to require a period of ~2 weeks of *in vivo* exposure to α-gal epitopes on WT lymphocytes in the absence of T cell help (Figs. 1 and 3A). As further shown in Fig. 3A, if T cells are activated by PKM immunization after a shorter period, they provided help to the memory anti-Gal B cells, which enables activation of these B cells and production of anti-Gal, instead of being tolerized. Accordingly, no tolerance induction was observed in GT-KO recipients of memory anti-Gal B cells and WT lymphocytes from semiallogeneic donor (H-2bxd). These recipients readily produced anti-Gal following PKM immunizations (Mohiuddin et al., 2003a). H-2d specific T cells in GT-KO mice were activated

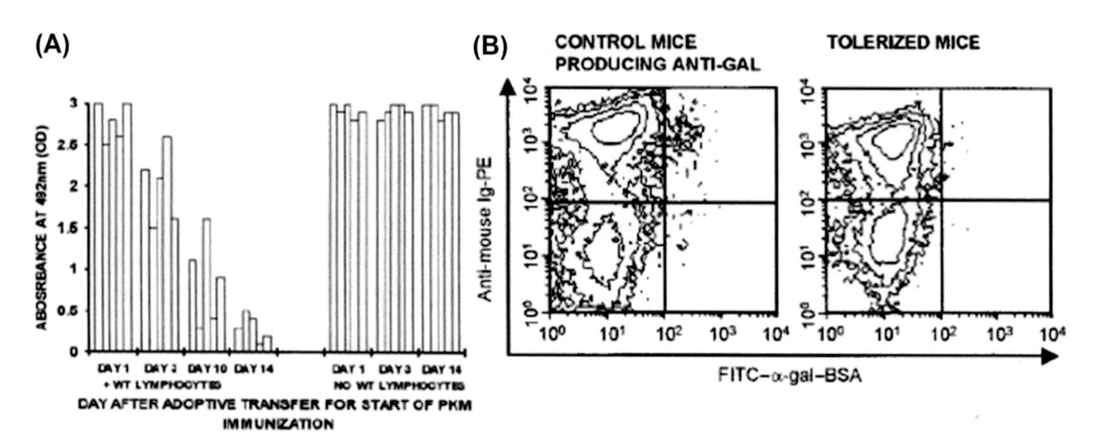

FIGURE 3 Tolerance induction in memory anti-Gal B cells by α-gal epitopes on wild-type (WT) lymphocytes. (A) Time required for tolerance induction. Irradiated α1,3galactosyltransferase knockout (GT-KO) mice received 20×10^6 lymphocytes that include memory anti-Gal B cells, 20×10^6 naïve GT-KO bone marrow cells, and 2×10^6 WT lymphocytes or no WT lymphocytes. The mice further received two pig kidney membranes (PKM) immunizations, the first of which was delivered at various time points following the adoptive transfer and the second, 1 week after the first immunization. Anti-Gal IgG production was measured by ELISA 1 week after the second PKM immunization. Absorbance values at serum dilution of 1:100 indicate that tolerance in all mice is observed when PKM immunizations start 14 days post adoptive transfer that includes WT lymphocytes. In the absence of WT lymphocytes, no tolerance is induced at any time point. (B) Flow cytometry identification of anti-Gal B cells by double staining with fluorescein (FITC)-α-gal bovine serum albumin (BSA) (green) and phycoerythrin (PE) anti-mouse Ig (red-staining all B cells). Control and tolerized mice, as in (A) received PKM immunizations on days 14 and 21 post adoptive transfer. Spleens were harvested for immunostaining on day 28. Note that as many as 1% of B cells bound α-gal epitopes of α-gal BSA in the control mice (i.e., anti-Gal B cells), whereas almost no such B cells were detected in the tolerizes mice. Data from one representative out of four mice in each group. *Reprinted from Mohiuddin, M., Ogawa, H., Yin, D.P., Galili, U., 2003a. Tolerance induction to a mammalian blood group like carbohydrate antigen by syngeneic lymphocytes expressing the antigen: II. Tolerance induction on memory B cells. Blood 102, 229–236, with permission.*

by the allo-antigen H-2d on the WT lymphocytes and provided help that prevented tolerance induction in anti-Gal B cells by α-gal epitopes on these WT lymphocytes. Tolerance induction by α-gal epitopes presented on syngeneic WT lymphocytes was found to be highly specific to memory anti-Gal B cells. This could be demonstrated by the activity of memory B cells capable of producing anti-blood group A antibodies in the same mice in which memory anti-Gal B cells were tolerized. Immunization of the mice with PKM yielded no anti-Gal production, whereas immunization of these mice with blood type A RBC resulted in extensive production of anti-A antibodies (Mohiuddin et al., 2003a).

In attempt to determine whether tolerized memory anti-Gal B cells are eliminated because of tolerance induction by α-gal epitopes on WT lymphocytes, specific immunostaining of memory anti-Gal B cells was performed with fluorescein (FITC)-labeled α-gal BSA. It was assumed that α-gal epitopes binding to anti-Gal BCR will label anti-Gal B cells by the FITC-α-gal BSA. Flow cytometry analysis demonstrated that anti-Gal B cells comprise ~1% of B cells in the control nontolerized recipients of B cells from PKM-immunized mice. However, almost no anti-Gal B cells were detected in the recipients that were tolerized by WT lymphocytes (Fig. 3B) (Mohiuddin et al., 2003a). Taken together, the studies on anti-Gal B cells in naïve and in PKM immunized GT-KO mice suggest that both naïve and memory anti-Gal B cells are tolerized after they engage the α-gal epitopes for long enough time in the absent of T cell help (Fig. 1).

The mechanisms tolerizing anti-Gal B cells were studied with immunoglobulin genes of monoclonal anti-Gal antibody M86 that was generated as hybridoma of an anti-Gal B cell from a GT-KO mouse (Galili et al., 1998). Knock-in studies in GT-KO mice of M86 anti-Gal IgH chain gene demonstrated production of the monoclonal antibody encoded by such transgene in GT-KO mice but not in WT mice or in heterozygotes of GT-KO and WT mice (i.e., mice having one copy of the disrupted α1,3GT gene and one copy of the intact gene) (Cretin and Iacomini, 2002). Similar observations were reported by Xu et al. (2002) with the immunoglobulin genes of another monoclonal anti-Gal antibody. These studies suggested deletion of the immature anti-Gal B cells in the bone marrow. In subsequent studies, Benatuil et al. (2008) cloned M86 VDJ_H and VJκ regions and targeted them into heavy and light chain immunoglobulin gene regions, correspondingly, in embryonic stem cells. The resulting IgH and IgL knock-in transgenes were readily expressed, producing monoclonal anti-Gal M86 antibody in GT-KO mice but not in heterozygote mice containing one copy of the intact α1,3GT gene. Analysis of the Ig transgenes in these heterozygotes suggested that the tolerance to the α-gal epitope was primarily achieved by receptor editing at the level of immature B cells, whereas negative selection for deletion of anti-Gal B cells was a secondary mechanism to receptor editing (Benatuil et al., 2008). The studies in this chapter demonstrate an additional tolerance induction mechanism in which mature naïve and memory anti-Gal B cells that left the bone marrow may also be tolerized by WT lymphocytes. This tolerance induction is feasible if no T cell help is provided to the anti-Gal B cells that engage the α-gal epitope via their BCR. As demonstrated below, this of B cell characteristic can be exploited for induction of tolerance to a given carbohydrate antigen by autologous lymphocytes engineered to present the tolerizing carbohydrate antigen.

TOLERANCE INDUCTION IN ANTI-GAL B CELLS BY AUTOLOGOUS LYMPHOCYTES ENGINEERED TO SYNTHESIZE α-GAL EPITOPES

A method for tolerance induction by lymphocytes presenting the tolerizing carbohydrate antigen may be of clinical significance in inducing tolerance to ABO incompatible blood group antigen prior to transplantation of the ABO incompatible organ. Thus, it was of interest to determine whether autologous lymphocytes of GT-KO mice can be engineered to present α-gal epitopes and tolerize anti-Gal B cells, like the tolerizing effects of WT lymphocytes (Ogawa et al., 2003; Mohiuddin et al., 2003a). Because lymphocytes are usually nondividing cells (unless they are activated), vectors such as retroviruses, which insert the transgene into the genome, may not be effective in introducing an intact α1,3GT transgene for the subsequent synthesis of α-gal epitopes. Therefore, transduction of this gene was performed with a replication defective adenovirus vector carrying the intact mouse α1,3GT gene, designated AdαGT (Deriy et al., 2002). Transduction of human or mouse cells with AdαGT was found to introduce 10–20 copies of the vector into the cells, resulting in effective synthesis and presentation of 0.5×10^6–1.0×10^6 α-gal epitopes/cell within 12–24 h (Deriy et al., 2002, 2005).

GT-KO lymphocytes were engineered to express α-gal epitopes by incubation with AdαGT for 4h. Subsequently, the transduced lymphocytes were administered into naive GT-KO mice via the tail vein, as 20×10^6 lymphocytes/mouse. Control lymphocytes were transduced with the parental "empty" adenovirus vector (Ogawa et al., 2004b) lacking the α1,3GT insert. This administration of transduced GT-KO mouse lymphocytes was repeated after 4 and 9days. The mice received 4 weekly immunizations with PKM, starting on day 14, and studied for anti-Gal production. Although all control mice produced anti-Gal antibodies at high titers (~1:800), half of the mice receiving lymphocytes transduced with AdαGT produced no anti-Gal, whereas the rest produced the antibody at low titers of <1:50 (Ogawa et al., 2004b).

Analysis of tolerance induction in memory anti-Gal B cells was performed by adoptive transfer into lethally irradiated GT-KO mice of 20×10^6 cells of the following populations: Lymphocytes from GT-KO mice immunized four times with PKM (i.e., lymphocytes including memory anti-Gal B cells), AdαGT-transduced GT-KO mouse lymphocytes (or lymphocytes transduce with "empty" vector), and GT-KO mouse bone marrow cells. Administration of transduced GT-KO mouse lymphocytes was repeated on days 4 and 9. As with tolerizing WT lymphocytes, this adoptive transfer of the various cell populations was required because lymphocytes presenting α-gal epitopes are destroyed in the presence of circulating anti-Gal. The mice were immunized with PKM, 2 and 3weeks post adoptive transfer. As with the naïve mice, the recipients of memory anti-Gal B cells and control lymphocyte transduced with the empty vector produced anti-Gal IgG at high titers. However, half of the recipients of memory anti-Gal B cells and AdαGT-transduced lymphocytes produced no significant amounts of anti-Gal, and the rest produced this antibody at very low levels (Fig. 4A). All tolerized mice produced no anti-Gal IgM antibodies (Ogawa et al., 2004b).

The tolerized mice that did not produce anti-Gal following PKM immunizations were further transplanted heterotopically with WT heart presenting α-gal epitopes. As many as 65% of the transplanted hearts continued to function for 45 or 100days before subjected to histological studies (Ogawa et al., 2004b). The other 35% of transplanted mice died after 62–64days for unknown reasons. All these mice continued to display inability for production of the anti-Gal antibody, despite additional 4 PKM immunizations (Fig. 4B). In contrast, GT-KO mouse recipients of GT-KO memory anti-Gal B cells and lymphocytes transduced with empty adenovirus vector, all produced anti-Gal at high titers following 2 PKM immunizations and rejected WT heart grafts within a period of 30min–18h.

The mechanism inducing tolerance to the α-gal epitope in recipients of memory anti-Gal B cells is not fully understood. It is possible that the encounter of memory anti-Gal B cells with α-gal epitopes on the AdαGT transduced GT-KO lymphocytes, in the absence of T cell help, results in induction of apoptotic signals that cause the deletion of these B cells. Alternatively, these memory anti-Gal B cells may have undergone receptor editing. It should be noted, however, that a literature search yielded no studies describing tolerance induction in memory B cells by the receptor editing mechanism. A secondary adoptive transfer of lymphocytes and bone marrow cells from these tolerized mice to naïve irradiated recipients followed by PKM immunizations of the recipients, resulted in no anti-Gal production. The transferred lymphocytes included no transducing AdαGT (determined by PCR), suggesting that the tolerance is not the result of anergy of the memory anti-Gal B cells, but that these B cell clones were eliminated either by deletion or by receptor editing (Ogawa et al., 2004b).

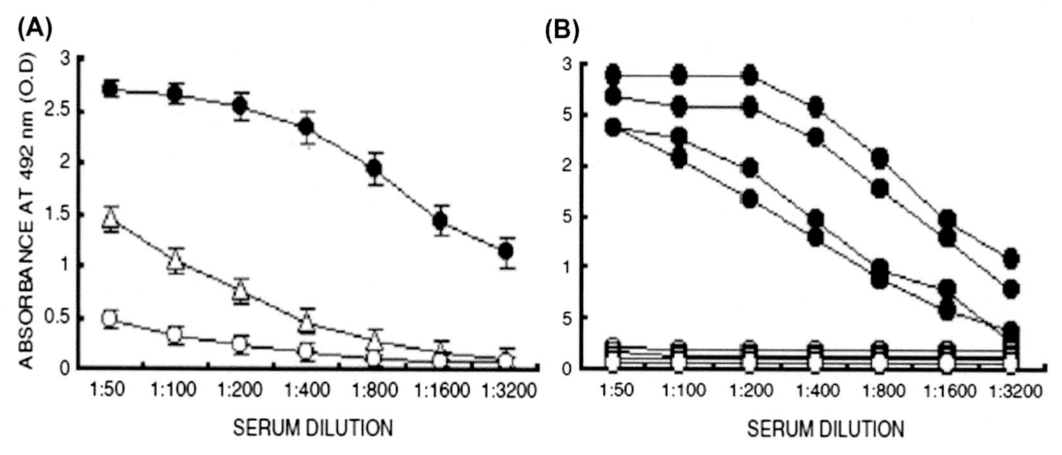

FIGURE 4 Induction of tolerance in memory anti-Gal B cells by α1,3galactosyltransferase knockout (GT-KO) lymphocytes transduced with AdαGT. (A) Anti-Gal activity in mice tolerized by AdαGT transduced lymphocytes and immunized with pig kidney membranes (PKM), 2 and 3 weeks postadoptive transfer. Control mice receiving lymphocytes transduced with "empty" adenovirus vector (*closed circles*, n = 18); mice receiving AdαGT-transduced lymphocytes and producing anti-Gal at a low level, which does not exceed 1.5 O.D., at serum dilution of 1:50 (*open triangles*, n = 12); mice receiving AdαGT-transduced lymphocytes and producing no significant amounts of anti-Gal (*open circles*, n = 16) (mean ± S.E.). (B) Perpetuation of the tolerance induced by AdαGT transduced GT-KO lymphocytes, followed by heterotopic transplantation of WT mouse heart on day 28 post adoptive transfer (n = 4). The mice were further immunized four times with PKM on weeks 8, 10, 12, and 14. Anti-Gal tested 2 weeks after the last immunization. All mice receiving lymphocytes including memory anti-Gal B cells and AdαGT-transduced lymphocytes (*open circles*) produced no anti-Gal antibody following PKM immunizations. Anti-Gal production was extensive in control mice lacking WT hearts and receiving lymphocytes including memory anti-Gal B cells from the same preparation as the tolerized mice and lymphocytes transduced with "empty" adenovirus vector (n = 4). These control mice were immunized twice with PKM (*closed circles*). *Reprinted from Ogawa, H., Yin, D.P., Galili, U., 2004b. Induction of immune tolerance to a transplantation carbohydrate antigen by gene therapy with autologous lymphocytes transduced with adenovirus containing the corresponding glycosyltransferase gene. Gene Ther. 11, 292–301, with permission.*

Overall, these studies on tolerance induction with AdαGT transduced autologous lymphocytes indicate that the ability of α-gal epitopes on the transduced lymphocytes to tolerize naïve and memory anti-Gal B cells, in the absence of T cell help, is similar to that of the α-gal epitope on WT lymphocytes. The one difference between the two type of tolerizing cells is that the activity of the α1,3GT gene in AdαGT transduced cells is limited in time to 2–4 days, probably because of AdαGT DNA destruction by nucleases of the transduced cells. For this reason, effective tolerance induction is achieved by repeated administration of AdαGT transduced lymphocytes also on days 4 and 9 after the initiation of the tolerizing process.

INDUCTION OF TOLERANCE AND ACCOMMODATION BY MOUSE HEART PRESENTING α-GAL EPITOPES

The studies above raised the question whether the ability of tolerance induction is limited to bone marrow cells and lymphocytes or can other WT syngeneic cells expressing α-gal epitopes tolerize anti-Gal B cells. This question was addressed by analysis of anti-Gal response in GT-KO mice transplanted with WT heart. Initial studies in GT-KO mice indicated that WT hearts presenting multiple α-gal epitopes, which were transplanted heterotopically in naïve GT-KO mice, were not rejected for months and did not induce production of anti-Gal (Pearse et al., 1998). However, if GT-KO mice were preimmunized and produced anti-Gal, the transplanted WT hearts were rejected within 30 min–24 h because of the interaction between this antibody and the multiple α-gal epitopes on endothelial cells of the graft, as shown in Fig. 2 (Pearse et al., 1998; Ogawa et al., 2003).

The α-gal epitopes on the transplanted syngeneic WT hearts can tolerize both naïve and memory anti-Gal B cells in the absence of the anti-Gal antibody and of T cell help. This was demonstrated in naïve GT-KO mice transplanted heterotopically with WT heart and lethally irradiated 2 weeks post transplantation. Subsequently, the mice received by adoptive transfer 20×10^6 lymphocytes from PKM immunized GT-KO mice (i.e., lymphocytes that include many memory anti-Gal B cells) and 20×10^6 bone marrow cells from naïve GT-KO mice. The memory anti-Gal B cells were exposed for different periods of time, ranging from 1 day to 28 days, to the α-gal epitopes on endothelial cells of the transplanted WT heart. At the end of each period, the GT-KO mouse recipients were immunized twice with PKM and anti-Gal activity, as well as WT heart function were evaluated. As indicated above, the immunogenic pig xenopeptides in the immunizing PKM effectively activate Th cells that provide help to anti-Gal B cells and enable their activation and production of the anti-Gal antibody. The outcomes of PKM immunizations differed in mice in which the memory anti-Gal B cell was exposed for different periods to the α-gal epitopes on the WT heart prior to PKM immunization (Fig. 5 and Table 1) (Mohiuddin et al., 2003b; Ogawa et al., 2004a).

1. *PKM immunization 24 h post adoptive transfer of memory anti-Gal B cells—In vivo* exposure of memory anti-Gal B cells for 24 h to α-gal epitopes on endothelial cells of the WT heart graft did not affect these B cells. Thus, immunization of the recipients with PKM 24 h post adoptive transfer resulted in activation of the memory anti-Gal B cells and their differentiation into plasma cells that produce anti-Gal IgM, IgG1, and IgG3 antibodies. These antibodies bound to the blood vessel walls of the WT heart and to cardiomyocytes causing edema and occlusion of blood vessels by blood clots, like the effects of anti-Gal shown in Fig. 2. These antibodies rejected the WT heart grafts within 3–6 days. Accordingly, *in vitro* incubation of mouse cells presenting α-gal epitopes with sera from theses mice diluted 1:100–1:1000 and with complement resulted in complement activation and extensive cytolysis.
2. *PKM immunization 7 days post adoptive transfer*—PKM immunization after 7 days exposure of the memory anti-Gal B cells to α-gal epitopes on the endothelial cells of the transplanted WT heart, resulted in induction of a state of accommodation. The mice in this group produced anti-Gal antibodies, which were comprised of IgM, IgG1, and IgG2b isotypes. These antibodies readily bound to the blood vessel walls of the WT heart; however, they did not induce rejection of the graft for >100 days, despite multiple (>4) repeated immunizations with PKM. The transplanted hearts displayed normal structure of the blood vessels and of cardiomyocytes, despite anti-Gal binding to them. A second WT heart transplanted into the accommodating mice producing anti-Gal at high titers continued to function for >2 months (Mohiuddin et al., 2003b). This suggests that accommodation in these mice was associated with the characteristics of anti-Gal antibodies produced by the activated anti-Gal B cells, rather than with unknown

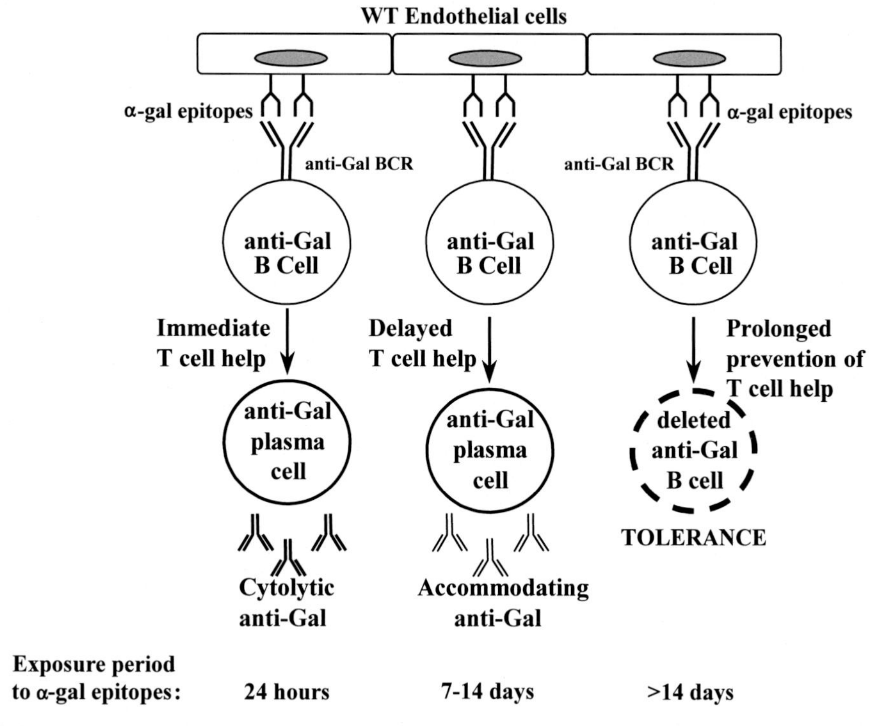

FIGURE 5 Association between the period of anti-Gal B cell exposure to α-gal epitopes on endothelial cells of wild-type (WT) heart transplanted into GT-KO mice, in the absence of T cell help, and production of anti-Gal. T cell help provided within 24-h exposure to α-gal epitopes activates anti-Gal B cells into plasma cells producing cytolytic antibodies. Exposure of anti-Gal B cells for 7–14 days prior to T cell help results in activation of anti-Gal B cells into plasma cells producing accommodating antibodies. In contrast, exposure of these B cells for more than 14 days in the absence of T cell help, tolerizes the anti-Gal B cells. *BCR*, B cell receptor. *Reprinted from Galili, U., 2004. Immune response, accommodation and tolerance to transplantation carbohydrate antigens. Transplantation 78, 1093–1098, with permission.*

changes in the cells of the WT heart graft. Accordingly, in the presence of complement, anti-Gal antibodies produced in the accommodating mice were noncytolytic to mouse cells presenting α-gal epitopes. Anti-Gal produced in the accommodating mice differed from the antibody produced in rejecting mice in that IgG3 anti-Gal antibodies were very low, whereas IgG2b anti-Gal antibodies were high. Combination of the accommodating anti-Gal antibodies with cytolytic anti-Gal antibodies from graft rejecting mice resulted in 50%–90% decrease in complement-mediated cytolysis of mouse cells presenting α-gal epitopes (Mohiuddin et al., 2003b). This observation suggests that the accommodating anti-Gal antibodies can protect cells from lysis by the cytolytic anti-Gal antibodies.

3. *PKM immunization 14 days post adoptive transfer*—PKM immunization of GT-KO mice transplanted with WT heart, starting 14 days post adoptive transfer of the lymphocytes which include memory anti-Gal B cells, resulted in a state of accommodation in 60% of the mice, whereas the remaining 40% became tolerant to the α-gal epitope and did not produce anti-Gal, like all mice receiving PKM immunizations 28 days post adoptive transfer (described below).

TABLE 1 Association between the type of immune response to syngeneic grafts presenting α-gal epitopes in α1,3galactosyltransferase knockout (GT-KO) mice and time of T cell activation[a]

Type of immune response	Period of absence of T cell help	Characteristics of the immune response
1. Rejection	T cell help is available to anti-Gal B cells that bind (engage) the α-gal epitope because no effective prevention of T cell activation is achieved.	Anti-Gal B cells engaging α-gal epitopes produce anti-Gal antibodies that can activate complement following binding to α-gal epitopes on the mouse wild type graft and induce rejection of the graft.
2. Accommodation	T cell help is available to anti-Gal B cells only after a period of several days, during which no T cell help is provided.	Naïve and memory anti-Gal B cells engaging α-gal epitopes for several days in the absence of T cell help differentiate into B cells capable of producing accommodating antibodies. These antibodies, which are produced when T cell help is provided, bind to the α-gal epitopes on the graft but do not activate complement. Therefore, the graft binding the accommodating anti-Gal antibodies is not rejected.
3. Tolerance	T cell help is effectively prevented for prolonged periods (2–4 weeks)	Naïve and memory anti-Gal B cells that are repeatedly engaging α-gal epitopes for prolonged periods in the absence of T cell help, are deleted. This results in induction of long-term tolerance as newly emerging anti-Gal B cells are deleted.

[a]Analogous responses may occur in recipients of ABO incompatible grafts.

Reprinted from Galili, U., 2006. Xenotransplantation and ABO incompatible transplantation: the similarities they share. Transfus. Apher. Sci. 35, 45–58, with permission.

4. *PKM immunization 28 days post adoptive transfer*—Prolonged exposure of memory anti-Gal B cells to α-gal epitopes on WT heart, for a period of 28 days, resulted in tolerization of these B cells. This was indicated by the inability of the mice to produce anti-Gal at significant levels, even after several immunizations with PKM. The WT heart grafts in these mice continued to function for >100 days, despite repeated PKM immunizations. Immunostaining of the heart histological sections revealed no binding of immunoglobulin to the blood vessel walls of the WT hearts and normal structure of the myocardium (Ogawa et al., 2004a).

Anergy of B cells is considered to be an important mechanism of immune unresponsiveness and tolerance to autologous antigens (Goodnow, 1996; Yarkoni et al., 2010). Thus, it was of interest to determine whether the memory anti-Gal B cells in these tolerized mice were anergized or deleted. This was studied by a second adoptive transfer of lymphocytes from the tolerized mice to naïve mice. The secondary recipients were lethally irradiated prior to the adoptive transfer of the lymphocytes from the tolerized mice and of naïve bone marrow cells. The transferred lymphocytes circulated in the secondary recipients for 14 days in the absence of α-gal epitopes, prior to immunization with PKM. It was assumed that if the memory anti-Gal B cells are anergized by α-gal epitopes of the WT heart, then in the absence of these epitopes for a period of 2 weeks in the secondary recipient, memory anti-Gal B cells should cease to be anergic and revert to a normal state of memory B cells. These secondary recipients failed, however, to produce anti-Gal, despite repeated PKM immunizations (Ogawa et al., 2004a). This suggested that memory anti-Gal B cells were deleted, rather than anergized, by 28 days exposure to α-gal epitopes of the WT heart in the absence of T cell help. ELISPOT studies for identification of antibody secreting anti-Gal B cells also demonstrated the complete absence of these B cells in the mice tolerized by α-gal epitopes on the WT heart graft versus high number of B cells secreting anti-Gal in control recipients of memory anti-Gal B cells that lack WT heart. This observation further suggests that tolerance induction by WT heart is the result of anti-Gal B cell deletion following repeated exposures to the α-gal epitopes on the WT heart graft in the absence of T cell help. Nevertheless, receptor editing cannot be excluded as a mechanism causing the observed tolerance of memory anti-Gal B cells.

The observations above on the transition in PKM-immunized mice from production of cytolytic anti-Gal antibodies to production of accommodating antibodies and ultimately to a state of tolerance suggest that the length of the period during which memory anti-Gal B cells repeatedly engage via their BCR α-gal epitopes on endothelial cells of the WT heart graft, in the absence of Th cell help, determines the fate of these B cells (Fig. 5). If that period is short (e.g., 24 h), activation of these anti-Gal B cells in presence of T cell help results in production of cytolytic antibodies. Exposure of these B cells to α-gal epitopes for 1–2 weeks in the absence of Th cell help results in the induction of class (isotype) switch that leads to production of accommodating anti-Gal antibodies, when activation of memory anti-Gal B cells includes Th cell help. If the exposure of these B cells to α-gal epitopes on WT heart in the absence of T cell help continues for a period of 2–4 weeks, the repeated engagement of their BCRs with the α-gal epitopes is likely to induce elimination of anti-Gal B cells, resulting in a state of immune tolerance and prevention of anti-Gal production. Because newly generated anti-Gal B cells leave the bone marrow and repeatedly engage α-gal epitopes of the graft without T cell help, these cells are deleted, and the state of tolerance is perpetuated. Thus, the tolerized mice do not produce anti-Gal antibodies, and the WT hearts continue to function for months, despite repeated immunizations with PKM.

Overall, these findings indicate that tolerance to α-gal epitopes can be induced not only by chimerism with bone marrow cells presenting these epitopes, but also with mature lymphocytes, as well as with endothelial cells presenting α-gal epitopes. The tolerance-inducing process is time-dependent and requires the BCRs on anti-Gal B cells to engage the α-gal epitopes for at least a period of 14 days in the complete absence of T cell help. Once the tolerance is induced, it is further maintained by α-gal epitopes on the graft, which continue to induce deletion of new anti-Gal B cells, emerging from the bone marrow.

SUGGESTED TOLERANCE INDUCTION TO ABO INCOMPATIBLE ANTIGENS IN ALLOGRAFT RECIPIENTS

Possible reasons for rejection of ABO incompatible allografts

The observations on tolerance induction by α-gal epitopes in GT-KO mice may help to understand the principles of tolerance and accommodation inductions to blood group A or B (A/B) antigens, observed in a proportion of the patients transplanted with ABO incompatible kidney allograft (Tanabe et al., 1998; Takahashi et al., 2010) and in infants transplanted with ABO incompatible heart (West et al., 2001; Urschel et al., 2013). The findings described above in GT-KO mice suggest that in the group of patients transplanted with ABO incompatible kidney and displaying tolerance toward the incompatible blood group antigen (i.e., do not produce antibodies against the incompatible A/B antigen), at least two factors contribute to the successful induction of this tolerance: (1) Effective removal of anti-blood group A/B antibody shortly prior to transplantation prevents the subsequent masking of the incompatible blood group antigen on the allograft, thus enabling the BCRs on B cells to engage the incompatible antigen on endothelial cells of the graft. (2) Effective prevention of help provided by Th cells by immunosuppression, which prevents Th cell activation by alloantigens of the graft. Under such conditions, B cells specific to the incompatible blood group antigen are exposed to it for a long enough period, in the absence of Th cells help, resulting in the deletion of these B cells. Conversely, if removal of the antibody to the incompatible blood group antigen is insufficient, the remaining antibody either kills the cells presenting the incompatible blood group antigen or masks this antigen on the graft and thus preventing the BCR of corresponding B cells from engaging the antigen. Without such BCR engaging, the B cells capable of producing antibodies to the incompatible blood group antigen will not be eliminated. If immunosuppression is insufficient to prevent Th activation, activated Th cells will "rescue" the B cells specific to the incompatible blood group antigen from being eliminated. Moreover, the Th help may enable these B cells to produce anti-A or anti-B antibodies that reject the allograft.

A demonstration of Th cell activation in allograft recipients to a level that enables activation of anti-carbohydrate B cells is that of diabetic patients transplanted with kidney allograft and with porcine fetal islet cell clusters xenograft (Groth et al., 1994). In these patients, the porcine islet cells were grafted either under the allograft kidney capsule or introduced into the portal vein. The standard immunosuppression treatment of these patients effectively prevented rejection of the kidney allograft. However, the multiple xenoproteins released from the porcine islet cells (many of which carrying α-gal epitopes) activated Th cells. These Th cells provided help to quiescent anti-Gal B cells that engaged the porcine α-gal epitopes and activated by them, ultimately resulting in increasing anti-Gal IgG titers by 30–80-fold in comparison with the pre-transplantation natural activity of the antibody (Galili et al., 1995). In view of these considerations, it would be of interest to closely monitor changes in antibody titers against the A/B incompatible blood group antigen in allograft recipients, if the level of T cell immunosuppression is decreased with time.

The studies on accommodation in GT-KO mice further describe a scenario of accommodation induction, which may occur in patients transplanted with ABO incompatible allograft. These studies suggest that B cells capable of producing antibodies to the incompatible blood group antigen undergo isotype switch to produce accommodating antibodies, if T cell help is accidentally provided after a period of several days post transplantation. The continuous activity of B cells producing such accommodating antibodies could result in the detection of these antibodies long after the transplantation, without detecting any immune damage to the allograft.

Suggested method for tolerance induction to ABO incompatible antigen in allograft recipients by transduced autologous lymphocytes presenting the incompatible A/B antigen

Transplantation of ABO mismatched allografts includes three immune intervention components: (1) immunosuppression of T cells, (2) removal of the anti-A/B antibodies either by plasmapheresis in early studies or by adsorption columns with the synthetic blood group A or B antigen (Tanabe et al., 1996, 1998; Ishida et al., 2000; Shishido et al., 2001; Takahashi, 2001; Tydén et al., 2005), and (3) suppression of B cell activity by splenectomy in early studies, or in more recent years by the administration of anti-CD20 (rituximab), which eliminates B cells (Tydén et al., 2005; Tanabe, 2007; Geyer et al., 2009; Fehr and Stussi, 2012; Melexopoulou et al., 2015). Some recipients of ABO incompatible allografts who were treated for suppressing B cell activity were reported to display higher risks of infectious and hemorrhagic complications than patients receiving ABO compatible kidney (i.e., patients that are not treated for B cell suppression) (Lentine et al., 2014).

The tolerance induction in GT-KO anti-Gal B by lymphocytes transduced with AdαGT (Ogawa et al., 2004b) raises the possibility that the need for nonspecific B cell suppression may be avoided by specific tolerance induction to the incompatible blood group antigen. Such tolerance induction may be studied using autologous lymphocytes transduced with adenovirus containing the corresponding A or B transferase gene. In such a hypothetical study, the natural antibody to the incompatible blood group A/B antigen is repeatedly removed by adsorption columns. This is followed by administration of autologous mononuclear cells transduced with an adenovirus vector containing blood group A or B transferase gene (Yamamoto et al., 1990), which corresponds to the incompatible blood group antigen (i.e., AdAT or AdBT, respectively). Because activity of the transduced AdAT or AdBT may be limited to several days, the administration of transduced lymphocytes might have to be repeated several times prior to transplantation. Subsequent transplantation of the kidney allograft under T cell immunosuppression may further block Th cell help to B cells specific to the incompatible blood group antigen, resulting in continuous deletion of such B cells, induced by the incompatible antigen on the allograft. Under such circumstances, no B cell clones with other antibody specificities are affected. Preclinical studies on tolerance induction to the α-gal epitope by AdαGT transduction of autologous mononuclear cells as described above, and transplantation of an allograft may be simulated in GT-KO pigs that are transplanted with a kidney from an allogeneic WT pig. Because the production of the natural anti-Gal antibody in GT-KO pigs is like that in humans (Dor et al., 2004; Fang et al., 2012; Galili, 2013), successful tolerization of anti-Gal B cells in GT-KO pigs and prevention of WT kidney allograft rejection will suggest that this method may be effective also in tolerance induction in recipients of ABO incompatible kidney allografts.

CONCLUSIONS

The immune response to large (complex) carbohydrate antigens such as the α-gal epitope differs from the immune response to protein antigens in that these carbohydrate antigens are not capable of activating T cells. Therefore, activation of both naïve and memory anti-Gal B cells requires Th cell help from T cells that are activated by immunogenic peptide antigens (e.g., xeno-peptides). However, *in vivo* exposure of naïve and memory anti-Gal B cells to α-gal epitopes without T cell help for prolonged periods (2–4 weeks) results in the deletion of these B cells and induction of immune tolerance to the α-gal epitope. Thus, administration into GT-KO mice of syngeneic WT bone marrow cells, WT lymphocytes, or WT heart grafts, all presenting multiple α-gal epitopes, results within 2–4 weeks in tolerance induction to this epitope. The tolerized mice do not produce anti-Gal even after several immunizations with PKM homogenates. Anti-Gal B cells exposed *in vivo* to α-gal epitopes for a period of ~7 days prior to T cell activation, are induced to produce noncytolytic accommodating anti-Gal antibodies that bind to α-gal epitopes on endothelial cells of WT syngeneic heart but do not reject the graft. The tolerance induction also can be achieved by autologous cells engineered to present α-gal epitopes. Transduction of GT-KO mouse lymphocytes with replication defective adenovirus containing the α1,3GT gene (AdαGT) results in presentation of α-gal epitopes on these lymphocytes. Administration of AdαGT transduced GT-KO lymphocytes into GT-KO mice tolerizes both naïve and memory anti-Gal B cells. It is suggested that this method may be applicable for tolerance induction to blood group antigens in patients receiving ABO incompatible allografts. Such tolerance induction may be performed with autologous mononuclear cells transduced by replication-defective adenovirus containing the gene of the glycosyltransferase synthesizing the corresponding incompatible blood group antigen. In such a hypothetical protocol, the transduced mononuclear cells may be administered several times to the patient only after effective removal of the antibody to the incompatible blood group antigen and under T cell suppression.

References

Alexandre, G.P., Squifflet, J.P., DeBruyere, B.M., Latinne, D., Reding, R., Gianello, P., et al., 1987. Present experiences in a series of 26 ABO-incompatible living donor renal allografts. Transplant. Proc. 19, 4538–4542.

Bach, F.H., Turman, M.A., Vercellotti, G.M., Platt, J.L., Dalmasso, A.P., 1991. Accommodation: a working paradigm for progressing toward clinical discordant xenografting. Transplant. Proc. 23, 205–207.

Bach, F.H., Ferran, C., Hechenleitner, P., Mark, W., Koyamada, N., Miyatake, T., et al., 1997. Accommodation of vascularized xenografts: expression of "protective genes" by donor endothelial cells in a host Th2 cytokine environment. Nat. Med. 3, 196–204.

Bannett, A.D., McAlack, R.F., Raja, R., Baquero, A., Morris, M., 1987. Experience with known ABO-mismatched renal transplants. Transplant. Proc. 19, 4543–4546.

Benatuil, L., Kaye, J., Cretin, N., Godwin, J.G., Cariappa, A., Pillai, S., et al., 2008. Ig knock-in mice producing anti-carbohydrate antibodies: breakthrough of B cells producing low affinity anti-self antibodies. J. Immunol. 180, 3839–3848.

Bracy, J.L., Sachs, D.H., Iacomini, J., 1998. Inhibition of xenoreactive natural antibody production by retroviral gene therapy. Science 281, 845–847.

Bracy, J.L., Iacomini, J., 2000. Induction of B-cell tolerance by retroviral gene therapy. Blood 96, 3008–3015.

Chopek, M.W., Simmons, R.L., Platt, J.L., 1987. ABO-incompatible kidney transplantation: initial immunopathologic evaluation. Transplant. Proc. 19, 4553–4557.

Cooper, D.K., 1990. Clinical survey of heart transplantation between ABO blood group-incompatible recipients and donors. J. Heart Transplant. 9, 376–381.

Cretin, N., Iacomini, J., 2002. Immunoglobulin heavy chain transgenic mice expressing Galα(1,3)Gal-reactive antibodies. Transplantation 73, 1558–1564.

Cretin, N., Bracy, J., Hanson, K., Iacomini, J., 2002. The role of T cell help in the production of antibodies specific for Galα1-3Gal. J. Immunol. 168, 1479–1483.

Deriy, L., Chen, Z.C., Gao, G.P., Galili, U., 2002. Expression of α-gal epitopes on HeLa cells transduced with adenovirus containing α1,3galactosyltransferase cDNA. Glycobiology 12, 135–144.

Deriy, L., Ogawa, H., Gao, G.P., Galili, U., 2005. In vivo targeting of vaccinating tumor cells to antigen-presenting cells by a gene therapy method with adenovirus containing the α1,3galactosyltransferase gene. Cancer Gene Ther. 12, 528–539.

Ding, J.W., Zhou, T., Ma, L., Yin, D., Shen, J., Ding, C.P., et al., 2008. Expression of complement regulatory proteins in accommodated xenografts induced by anti-α-Gal IgG1 in a rat-to-mouse model. Am. J. Transplant. 8, 32–40.

Dor, F.J., Tseng, Y.L., Cheng, J., Moran, K., Sanderson, T.M., Lancos, C.J., et al., 2004. α1,3-Galactosyltransferase gene-knockout miniature swine produce natural cytotoxic anti-Gal antibodies. Transplantation 78, 15–20.

Fang, J., Walters, A., Hara, H., Long, C., Yeh, P., Ayares, D., et al., 2012. Anti-gal antibodies in α1,3-galactosyltransferase gene-knockout pigs. Xenotransplantation 19, 305–310.

Fehr, T., Stussi, G., 2012. ABO-incompatible kidney transplantation. Curr. Opin. Organ Transplant. 17, 376–385.

Galili, U., 2004. Immune response, accommodation and tolerance to transplantation carbohydrate antigens. Transplantation 78, 1093–1098.

Galili, U., 2006. Xenotransplantation and ABO incompatible transplantation: the similarities they share. Transfus. Apher. Sci. 35, 45–58.

Galili, U., 2013. α1,3Galactosyltransferase knockout pigs produce the natural anti-Gal antibody and simulate the evolutionary appearance of this antibody in primates. Xenotransplantation 20, 267–276.

Galili, U., Rachmilewitz, E.A., Peleg, A., Flechner, I., 1984. A unique natural human IgG antibody with anti-α-galactosyl specificity. J. Exp. Med. 160, 1519–1531.

Galili, U., Tibell, A., Samuelsson, B., Rydberg, B., Groth, C.G., 1995. Increased anti-Gal activity in diabetic patients transplanted with fetal porcine islet cell clusters. Transplantation 59, 1549–1556.

Galili, U., LaTemple, D.C., Radic, M.Z., 1998. A sensitive assay for measuring α-gal epitope expression on cells by a monoclonal anti-Gal antibody. Transplantation 65, 1129–1132.

Geyer, M., Fischer, K.G., Drognitz, O., Walz, G., Pisarski, P., Wilpert, J., 2009. ABO-incompatible kidney transplantation with antigen-specific immunoadsorption and rituximab - insights and uncertainties. Contrib. Nephrol. 162, 47–60.

Goodnow, C.C., 1996. Balancing immunity and tolerance: deleting and tuning lymphocyte repertoires. Proc. Natl. Acad. Sci. U.S.A. 93, 2264–2271.

Groth, C.G., Korsgren, O., Tibell, A., Tollemar, J., Möller, E., Bolinder, J., et al., 1994. Transplantation of porcine fetal pancreas to diabetic patients. Lancet 344, 1402–1404.

Hasan, R., Van den Bogaerde, J., Forty, J., Wright, L., Wallwork, J., White, D.J., 1992. Xenograft adaptation is dependent on the presence of anti-species antibody, not prolonged residence in the recipient. Transplant. Proc. 24, 531–532.

Holgersson, J., Rydberg, L., Breimer, M.E., 2014. Molecular deciphering of the ABO system as a basis for novel diagnostics and therapeutics in ABO incompatible transplantation. Int. Rev. Immunol. 33, 174–194.

Ishida, H., Koyama, I., Sawada, T., Utsumi, K., Murakami, T., Sannomiya, A., et al., 2000. Anti-AB titer changes in patients with ABO incompatibility after living related kidney transplantations: survey of 101 cases to determine whether splenectomies are necessary for successful transplantation. Transplantation 70, 681–685.

Ishioka, G.Y., Lamont, A.G., Thomson, D., Bulbow, N., Gaeta, F.C., Sette, A., et al., 1993. Major histocompatibility complex class II association and induction of T cell responses by carbohydrates and glycopeptides. Springer Semin. Immunopathol. 15, 293–302.

Kobayashi, T., Yokoyama, I., Nagasaka, T., Liu, D., Kato, T., Tokoro, T., et al., 2000. Comparative study of antibody removal before pig-to-baboon and human ABO-incompatible renal transplantation. Transplant. Proc. 32, 1097.

LaTemple, D.C., Galili, U., 1998. Adult and neonatal anti-Gal response in knock-out mice for α1,3galactosyltransferase. Xenotransplantation 5, 191–196.

Latinne, D., Squifflet, J.P., De Bruyere, M., Pirson, Y., Gianello, P., Sokal, G., et al., 1989. Subclasses of ABO isoagglutinins in ABO-incompatible kidney transplantation. Transpl. Proc. 21, 641–642.

Lee, L.A., Sergio, J.J., Sachs, D.H., Sykes, M., 1994. Mechanism of tolerance in mixed xenogeneic chimeras prepared with a nonmyeloablative conditioning regimen. Transplant. Proc. 26, 1197–1198.

Lentine, K.L., Axelrod, D., Klein, C., Simpkins, C., Xiao, H., Schnitzler, M.A., et al., 2014. Early clinical complications after ABO-incompatible live-donor kidney transplantation: a national study of Medicare-insured recipients. Transplantation 98, 54–65.

Mayumi, H., Good, R.A., 1989. The necessity of both allogeneic antigens and stem cells for cyclophosphamide-induced skin allograft tolerance in mice. Immunobiology 178, 287–304.

Melexopoulou, C., Marinaki, S., Liapis, G., Skalioti, C., Gavalaki, M., Zavos, G., et al., 2015. Excellent long term patient and renal allograft survival after ABO-incompatible kidney transplantation: experience of one center. World J. Transplant. 5, 329–337.

Mendez, R., Sakhrani, L., Aswad, S., Minasian, R., Obispo, E., Mendez, R.G., 1992. Successful living-related ABO incompatible renal transplant using the BIOSYNSORB Immunoadsorption Column. Transplant. Proc. 24, 1738–1746.

Mohiuddin, M., Ogawa, H., Yin, D.P., Galili, U., 2003a. Tolerance induction to a mammalian blood group like carbohydrate antigen by syngeneic lymphocytes expressing the antigen: II. Tolerance induction on memory B cells. Blood 102, 229–236.

Mohiuddin, M.M., Ogawa, H., Yin, D.P., Shen, J., Galili, U., 2003b. Antibody-mediated accommodation of heart grafts expressing an incompatible carbohydrate antigen. Transplantation 75, 258–262.

Nemazee, D., 2000. Receptor editing in B cells. Adv. Immunol. 74, 89–126.

Nossal, G.J., 1994. Negative selection of lymphocytes. Cell 76, 229–239.

Nydegger, U., Mohacsi, P., Koestner, S., Kappeler, A., Schaffner, T., Carrel, T., 2005. ABO histo-blood group system-incompatible allografting. Int. Immunopharmacol. 5, 147–153.

Ogawa, H., Yin, D.P., Shen, J., Galili, U., 2003. Tolerance induction to a mammalian blood group-like carbohydrate antigen by syngeneic lymphocytes expressing the antigen. Blood 101, 2318–2320.

Ogawa, H., Mohiuddin, M.M., Yin, D.P., Shen, J., Chong, A.S., Galili, U., 2004a. Mouse-heart grafts expressing an incompatible carbohydrate antigen. II. Transition from accommodation to tolerance. Transplantation 77, 366–373.

Ogawa, H., Yin, D.P., Galili, U., 2004b. Induction of immune tolerance to a transplantation carbohydrate antigen by gene therapy with autologous lymphocytes transduced with adenovirus containing the corresponding glycosyltransferase gene. Gene Ther. 11, 292–301.

Ohdan, H., Yang, Y.G., Shimizu, A., Swenson, K.G., Sykes, M., 1999. Mixed chimerism induced without lethal conditioning prevents T cell- and anti-Gal α1,3Gal-mediated graft rejection. J. Clin. Investig. 104, 281–290.

Park, W.D., Grande, J.P., Ninova, D., Nath, K.A., Platt, J.L., Gloor, J.M., et al., 2003. Accommodation in ABO-incompatible kidney allografts, a novel mechanism of self-protection against antibody-mediated injury. Am. J. Transplant. 3, 952–960.

Pearse, M.J., Witort, E., Mottram, P., Han, W., Murray-Segal, L., Romanella, M., et al., 1998. Anti-Gal antibody-mediated allograft rejection in α1,3-galactosyltransferase gene knockout mice: a model of delayed xenograft rejection. Transplantation 66, 748–754.

Platt, J., 1994. A perspective on xenograft rejection and accommodation. Immunol. Rev. 141, 127–149.

Platt, J.L., Bach, F.H., 1991. The barrier to xenotransplantation. Transplantation 52, 937–947.

Posekany, K.J., Pittman, H.K., Bradfield, J.F., Haisch, C.E., Verbanac, K.M., 2002. Induction of cytolytic anti-Gal antibodies in α-1,3-galactosyltransferase gene knockout mice by oral inoculation with *Escherichia coli O86:B7* bacteria. Infect. Immun. 70, 6215–6222.

Radic, M.Z., Zouali, M., 1996. Receptor editing, immune diversification, and self-tolerance. Immunity 5, 505–511.

Rydberg, L., 2001. ABO incompatibility in solid organ transplantation. Transfus. Med. 11, 325–342.

Sharabi, Y., Sachs, D.H., 1989. Mixed chimerism and permanent specific transplantation tolerance induced by a non-lethal preparative regimen. J. Exp. Med. 169, 493–502.

Shishido, S., Asanuma, H., Tajima, E., Hoshinaga, K., Ogawa, O., Hasegawa, A., et al., 2001. ABO-incompatible living-donor kidney transplantation in children. Transplantation 72, 1037–1042.

Slapak, M., Digard, N., Ahmed, M., Shell, T., Thompson, F., 1990. Renal transplantation across the ABO barrier-A 9-year experience. Transplant. Proc. 22, 1425–1428.

Soares, M.P., Lin, Y., Sato, K., Stuhlmeier, K.M., Bach, F.H., 1999. Accommodation. Immunol. Today 20, 434–437.

Speir, J.A., Abdel-Motal, U.M., Jondal, M., Wilson, I.A., 1999. Crystal structure of an MHC class I presented glycopeptide that generates carbohydrate-specific CTL. Immunity 10, 51–61.

Starzl, T.E., 1964. Renal homografts in patients with major donor recipient blood group incompatibilities. Surgery 55, 195–200.

Suhr, B.D., Guzman-Paz, M., Apasova, E.P., Matas, A.J., Dalmasso, A.P., 2000. Induction of accommodation in the hamster-to-rat model requires inhibition of the membrane attack complex of complement. Transplant. Proc. 32, 976.

Sykes, M., Shimizu, I., Kawahara, T., 2005. Mixed hematopoietic chimerism for the simultaneous induction of T and B cell tolerance. Transplantation 79 (3 Suppl.), S28–S29.

Takahashi, K., 2001. ABO incompatible kidney transplantation. Elsevier Science, Amsterdam.

Takahashi, K., Saito, K., Nakagawa, Y., Tasaki, M., Hara, N., Imai, N., 2010. Mechanism of acute antibody-mediated rejection in ABO- incompatible kidney transplantation: which anti-A/anti-B antibodies are responsible, natural or de novo? Transplantation 89, 635–637.

Tanabe, K., 2007. Japanese experience of ABO- incompatible living kidney transplantation. Transplantation 84 (Suppl.), S4–S7.

Tanabe, K., Takahashi, K., Agishi, T., Toma, H., Ota, K., 1996. Removal of anti-A/B antibodies for successful kidney transplantation between ABO blood type incompatible couples. Transfus. Sci. 17, 455–462.

Tanabe, K., Takahashi, K., Sonda, K., Tokumoto, T., Ishikawa, N., Kawai, T., et al., 1998. Long-term results of ABO-incompatible living kidney transplantation: a single-center experience. Transplantation 65, 224–228.

Tanemura, M., Yin, D., Chong, A.S., Galili, U., 2000a. Differential immune response to α-gal epitopes on xenografts and allografts: implications for accommodation in xenotransplantation. J. Clin. Investig. 105, 301–310.

Tanemura, M., Maruyama, S., Galili, U., 2000b. Differential expression of α-gal epitopes (Galα1-3Galß1-4GlcNAc-R) on pig and mouse organs. Transplantation 69, 187–190.

Thall, A.D., Maly, P., Lowe, J.B., 1995. Oocyte Galα1,3Gal epitopes implicated in sperm adhesion to the zona pellucida glycoprotein ZP3 are not required for fertilization in the mouse. J. Biol. Chem. 270, 21437–21440.

Tydén, G., Kumlien, G., Genberg, H., Sandberg, J., Lundgren, T., Fehrman, I., 2005. ABO incompatible kidney transplantations without splenectomy, using antigen-specific immunoadsorption and rituximab. Am. J. Transplant. 5, 145–148.

Ulfvin, A., Backer, A.E., Clausen, H., Hakomori, S., Rydberg, L., Samuelsson, B.E., et al., 1993. Expression of glycolipid blood group antigens in single human kidneys: change in antigen expression of rejected ABO incompatible kidney grafts. Kidney Int. 44, 1289–1297.

Urschel, S., Larsen, I.M., Kirk, R., Flett, J., Burch, M., Shaw, N., et al., 2013. ABO-incompatible heart transplantation in early childhood: an international multicenter study of clinical experiences and limits. J. Heart Lung Transplant. 32, 285–292.

West, L.J., Pollock-Barziv, S.M., Dipchand, A.I., Lee, K.J., Cardella, C.J., Benson, L.N., et al., 2001. ABO-incompatible heart transplantation in infants. N. Engl. J. Med. 344, 793–800.

Whalen, G.F., Sullivan, M., Piperdi, B., Wasseff, W., Galili, U., 2012. Cancer immunotherapy by intratumoral injection of α-gal glycolipids. Anticancer Res. 32, 3861–3868.

Wilbrandt, R., Tung, K.S., Deodhar, S.D., Nakamoto, S., Kolff, W.J., 1969. ABO blood group incompatibility in human renal homotransplantation. Am. J. Clin. Pathol. 51, 15–23.

Xu, H., Sharma, A., Lei, Y., Okabe, J., Wan, H., Chong, A.S., et al., 2002. Development and characterization of anti-Gal B cell receptor transgenic Gal-/- mice. Transplantation 73, 1549–1557.

Yamamoto, F., Clausen, H., White, T., Marken, J., Hakomori, S.-I., 1990. Molecular genetic basis of the histo-blood group ABO system. Nature 345, 229–233.

Yamashita, M., Aikawa, A., Ohara, T., Kawamura, S., Nagamine, K., Hasegawa, A., 1993. Local immune states in ABO-incompatible renal allografts. Transplant. Proc. 25, 274–276.

Yang, Y.G., deGoma, E., Ohdan, H., Bracy, J.L., Xu, Y., Iacomini, J., et al., 1998. Tolerization of anti-Galα1-3Gal natural antibody-forming B cells by induction of mixed chimerism. J. Exp. Med. 187, 1335–1342.

Yarkoni, Y., Getahun, A., Cambier, J.C., 2010. Molecular underpinning of B-cell anergy. Immunol. Rev. 237, 249–263.

Yuzawa, Y., Brett, J., Fukatsu, A., Matsuo, S., Caldwell, P.R., Niesen, N., et al., 1995. Interaction of antibody with Forssman antigen in guinea pigs. A mechanism of adaptation to antibody- and complement-mediated injury. Am. J. Pathol. 146, 1260–1272.

Zschiedrich, S., Kramer-Zucker, A., Jänigen, B., Seidl, M., Emmerich, F., Pisarski, P., et al., 2015. An update on ABO-incompatible kidney transplantation. Transpl. Int. 28, 387–397.

SECTION 2

ANTI-GAL AS FOE

CHAPTER

6

Anti-Gal and Other Immune Barriers in Xenotransplantation

OUTLINE

INTRODUCTION

The interest in the natural anti-Gal antibody and the α-gal epitope was limited during the first decade following the discovery of this antibody and identification of its mammalian antigen. This interest increased with the realization that anti-Gal forms a barrier to transplantation of organs and tissues from other species into humans—xenotransplantation. The success in allotransplantation as a standard clinical treatment, due to development of effective immunosuppressive regimens, resulted already in the 1970s in a general lack of sufficient human allograft donors. Thus, in the recent 3–4 decades, a large proportion of patients in need of a heart graft die without being transplanted with a heart allograft. The number of patients in need of a kidney transplant, because of impaired kidneys function, is also much higher than the number of kidney allografts available. In addition, many patients with fulminant liver failure die without transplantation of a liver allograft. Thus, the use of organs from other animals has been considered as potential option for providing functioning organs to patients

http://dx.doi.org/10.1016/B978-0-12-813362-0.00006-3

who have no available allografts and are in risk of death (cf., Cooper et al., 1997; Taniguchi and Cooper, 1997).

Xenografts such as kidney, liver, and heart from pig, sheep, goat, baboon, and chimpanzee have been sporadically transplanted in patients at advanced stages of their disease throughout the 20th century (cf., Geller et al., 1992; Taniguchi and Cooper, 1997). In most recipients, rejection of the xenograft was observed within <1 day to <3 weeks with organs from nonprimate mammals and some of the primates, whereas in recipients of organs from chimpanzee or baboon, survival of the few xenografts was reported to be for several months and even up to 8 months (Reemtsma et al., 1964; Starzl et al., 1964; Rose et al., 1991). Thus, prolonged survival of xenografts was observed in donor/recipient pairs that are evolutionarily close (concordant) to each other (e.g., mouse/rat and human/chimpanzee), whereas rapid rejection of xenografts occurred in donor/recipient pairs that are phylogenetically far from each other (discordant) (Calne, 1970). These differences in graft survival led to the suggestion that the survival time of xenografts is inversely proportional to the phylogenetic distance between donor and recipient species. However, several considerations, including similarity in size of organ, availability of donors, and the risk of transmission of unknown infections by primate viruses, led most investigators in xenotransplantation to choose pig as a possible xenograft donor (Cooper et al., 1997).

The rapid discordant rejection of pig organs in humans or monkeys may occur within less than an hour to several hours. This rejection is referred to as "hyperacute" rejection because it displays similar characteristics as the hyperacute rejection observed following transplantation of ABO incompatible kidney allograft in patients in whom the corresponding anti-blood group A or B antibodies were not removed prior to the transplantation. In incompatible ABO allograft hyperacute rejection, such as incompatible kidney allografts, preexisting natural anti-blood group A or anti-B antibodies bind to the incompatible blood group antigens on endothelial cells lining the vascular system of the graft. These antibodies induce complement-mediated lysis of the cells, platelet aggregation, intravascular thrombosis, and thus, collapse of the vascular bed (Starzl, 1964; Wilbrandt et al., 1969; Chopek et al., 1987; Cooper, 1990). Similar rapid binding of antibodies to the endothelial cells of xenografts, activation of complement and cytolysis of the endothelial cells, as well as platelet aggregation, and collapse of the vascular bed were observed in hyperacute rejection of porcine xenografts in monkeys, within minutes to hours (Lexer et al., 1986; Cooper et al., 1988, 1997; Geller et al., 1992). The rejection was slowed, if serum immunoglobulins were removed by plasmapheresis. Similarly, a rejection damage to the endothelial cells of the xenograft was observed within 3 h post transplantation in a patient with fulminant liver failure transplanted for temporary support with porcine liver, despite the removal of a large proportion of serum immunoglobulins by plasmapheresis (Makowka et al., 1995). These observations suggested that there are natural antibodies that mediate the hyperacute rejection of discordant xenografts, such as porcine organs. As described in this chapter, the primary natural antibody mediating hyperacute rejection is the natural anti-Gal antibody that interacts with the millions of α-gal epitopes presented on each endothelial cell of the porcine xenograft. This chapter describes additional natural and induced antibodies that form immune barriers against xenografts in humans. The chapter also describes methods that have enabled elimination of some of these barriers and possible methods for confronting barriers that have not been overcome as yet.

ANTI-GAL MEDIATES HYPERACUTE REJECTION OF XENOGRAFTS

As discussed in Chapters 1 and 2, α-gal epitopes are abundant in nonprimate mammals, prosimians, and New World monkeys, whereas the natural anti-Gal antibody is produced in Old World monkeys, apes, and humans, all of which lack the α-gal epitope (Galili et al., 1984, 1987, 1988a; Oriol et al., 1999; Teranishi et al., 2002). In 1987, the α-gal epitope was reported to be present in high numbers on red blood cells (RBC) of various mammals, including pigs, and to readily bind the anti-Gal antibody purified from serum of human blood type AB donors (Galili et al., 1987). This human natural anti-Gal antibody was further shown to bind to nucleated cells of nonprimate mammals, including porcine endothelial cells and epithelial cells, as well as to nucleated cells from lemurs and New World monkeys (Galili et al., 1988a). That study further measured the binding to various cells of the α-gal epitope specific lectin *Bandeiraea (Griffonia) simplicifolia* IB4 radiolabeled with ^{125}I. Based on these measurements and calculations performed by Scatchard plotting analysis, the number of lectin binding sites/cell was determined. It was estimated that porcine endothelial cells present at least ~10^7 α-gal epitopes per cell and porcine epithelial cells ~3×10^7 α-gal epitopes per cell (Galili et al., 1988a).

In vitro xenotransplantation studies with anti-Gal

The first xenotransplantation associated studies demonstrating that most of human natural antibodies binding to the porcine kidney cell line PK15 are anti-Gal were those of Cooper and colleague (Good et al., 1992; Cooper et al., 1993). These studies demonstrated that the antibodies binding to pig cells could be removed from normal human serum by adsorption on a column of synthetic α-gal epitopes linked to silica beads and to a much lesser extent by adsorption on silica beads presenting other carbohydrate structures. Subsequent studies demonstrated antibody-dependent cell cytotoxicity (ADCC) of pig endothelial cells, smooth muscle cells and fibroblasts, following binding of purified human anti-Gal to these cells and incubation with human mononuclear cells (Galili, 1993). In addition, binding of human anti-Gal to de novo synthesized α-gal epitopes on monkey cells was shown with Old World monkey cells that were transfected with mouse α1,3galactosyltransferase gene (α1,3GT gene also called *GGTA1*) (Sandrin et al., 1993). These initial xenotransplantation studies suggested that the anti-Gal/α-gal epitopes interaction is likely to be the main cause for hyperacute rejection of porcine xenografts in Old World monkeys and humans. This interaction induces rapid activation of the complement system, resulting in complement-mediated cytolysis of the xenograft endothelial cells. It is further suggested that the destruction of endothelial cells leads to induction of platelet aggregation and thrombosis by substances released from the lysed cells and by the exposed blood vessels wall (Collins et al., 1995). Indeed, *in vitro* neutralization of anti-Gal in human and baboon serum with free oligosaccharides indicated that the α-gal epitope free trisaccharide (Galα1-3Galβ1-4GlcNAc) is the most effective oligosaccharide to inhibit complement-mediated cytolysis of PK15 pig kidney cells (Neethling et al., 1996). This trisaccharide was >10-fold more effective than the Galα1-3Gal disaccharide, whereas oligosaccharides with terminal carbohydrates other than α-galactosyl lacked inhibitory activity even at 500-fold higher concentrations.

In vivo xenotransplantation studies with anti-Gal

The significance of the anti-Gal/α-gal epitope interaction in hyperacute rejection of xenografts, rather than the evolutionary distance between donor and recipient, was demonstrated in studies of New World monkey heart xenografts transplanted in Old World monkey recipients (Collins et al., 1995). As indicated above, New World monkeys may be considered concordant to Old World monkeys on an evolutionary scale. Nevertheless, the former synthesize α-gal epitopes on their cells, whereas the latter produce the anti-Gal antibody. Accordingly, squirrel monkey (New World monkey) heart xenografts underwent hyperacute rejection in baboon (Old World monkey) recipients within a period of <1 h (Collins et al., 1995).

In vivo studies, in which anti-Gal activity was inhibited, further validated the role of anti-Gal in hyperacute rejection of porcine xenografts in Old World monkeys. Infusion of the disaccharide Galα1-3Gal into baboon was found to delay hyperacute rejection of porcine heart xenograft (Simon et al., 1998). The delay was for a relatively short time because this disaccharide is rapidly removed from the circulation by the kidneys. A longer delay in rejection of kidney xenografts in baboons was observed in recipients in which anti-Gal was removed from the circulation by adsorption as a result of passing their blood through a column presenting synthetic α-gal epitopes (Xu et al., 1998). The kidney xenografts were rejected, on average, only after 7 days, when anti-Gal reappeared in the blood because of the continuous production of this antibody as part of the immune response to bacteria of the natural flora in the recipients (Galili et al., 1988b). These *in vitro* and *in vivo* studies were validated in the late 1990s in additional research centers and indicated the need for elimination of the α-gal epitopes from porcine xenografts.

The last stage in synthesis of α-gal epitopes in mammalian cells is the linking of the terminal galactose in an α1,3 link to the penultimate galactose in the nascent carbohydrate chain to generate the Galα1-3Galβ1-3GlcNAc-R structure (Fig. 2A in Chapter 1). This step occurs in the trans-Golgi, a cellular compartment that includes in addition to α1,3GT, other glycosyltransferases that compete with α1,3GT to cap the nascent carbohydrate chain with carbohydrate units such as fucose or sialic acid (Smith et al., 1990). Therefore, initial attempts to eliminate the α-gal epitope included introduction into mouse or pig transgenes such as fucosyltransferase (Sandrin et al., 1995; Costa et al., 1999), *N*-acetylglucosaminyltransferase (Tanemura et al., 1997; Koma et al., 2000; Miyagawa et al., 2001), or sialyltransferase (Tanemura et al., 1998) to compete with α1,3GT or reduce its ability to link the terminal galactose. Such transgenes indeed decreased α-gal epitope expression on mouse cells but did not eliminate it completely. Because as many as ~1% of circulating B cells in humans can produce anti-Gal (Chapter 1 and Galili et al., 1993), even reduced numbers of α-gal epitopes on xenografts suffice to activate the quiescent anti-Gal B cells. The resulting marked increase in production of anti-Gal exacerbates the rejection of xenografts. Successful elimination of the α-gal epitope in pigs was achieved in the early 2000s by disruption of the α1,3GT gene, as described below.

PIGS LACKING α-GAL EPITOPES AS DONORS IN XENOTRANSPLANTATION

The observations regarding the synthesis of α-gal epitopes as hundreds of thousands to millions of epitopes per cell in nonprimate mammals (Galili et al., 1988a) raised the question whether mammals, such as pigs, can develop properly and survive in the absence of these epitopes. The very large number of these epitopes on mouse and pig cells and in basement membranes in nonprimate mammals suggested that the α-gal epitope may fulfill an important biological function. Indeed, studies of the process of fertilization in mice indicated that terminal α-galactosyls on glycoproteins of the egg zona pellucida serve as "docking receptors" for spermatozoa, which have a protein binding to the α-gal epitope (Bleil and Wassarman, 1988). Carbohydrates of glycoproteins and glycolipids (glycans) were also found to be essential in cell–cell interaction in fetal development. One example is the disruption (knockout) of an *N*-acetylglucosaminyltransferase gene encoding the enzyme that links of *N*-acetylglucosamine β1,3 to the nascent core structure of *N*-linked carbohydrate chains on glycans in the Golgi apparatus (see Fig. 2 in Chapter 1). Mice lacking this enzyme could not complete the synthesis of their carbohydrate chains on glycans and died *in utero* (Shafi et al., 2000). However, the findings suggesting that ancestral Old World monkeys and apes lost their ability to synthesize the α-gal epitope 20–30 million years ago, suggested that there are other carbohydrate epitopes, which may serve as backup molecules, if α-gal epitopes are eliminated (see Chapter 2). Thus, it was of interest to determine whether nonprimate mammals expressing α-gal epitopes, such a mice and pigs, can develop and survive in the absence of this epitope.

α1,3galactosyltransferase knockout mice

Two groups succeeded in the mid-1990s to generate mice lacking α-gal epitopes (i.e., knockout mice for the α1,3GT gene—"GT-KO mice") by the method of nuclear transfer. Thall et al. (1995) generated these mice from parental C57BL/6 × BALB/c wild-type (WT) mice and Tearle et al. (1996) from parental BALB/c mice. In most animal facilities, these mice do not produce anti-Gal as a natural antibody because the sterile conditions in which they are kept, and the sterile food they receive. The sterile environment and food do not enable the development of a gastrointestinal flora presenting antigens that stimulate the mice to produce the anti-Gal antibody. However, immunization of the mice with rabbit RBC membranes presenting multiple α-gal epitopes (LaTemple et al., 1999), or with pig kidney membrane homogenates containing many of these epitopes (Tanemura et al., 2000a), results in production of anti-Gal in these mice. Feeding of GT-KO mice with *Escherichia coli O86* also induces anti-Gal IgM production in these mice (Posekany et al., 2002). In early studies, this *E. coli* was found to bind effectively the human anti-Gal antibody, suggesting that α-gal-like carbohydrate epitopes on its wall induce anti-Gal production (Galili et al., 1988b). Interestingly, GT-KO mice develop cataract at the age of 6–8 weeks (Thall, 1999), suggesting that the α-gal epitope has a biological function in the eye of mice, which cannot be replaced by other carbohydrates. However, this function is not vital to mice kept in animal facilities, and thus, the mice display normal life span and normal fecundity.

α1,3galactosyltransferase knockout pigs

Initial attempts to genetically engineer pigs to overcome hyperacute rejection aimed to prevent activation of the complement system. This was based on the observations that natural antibodies in human serum induce complement-mediated cytolysis of pig cells (Dalmasso et al., 1991; Zhao et al., 1994) and that treatment of pig hearts with cobra venom factor, which inactivates complement, delays rejection of these xenografts in baboons

(Kobayashi et al., 1997). Transgenic pigs containing genes coding human complement-regulatory proteins, such as human decay accelerating factor [CD55] (Cozzi and White, 1995; White and Yannoutsos, 1996; Cozzi et al., 1997), membrane cofactor protein—CD46 (Diamond et al., 2001; Loveland et al., 2004), and MAC-inhibitory protein—CD59 (Fodor et al., 1994), or double transgenic pigs for both CD55 and CD59 (Byrne et al., 1997), were generated. However, as discussed in Chapter 1 and shown in Fig. 3 in that chapter, α-gal epitopes, even in small numbers, induce extensive production of elicited anti-Gal antibodies. These antibodies are very effective in killing pig cells by ADCC (Galili, 1993; Watier et al., 1996; Galili et al., 2001), even in the absence of complement activation. Thus, it became evident that complete elimination of α-gal epitopes from pig tissues is required to overcome the immune barrier of the anti-Gal antibody. The reports on the successful disruption of α1,3GT gene in mice (Thall et al., 1995; Tearle et al., 1996) prompted researchers to determine whether generation of GT-KO pigs lacking α-gal epitopes is feasible, as well. In view of the 10–500-fold higher concentration of α-gal epitopes in pig tissues in comparison with that in mouse tissues (Tanemura et al., 2000b), it was not clear whether pigs may develop normally in the absence of α-gal epitopes. However, by using animal cloning and nuclear transfer technologies in pigs, the α1,3GT gene was inactivated and knockout pigs for this gene were successfully generated (Lai et al., 2002; Phelps et al., 2003; Kolber-Simonds et al., 2004; Takahagi et al., 2005). These knockout pigs are referred to here as "GT-KO pigs." Generation of such pigs was repeated by additional groups in subsequent years including GT-KO pigs that are also transgenic for one or more complement-regulatory genes (McGregor et al., 2012; Mohiuddin et al., 2012, 2016; Azimzadeh et al., 2015; Le Bas-Bernardet et al., 2015), double knockouts for α1,3GT, and for genes encoding other enzymes participating in synthesis of carbohydrate antigens (Lutz et al., 2013; Butler et al., 2016), or GT-KO pigs that have *N*-acetylglucosaminyltransferase III transgene gene, as well as a transgene for decay accelerating factor (Takahagi et al., 2005).

GT-KO pigs naturally produce the anti-Gal antibody (Dor et al., 2004; Fang et al., 2012; Galili, 2013), probably because of the growth of bacteria in their gastrointestinal tract, which stimulate the immune system to produce this antibody. The natural anti-Gal antibody in GT-KO pigs displays characteristics like those of human anti-Gal antibody, implying that GT-KO pigs completely lack α-gal epitopes, which otherwise, may tolerize their immune system to this epitope (Galili, 2013). Accordingly, baboons transplanted with GT-KO pig heart display no increase in anti-Gal titers because there are no immunizing α-gal epitopes on the xenograft (Ezzelarab et al., 2006; Yeh et al., 2010). In contrast, baboons and other Old World monkeys transplanted with WT porcine tissue display marked increase in anti-Gal activity, which may reach titers that are 1000-fold higher than the natural titer of the antibody because of the activation of quiescent anti-Gal B cells by xenogeneic α-gal epitopes (Galili et al., 1997; Ezzelarab et al., 2006; Stone et al., 2007b; Yeh et al., 2010). Several studies with heart or kidney xenografts from GT-KO pigs indeed indicated that the xenografts did not undergo hyperacute rejection but were rejected after several weeks to several months (Kuwaki et al., 2005; Chen et al., 2005; Tseng et al., 2005; Ezzelarab et al., 2006; Hisashi et al., 2008). The recipient monkeys rejecting the xenografts were found to produce antibodies that could be detected *in vitro* as antibodies binding to GT-KO pig cells (Ezzelarab et al., 2006; Yeh et al., 2010). Evidently, these antibodies are not anti-Gal because the GT-KO pig xenografts completely lack α-gal epitopes. Thus, these antibodies are in general referred to as anti-non-gal antibodies. Although not directly proven, it is probable that these anti-non-gal antibodies contribute significantly to the observed rejection of GT-KO pig xenografts.

GT-KO pigs, which were further engineered for introduction of transgenes, or in which additional genes were knocked out, were found to serve also as xenograft donors for pancreatic islet cells that may provide long-term independence from insulin injections in diabetic patients. In some studies, xenograft islet cells were obtained from pig fetal pancreatic islet cells that were cultured *in vitro* to form cell clusters (Korsgren et al., 1988). These fetal islet cell clusters from WT pig donors grafted into diabetic patients were found to last only for few weeks (Groth et al., 1994). However, transplantation in cynomolgus monkey of islet cells from GT-KO pig donors that also are transgenic for the complement-regulatory molecule CD46, resulted in function of the transplanted islets for at least 3 months, and in one monkey for >1 year (van der Windt et al., 2009). A marked increased in survival and function of GT-KO newborn pig islet cells in rhesus monkeys in comparison with survival of islet cells from WT newborn pigs was also observed by Thompson et al. (2011). All these studied have implied that the GT-KO pig should serve the starting point for additional genetic engineering to achieve improved survival of pig xenografts in humans.

ANTI-NON-GAL ANTIBODIES

There are two types of anti-non-gal antibodies produced in humans against porcine xenografts or against xenografts from most other mammalian species: (1) anti-carbohydrate non-gal antibodies, and (2) anti-protein non-gal antibodies.

Anti-carbohydrate non-gal antibodies

A natural anti-carbohydrate antibody that was reported to be produced in humans, in addition to anti-Gal, is the anti-Neu5Gc antibody, which binds to the sialic acid (Sia) *N*-glycolyl-5-neuraminic acid (Neu5Gc). Two of the common forms of Sia found in mammals, including pigs, are *N*-acetyl-5-neuraminic acid (Neu5Ac) and Neu5Gc (Bouhours et al., 1996; Padler-Karavani and Varki, 2011). Humans and New World monkeys are the exception, as they synthesize only Neu5Ac and not Neu5Gc (Muchmore et al., 1998; Springer et al., 2014). Instead, humans were found to produce a natural antibody against the Neu5Gc epitope, the anti-Neu5Gc antibody, also called Hanganutziu-Deicher antibody (Higashi et al., 1977; Merrick et al., 1978; Zhu and Hurst, 2002; Padler-Karavani et al., 2008; Varki, 2010). The lack of Neu5Gc in humans is the result of evolutionary inactivation of the gene coding cytidine-monophosphate-*N*-acetyl-neuraminic acid hydroxylase (*CMAH*) in hominins following the divergence from ancestors of chimpanzee (Varki, 2010). Anti-Neu5Gc antibodies were found to bind *in vitro* to Neu5Gc on pig cells and destroy these cells, thereby acting as an immunological obstacle in xenotransplantation (Zhu and Hurst, 2002; Tahara et al., 2010; Padler-Karavani and Varki, 2011). An elevated activity of anti-Neu5Gc antibody was reported in patients in whom burns were covered with pig skin (Scobie et al., 2013), in patients implanted with porcine heart valve (Reuven et al., 2016) and in patients treated with rabbit antithymocyte globulin (ATG), i.e., with antibodies presenting Neu5Gc (Salama et al., 2017). These observations suggest that anti-Neu5Gc antibodies may be deleterious in human xenograft recipients (Padler-Karavani and Varki, 2011; Salama et al., 2015). The availability of double knockout pigs for both α1,3GT gene and for the *CMAH* gene, i.e., GT-KO pigs, in which the *CMAH* gene (the gene encoding hydroxylase that converts *N*-Neu5Ac into Neu5Gc) was inactivated as well (Lutz et al., 2013), may be a safe approach for overcoming the potential immunological barrier of anti-Neu5Gc antibodies.

Additional anti-carbohydrate antibodies are being studied by arrays of carbohydrates binding antibodies from monkeys and humans exposed to pig tissues (Byrne et al., 2014, 2015). By such methodology, antibodies to the biosynthetic product of the enzyme β(1,4)*N*-acetylgalactosaminyl transferase have been identified (Byrne et al., 2014). The significance of these antibodies in preventing xenotransplantation in humans awaits further elucidation.

Anti-protein non-gal antibodies and the immunosuppression conundrum

Most homologous proteins in humans and pigs (or other nonprimate mammals) differ from each other in their amino acid sequence by ~3%–40%. These differences, which are the result of random mutations that have accumulated independently in the various mammalian lineages for at least 75 million years, are likely to result in immunogenicity (i.e., in eliciting antibody production) of >99% of porcine proteins in humans. The proportion of such mutations varies in different regions of a given protein based on functional constraints. Thus, in a membrane-bound receptor there are more mutations in the tether region than in the ligand-binding region (Galili, 2012). Regardless of their location, if they are absent in humans, these mutations form immunogenic amino acid sequences in pig proteins. The production of these anti-protein non-gal antibodies in human recipients of xenografts was suggested in studies with sera of diabetic patients transplanted with a kidney allograft and with fetal porcine islet cell clusters xenograft (Groth et al., 1994; Galili et al., 1995). In addition to the marked increase in anti-Gal activity as a result of anti-Gal B cell response to α-gal epitopes on the xenograft, the patients displayed post transplantation production of antibodies capable of binding to pig endothelial cells, despite the removal of anti-Gal antibodies by columns with synthetic α-gal epitopes. Similar studies on antibody response were performed in a recipient of mouse fibroblasts as a packaging cell line for a virus vector in a gene therapy study (Galili et al., 2001). These studies demonstrated marked production of elicited anti-Gal antibodies (Fig. 3A in Chapter 1) and anti-non-gal antibodies. Whereas the elicited anti-Gal antibodies among IgG subclasses were measured as IgG2 > IgG1 > IgG3 > IgG4, the relative activity of IgG subclasses of elicited anti-non-gal antibodies against mouse protein xenoantigens was IgG1 > IgG2 > IgG3 > IgG4 (Galili et al., 2001).

The extensive diversity in specificity of anti-non-gal antibodies produced in porcine xenograft recipients could be demonstrated in patients with ruptured anterior cruciate ligament (ACL). These patients were implanted with porcine tendon depleted of α-gal epitopes (by incubation with recombinant α-galactosidase) and partially cross-linked with glutaraldehyde to enable gradual remodeling of the graft into autologous ACL (Stone et al., 2007a). The specificity of elicited antibodies in the sera of the patients was determined by Western blots of porcine tendon proteins separated by sodium dodecyl sulfate-polyacrylamide gel electrophoresis (PAGE) and immunostained with sera of these patients. Prior to the assay, these sera were adsorbed on an equal volume of packed glutaraldehyde fixed rabbit RBC.

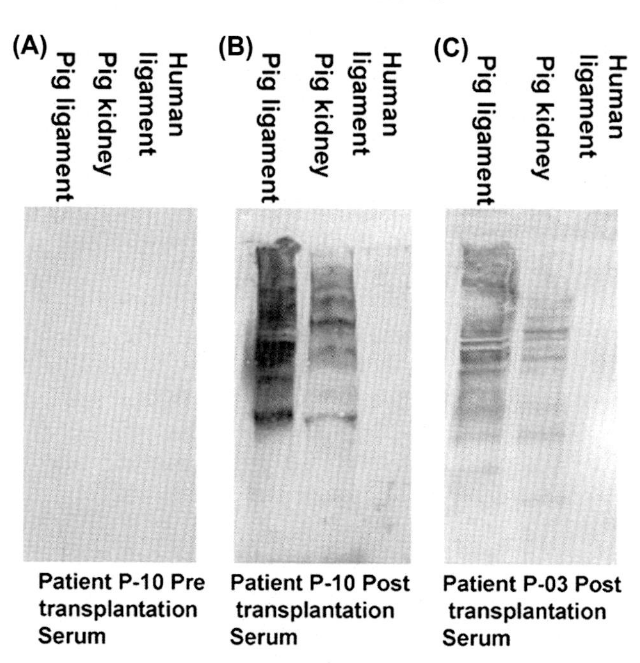

FIGURE 1 Anti-non-gal antibody activity in the serum of patients with ruptured anterior cruciate ligament who were grafted with porcine patellar tendon. The porcine tendon was enzymatically treated with α-galactosidase for elimination of α-gal epitopes and partially cross-linked with glutaraldehyde. Proteins extracted from porcine patellar tendon, porcine kidney, and human tendon were separated by sodium dodecyl sulfate-polyacrylamide gel electrophoresis, blotted on nitrocellulose paper and studied for binding of serum antibodies in Western blot analysis. Peroxidase-coupled anti-human IgG antibody served as secondary antibody. (A) Preimplantation serum of patient P-10. (B) Serum of patient P-10, 6 months post grafting. (C) Serum of patient P-03, 6 months post implantation. Prior to this analysis, the sera were adsorbed on equal volume of packed, glutaraldehyde fixed rabbit RBC and were diluted 1:10. *Adapted from Stone, K.R., Abdel-Motal, U.M., Walgenbach, A.W., Turek, T.J., Galili, U., 2007a. Replacement of human anterior cruciate ligaments with pig ligaments: a model for anti-non-gal antibody response in long-term xenotransplantation. Transplantation 83, 211–219, with permission.*

Rabbit RBC present multiple α-gal epitopes that remove anti-Gal antibodies (Galili et al., 1984, 2007) and Neu5Gc epitopes binding anti-Neu5Gc antibodies (Takemae et al., 2010). Sera obtained prior to grafting of the pig tendon implant contained no IgG antibodies that bound to the separated pig proteins (Fig. 1), implying that these sera do not contain antibodies that bind to porcine tendon proteins. However, sera obtained 6 months post-implantation displayed extensive production of anti-non-gal antibodies. These antibodies bound to a very large number of porcine tendon proteins, implying that these proteins are immunogenic in humans (Fig. 1) (Stone et al., 2007a). Because of the high number of the proteins-binding anti-non-gal antibodies, the binding pattern is a "smear" of partially overlapping bands rather than of individual bands of proteins. Some of the anti-non-gal antibody–binding proteins were also found in porcine kidney, suggesting that these are proteins common to different tissues. No antibodies were found to bind to human ligament proteins, indicating that these anti-non-gal antibodies are specific to porcine proteins (Fig. 1). This negative observation further suggests that an effective immune tolerance mechanism in humans prevents production of autoantibodies to homologous proteins in the recipients. A similar production of elicited anti-non-gal antibodies following transplantation of xenografts from GT-KO pigs was observed in baboons transplanted with GT-KO pig heart (Tseng et al., 2006) or kidney (Chen et al., 2005). WT pig heart xenografts elicited such anti-non-gal antibodies in baboons, as well (Lam et al., 2004).

The contribution of elicited anti-non-gal antibodies in rejection of WT pig xenografts, during the first months post transplantation is much lower than that of anti-Gal. This could be demonstrated in cynomolgus monkeys implanted in the suprapatellar pouch with porcine meniscus cartilage or with the same cartilage depleted of α-gal epitopes by treatment with recombinant α-galactosidase (Stone et al., 1998). Cartilage grafts explanted after 2 months displayed 90%–95% lower inflammatory response in the absence of α-gal epitopes in comparison with cartilage grafts containing α-gal epitopes. However, anti-non-gal antibody response against porcine protein antigens continues as long as there is pig tissue in the recipient. This could be shown in patients grafted with processed pig tendon. Anti-non-gal antibody production continued in these patients for almost 2 years, until all the porcine tissue was replaced by autologous human ACL tissue produced by human fibroblasts that align with the porcine collagen fibers scaffold (Figs. 2 and 3 in Chapter 13; Stone et al., 2007a).

The initial number of B cells in of each of the multiple B cell clones capable of producing anti-non-gal antibodies is very small prior to grafting of porcine tissue. Thus, it takes a period of several weeks post transplantation for these clones to expand and produce their antibodies at a measurable level (Galili et al., 2001). However, the expanded clones continuously produce these antibodies that may mediate chronic rejection of the xenograft. The antibody deposits in rejected heart and kidney of GT-KO pigs transplanted in monkeys are likely to be of these anti-non-gal antibodies because monkeys lack the ability to produce anti-Neu5Gc antibodies (Kuwaki et al., 2005; Chen et al., 2005; Tseng et al., 2005; Ezzelarab et al., 2006; Hisashi et al., 2008).

In a recent study, baboons were transplanted with porcine hearts from GT-KO pigs, which had also human CD46 and human thrombomodulin transgenes. The immunosuppression induction regimen included anti-CD20, anti-CD40, and ATG antibodies, and the immunosuppression maintenance regimen included anti-CD40, mycophenolate mofetil, and Solu-Medrol (Mohiuddin et al., 2016). This resulted in an unprecedented survival of the porcine heart for 1 year to almost 3 years. This protocol suggests that anti-non-gal antibody production was effectively suppressed, thus preventing rejection of the xenograft by these antibodies. However, this observation leads to an immunological conundrum inherent to immunosuppression of the wide range of anti-non-gal antibodies. The immunosuppression treatment preventing anti-non-gal antibody production includes drugs and antibodies affecting nonspecifically the whole immune system. This suppression of tens or hundreds of anti-non-gal antibody specificities may affect production of other antibodies as well, including antibodies protecting against various opportunistic infections. Therefore, prevention of anti-non-gal antibody production may expose the recipients to various infections, which are of no risk to individuals with a nonsuppressed immune system. Based on these considerations, it is suggested that development of methods for induction of immune tolerance specific to the multiple immunogenic porcine proteins may be beneficial to the success of porcine xenografts in humans.

INDUCTION OF IMMUNE TOLERANCE TO PORCINE PROTEIN ANTIGENS IN XENOGRAFT RECIPIENTS

Tolerance induction by thymokidney xenografts and mixed chimerism

Yamada et al. (2005) reported on one approach for induction of tolerance to porcine xenograft antigens by placing porcine thymus tissue under the porcine kidney capsule in GT-KO pigs. This procedure was performed in the kidney xenograft several weeks prior to transplantation into a monkey recipient to achieve vascularization of this thymus tissue. The thymus component of the xenograft is expected to tolerize the recipient against pig xenoantigens (Yamada et al., 2005). The thymokidney xenografts transplanted into monkey recipients survived for almost 3 months, a period that was much longer than the survival period of kidney xenografts lacking the thymus component (Yamada et al., 2005; Griesemer et al., 2009).

Mixed bone marrow chimerism was found to induce tolerance to alloantigens (Mayumi and Good, 1989; Sharabi and Sachs, 1989) and to the α-gal epitope (Yang et al., 1998). Thus, it was of interest to determine whether tolerance to GT-KO pig antigens can be achieved by generating bone marrow chimerism in monkeys with GT-KO pig bone marrow cells. Infusion of GT-KO bone marrow cells to baboons resulted in elimination of most of the bone marrow cells within 24 h (Griesemer et al., 2010; Tasaki et al., 2015). However, if the porcine bone marrow cells express human CD47 (a molecule preventing phagocytosis of the porcine cells by human macrophages) (Tena et al., 2014), or if the porcine bone marrow was injected into bone (Tasaki et al., 2015), the survival of porcine bone marrow cells could be extended to several weeks. Additional research in increasing survival of porcine bone marrow cells in monkeys may result in improving the efficacy of tolerance induction to porcine protein antigens in monkey recipients of GT-KO pig xenografts by mixed bone marrow chimerism.

Tolerance induction following elimination of B cells: theoretical considerations

One could hypothesize that immune tolerance preventing production of anti-non-gal antibodies may be achieved by pre-transplantation elimination of B cells in xenograft recipients. B cell tolerance involves deletion or receptor editing in newly generated B cell clones that emerge in the bone marrow, when these clones reach the differentiation stage of immature B cells. At this differentiation stage, B cells express B cell receptors which engage the many antigens that are self-antigens and are tolerized following this interaction (Radic and Zouali, 1996; Sandel and Monroe, 1999; Nemazee, 2000). Thus, it would be of interest to determine whether tolerance to porcine protein antigens may be achieved by elimination of the B cell system in the recipient, including the immature B cells, prior to transplantation of the xenograft. This may be achieved by anti-B cell antibodies that bind to B cells at the various differentiation stages. Shortly following such B cells elimination, the xenograft may be grafted and porcine bone marrow cells administered into the circulation. This would allow the immune system to regenerate in the presence of porcine cells producing the tolerizing antigens. The exposure of newly emerging B cells to porcine antigens provided by bone marrow cells and by the xenograft may result in "fooling" the immune system to "regard" these xenoproteins as self-antigens. This may prevent the maturation of B cells specific to the multiple porcine antigens. Inhibition of helper T cell activity in the first stages following transplantation may further contribute to the initial establishment of tolerance to a wide range of porcine antigens. It would be of interest to determine whether porcine mononuclear cells can provide such tolerizing antigens, as well. It would also be of interest to determine whether inclusion of the thymokidney approach to the tolerance induction process, described above, would improve the perpetuation of the state of tolerance toward the xenograft. It is possible that the long survival of heart xenograft in baboons following the use of anti-CD20 and anti-CD40 antibodies in the study of Mohiuddin et al. (2016) may reflect the induction of such B cell tolerance to porcine antigens. Studies on the tolerizing method by elimination of the B cell system shortly prior to transplantation may result in the subsequent perpetuation of specific tolerance to porcine antigens (i.e., prevention of anti-non-gal antibody production) by the xenograft antigens functioning as tolerizing antigens. This may be achieved without the need for long-term nonspecific immunosuppression of the B cell system of the xenograft recipient.

CONCLUSIONS

The anti-Gal antibody has been a major barrier for the possible use of pig organs as xenografts in humans. Studies in primates indicated that transplantation of porcine heart or kidney in anti-Gal producing monkeys results in rapid binding of the natural anti-Gal antibody to the millions of α-gal epitopes on each endothelial cell of the xenograft. This antigen/antibody interaction activates the complement system, which mediates lysis of these cells and collapse of the vascular bed, ultimately resulting in hyperacute rejection of the xenograft within 30 min to several hours. If complement activation is inhibited, anti-Gal IgG binding to the xenograft cells induces destruction of these cells by the ADCC mechanism. This obstacle of anti-Gal/α-gal epitope interaction was eliminated by generation of knockout pigs for the α1,3GT gene (*GGTA1*), designated GT-KO pigs. Generation of GT-KO pigs in which the *CMAH* gene was knocked out, as well, avoids the detrimental effects of the human anti-non-gal antibodies to the Neu5Gc carbohydrate epitope, which is synthesized by WT pigs. Another immunological barrier, which needs to be eliminated, is the anti-non-gal antibodies produced against most porcine protein antigens. These antigens are immunogenic in humans because they contain multiple amino acid sequences that are absent in homologous human proteins. It is suggested that temporary elimination of the B cell system prior to transplantation of the xenograft may result in specific tolerance induction to porcine xenoproteins, which may be "regarded" by the tolerized immune system as self-antigens. Prevention of T cell help in the initial period post-transplantation may further help in specific tolerance induction in B cell clones reacting against porcine protein xenoantigens, while enabling regeneration of the rest of the immunoglobulin repertoire.

References

Azimzadeh, A.M., Kelishadi, S.S., Ezzelarab, M.B., Singh, A.K., Stoddard, T., Iwase, H., et al., 2015. Early graft failure of GalTKO pig organs in baboons is reduced by expression of a human complement pathway-regulatory protein. Xenotransplantation 22, 310–316.

Bleil, J.D., Wassarman, P.M., 1988. Galactose at the nonreducing terminus of O-linked oligosaccharides of mouse egg zona pellucida glycoprotein ZP3 is essential for the glycoprotein's sperm receptor activity. Proc. Natl. Acad. Sci. U.S.A. 85, 6778–6782.

Bouhours, D., Pourcel, C., Bouhours, J.E., 1996. Simultaneous expression by porcine aorta endothelial cells of glycosphingolipids bearing the major epitope for human xenoreactive antibodies (Galα1-3Gal), blood group H determinant and *N*-glycolylneuraminic acid. Glycoconj. J. 13, 947–953.

Butler, J.R., Paris, L.L., Blankenship, R.L., Sidner, R.A., Martens, G.R., Ladowski, J.M., et al., 2016. Silencing porcine CMAH and GGTA1 genes significantly reduces xenogeneic consumption of human platelets by porcine livers. Transplantation 100, 571–576.

Byrne, G.W., McCurry, K.R., Martin, M.J., McClellan, S.M., Platt, J.L., Logan, J.S., 1997. Transgenic pigs expressing human CD59 and decay-accelerating factor produce an intrinsic barrier to complement mediated damage. Transplantation 63, 149–155.

Byrne, G.W., Du, Z., Stalboerger, P., Kogelberg, H., McGregor, C.G., 2014. Cloning and expression of porcine β1,4 *N*-acetylgalactosaminyl transferase encoding a new xenoreactive antigen. Xenotransplantation 21, 543–555.

Byrne, G.W., McGregor, C.G., Breimer, M.E., 2015. Recent investigations into pig antigen and anti-pig antibody expression. Int. J. Surg. 23, 223–228.

Calne, R.Y., 1970. Organ transplantation between widely disparate species. Transplant. Proc. 2, 550–556.

Chen, G., Qian, H., Starzl, T., Sun, H., Garcia, B., Wang, X., et al., 2005. Acute rejection is associated with antibodies to non-Gal antigens in baboons using Gal-knockout pig kidneys. Nat. Med. 11, 1295–1298.

Chopek, M.W., Simmons, R.L., Platt, J.L., 1987. ABO-incompatible kidney transplantation: initial immunopathologic evaluation. Transplant. Proc. 19, 4553–4557.

Collins, B.H., Cotterell, A.H., McCurry, K.R., Alvarado, C.G., Magee, J.C., Parker, W., et al., 1995. Cardiac xenografts between primate species provide evidence for the importance of the α-galactosyl determinant in hyperacute rejection. J. Immunol. 154, 5500–5510.

Cooper, D.K., 1990. Clinical survey of heart transplantation between ABO blood group-incompatible recipients and donors. J. Heart Transplant. 9, 376–381.

Cooper, D.K., Human, P.A., Lexer, G., Rose, A.G., Rees, J., Keraan, M., et al., 1988. Effects of cyclosporine and antibody adsorption on pig cardiac xenograft survival in the baboon. J. Heart Transplant. 7, 238–246.

Cooper, D.K.C., Good, A.H., Koren, E., Oriol, R., Malcolm, A.J., Ippolito, R.M., et al., 1993. Identification of α-galactosyl and other carbohydrate epitopes that are bound by human anti-pig antibodies: relevance to discordant xenografting in man. Transpl. Immunol. 1, 198–205.

Cooper, D.K.C., Kemp, E., Reemtsma, K., White, D.J.G. (Eds.), 1997. Xenotransplantation. The Transplantation of Organs and Tissues between Species. Springer, Heidelberg.

Costa, C., Zhao, L., Burton, W.V., Bondioli, K.R., Williams, B.L., Hoagland, T.A., et al., 1999. Expression of the human αl,2-fucosyltransferase in transgenic pigs modifies the cell surface carbohydrate phenotype and confers resistance to human serum-mediated cytolysis. FASEB J. 13, 1762–1773.

Cozzi, E., White, D.J., 1995. The generation of transgenic pigs as potential organ donors for humans. Nat. Med. 1, 964–966.

Cozzi, E., Tucker, A.W., Langford, G.A., Pino-Chavez, G., Wright, L., O'Connell, M.J., et al., 1997. Characterization of pigs transgenic for human decay-accelerating factor. Transplantation 64, 1383–1392.

Dalmasso, A.P., Platt, J.L., Bach, F.H., 1991. Reaction of complement with endothelial cells in a model of xenotransplantation. Clin. Exp. Immunol. 86 (Suppl. 1), 31–35.

Diamond, L.E., Quinn, C.M., Martin, M.J., Lawson, J., Platt, J.L., Logan, J.S., 2001. A human CD46 transgenic pig model system for the study of discordant xenotransplantation. Transplantation 71, 132–142.

Dor, F.J., Tseng, Y.L., Cheng, J., Moran, K., Sanderson, T.M., Lancos, C.J., et al., 2004. α1,3-Galactosyltransferase geneknockout miniature swine produce natural cytotoxic anti-Gal antibodies. Transplantation 78, 15–20.

Ezzelarab, M., Hara, H., Busch, J., Rood, P.P., Zhu, X., Ibrahim, Z., et al., 2006. Antibodies directed to pig non-Gal antigens in naïve and sensitized baboons. Xenotransplantation 13, 400–407.

Fang, J., Walters, A., Hara, H., Long, C., Yeh, P., Ayares, D., et al., 2012. Anti-gal antibodies in α1,3-galactosyltransferase gene-knockout pigs. Xenotransplantation 19, 305–310.

Fodor, W.L., Williams, B.L., Matis, L.A., Madri, J.A., Rollins, S.A., Knight, J.W., et al., 1994. Expression of a functional human complement inhibitor in a transgenic pig as a model for the prevention of xenogeneic hyperacute organ rejection. Proc. Natl. Acad. Sci. U.S.A. 91, 11153–11157.

Galili, U., 1993. Interaction of the natural anti-Gal antibody with α-galactosyl epitopes: a major obstacle for xenotransplantation in humans. Immunol. Today 14, 480–482.

Galili, U., 2012. Induced anti-non gal antibodies in human xenograft recipients. Transplantation 93, 11–16.

Galili, U., 2013. α1,3Galactosyltransferase knockout pigs produce the natural anti-Gal antibody and simulate the evolutionary appearance of this antibody in primates. Xenotransplantation 20, 267–276.

Galili, U., Rachmilewitz, E.A., Peleg, A., Flechner, I., 1984. A unique natural human IgG antibody with anti-α-galactosyl specificity. J. Exp. Med. 160, 1519–1531.

Galili, U., Clark, M.R., Shohet, S.B., Buehler, J., Macher, B.A., 1987. Evolutionary relationship between the natural anti-Gal antibody and the Galα1-3Gal epitope in primates. Proc. Natl. Acad. Sci. U.S.A. 84, 1369–1373.

Galili, U., Shohet, S.B., Kobrin, E., Stults, C.L.M., Macher, B.A., 1988a. Man, apes, and Old World monkeys differ from other mammals in the expression of α-galactosyl epitopes on nucleated cells. J. Biol. Chem. 263, 17755–17762.

Galili, U., Mandrell, R.E., Hamadeh, R.M., Shohet, S.B., Griffiss, J.M., 1988b. Interaction between human natural anti-α-galactosyl immunoglobulin G and bacteria of the human flora. Infect. Immun. 56, 1730–1737.

Galili, U., Tibell, A., Samuelsson, B., Rydberg, L., Groth, C.G., 1995. Increased anti-Gal activity in diabetic patients transplanted with fetal porcine islet cell clusters. Transplantation 59, 1549–1556.

Galili, U., Anaraki, F., Thall, A., Hill-Black, C., Radic, M., 1993. One percent of circulating B lymphocytes are capable of producing the natural anti-Gal antibody. Blood 82, 2485–2493.

Galili, U., Chen, Z.C., Tanemura, M., Seregina, T., Link, C.J., 2001. Understanding the induced antibody response (in xenograft recipients). Graft 4, 32–35.

Galili, U., LaTemple, D.C., Walgenbach, A.W., Stone, K.R., 1997. Porcine and bovine cartilage transplants in cynomolgus monkey: II. Changes in anti-Gal response during chronic rejection. Transplantation 63, 646–651.

Galili, U., Wigglesworth, K., Abdel-Motal, U.M., 2007. Intratumoral injection of α-gal glycolipids induces xenograft-like destruction and conversion of lesions into endogenous vaccines. J. Immunol. 178, 4676–4687.

Geller, R.L., Turman, M.A., Dalmasso, A.P., Platt, J.L., 1992. The natural immune barrier to xenotransplantation. J. Am. Soc. Nephrol. 3, 1189–1200.

Good, A.H., Cooper, D.C.K., Malcolm, A.J., Ippolito, R.M., Koren, E., Neethling, F.A., et al., 1992. Identification of carbohydrate structures which bind human anti-porcine antibodies: implication for discordant xenografting in man. Transplant. Proc. 24, 559–562.

Griesemer, A.D., Hirakata, A., Shimizu, A., Moran, S., Tena, A., Iwaki, H., et al., 2009. Results of gal-knockout porcine thymo-kidney xenografts. Am. J. Transplant. 9, 2669–2678.

Griesemer, A., Liang, F., Hirakata, A., Hirsh, E., Lo, D., Okumi, M., et al., 2010. Occurrence of specific humoral non-responsiveness to swine antigens following administration of GalT-KO bone marrow to baboons. Xenotransplantation 17, 300–312.

Groth, C.G., Korsgren, O., Tibell, A., Tollerman, J., Möller, E., Bolinder, J., et al., 1994. Transplantation of fetal porcine pancreas to diabetic patients: biochemical and histological evidence for graft survival. Lancet 344, 1402–1404.

Higashi, H., Naiki, M., Matuo, S., Okouchi, K., 1977. Antigen of "serum sickness" type of heterophile antibodies in human sera: Identification as gangliosides with *N*-glycolylneuraminic acid. Biochem. Biophys. Res. Commun. 79, 388–395.

Hisashi, Y., Yamada, K., Kuwaki, K., Tseng, Y.L., Dor, F.J., Houser, S.L., et al., 2008. Rejection of cardiac xenografts transplanted from α1,3-galactosyltransferase gene-knockout (GalT-KO) pigs to baboons. Am. J. Transplant. 8, 2516–2526.

Kobayashi, T., Taniguchi, S., Neethling, F.A., Rose, A.G., Hancock, W.W., Ye, Y., et al., 1997. Delayed xenograft rejection of pig-to-baboon cardiac transplants after cobra venom factor therapy. Transplantation 64, 1255–1261.

Kolber-Simonds, D., Lai, L., Watt, S.R., Denaro, M., Arn, S., Augenstein, M.L., et al., 2004. Production of α1,3-galactosyltransferase null pigs by means of nuclear transfer with fibroblasts bearing loss of heterozygosity mutations. Proc. Natl. Acad. Sci. U.S.A. 101, 7335–7340.

Koma, M., Miyagawa, S., Honke, K., Ikeda, Y., Koyota, S., Miyoshi, S., et al., 2000. Reduction of the major xenoantigen on glycosphingolipids of swine endothelial cells by various glycosyltransferases. Glycobiology 10, 745–751.

Korsgren, O., Sandler, S., Landström, A.S., Jansson, L., Andersson, A., 1988. Large-scale production of fetal porcine pancreatic islet-like cell clusters. An experimental tool for studies of islet cell differentiation, and xenotransplantation. Transplantation 45, 509–514.

Kuwaki, K., Tseng, Y.L., Dor, F.J., Shimizu, A., Houser, S.L., Sanderson, T.M., et al., 2005. Heart transplantation in baboons using α1,3galactosyltransferase gene-knockout pigs as donors: initial experience. Nat. Med. 11, 29–31.

Lai, L., Kolber-Simonds, D., Park, K.W., Cheong, H.T., Greenstein, J.L., Im, G.S., et al., 2002. Production of α-1,3-galactosyltransferase knockout pigs by nuclear transfer cloning. Science 295, 1089–1092.

Lam, T.T., Paniagua, R., Shivaram, G., Schuurman, H.J., Borie, D.C., Morris, R.E., 2004. Anti-non- Gal porcine endothelial cell antibodies in acute humoral xenograft rejection of hDAF-transgenic porcine hearts in cynomolgus monkeys. Xenotransplantation 11, 531–535.

LaTemple, D.C., Abrams, J.T., Zhang, S.U., Galili, U., 1999. Increased immunogenicity of tumor vaccines complexed with anti-Gal: studies in knock out mice for α1,3galactosyltranferase. Cancer Res. 59, 3417–3423.

Le Bas-Bernardet, S., Tillou, X., Branchereau, J., Dilek, N., Poirier, N., Châtelais, M., et al., 2015. Bortezomib, C1-inhibitor and plasma exchange do not prolong the survival of multi-transgenic GalT-KO pig kidney xenografts in baboons. Am. J. Transplant. 15, 358–370.

Lexer, G., Cooper, D.K., Rose, A.G., Wicomb, W.N., Rees, J., Keraan, M., et al., 1986. Hyperacute rejection in a discordant (pig to baboon) cardiac xenograft model. J. Heart Transplant. 5, 411–418.

Loveland, B.E., Milland, J., Kyriakou, P., Thorley, B.R., Christiansen, D., Lanteri, M.B., et al., 2004. Characterization of a CD46 transgenic pig and protection of transgenic kidneys against hyperacute rejection in non-immunosuppressed baboons. Xenotransplantation 11, 171–183.

Lutz, A.J., Li, P., Estrada, J.L., Sidner, R.A., Chihara, R.K., Downey, S.M., Burlak, C., et al., 2013. Double knockout pigs deficient in *N*-glycolylneuraminic acid and galactose α-1,3-galactose reduce the humoral barrier to xenotransplantation. Xenotransplantation 20, 27–35.

Makowka, L., Cramer, D.V., Hoffman, A., Breda, M., Sher, L., Eiras-Hreha, G., et al., 1995. The use of a pig liver xenograft for temporary support of a patient with fulminant hepatic failure. Transplantation 59, 1654–1659.

Mayumi, H., Good, R.A., 1989. The necessity of both allogeneic antigens and stem cells for cyclophosphamide-induced skin allograft tolerance in mice. Immunobiology 178, 287–304.

McGregor, C.G., Ricci, D., Miyagi, N., Stalboerger, P.G., Du, Z., Oehler, E.A., Tazelaar, H.D., et al., 2012. Human CD55 expression blocks hyperacute rejection and restricts complement activation in Gal knockout cardiac xenografts. Transplantation 93, 686–692.

Merrick, J.M., Zadarlik, K., Milgrom, F., 1978. Characterization of the Hanganutziu-Deicher (serum-sickness) antigen as gangliosides containing *N*-glycolylneuraminic acid. Int. Arch. Allergy Appl. Immunol. 57, 477–480.

Miyagawa, S., Murakami, H., Takahagi, Y., Nakai, R., Yamada, M., Murase, A., et al., 2001. Remodeling of the major pig xenoantigen by *N*-acetylglucosaminyltransferase III in transgenic pig. J. Biol. Chem. 276, 39310–39319.

Mohiuddin, M.M., Corcoran, P.C., Singh, A.K., Azimzadeh, A., Hoyt Jr., R.F., Thomas, M.L., 2012. B-cell depletion extends the survival of GTKO.hCD46Tg pig heart xenografts in baboons for up to 8 months. Am. J. Transplant. 12, 763–771.

Mohiuddin, M.M., Singh, A.K., Corcoran, P.C., Thomas 3rd, M.L., Clark, T., Lewis, B.G., 2016. Chimeric 2C10R4 anti-CD40 antibody therapy is critical for long-term survival of GTKO.hCD46.hTBM pig-to-primate cardiac xenograft. Nat. Commun. 7, 11138–11148.

Muchmore, E.A., Diaz, S., Varki, A., 1998. A structural difference between the cell surfaces of humans and the great apes. Am. Phys. Anthropol. 107, 187–198.

Neethling, F.A., Joziasse, D., Bovin, N., Cooper, D.K., Oriol, R., 1996. The reducing end of α-Gal oligosaccharides contributes to their efficiency in blocking natural antibodies of human and baboon sera. Transpl. Int. 9, 98–101.

Nemazee, D., 2000. Receptor editing in B cells. Adv. Immunol. 74, 89–126.

Oriol, R., Candelier, J.J., Taniguchi, S., Balanzino, L., Peters, L., Niekrasz, M., et al., 1999. Major carbohydrate epitopes in tissues of domestic and African wild animals of potential interest for xenotransplantation research. Xenotransplantation 6, 79–89.

Padler-Karavani, V., Yu, H., Cao, H., Karp, F., Chokhawala, H., Varki, N., et al., 2008. Diversity in specificity, abundance, and composition of anti-Neu5Gc antibodies in normal humans: potential implications for disease. Glycobiology 18, 818–830.

Padler-Karavani, V., Varki, A., 2011. Potential impact of the non-human sialic acid *N*-glycolylneuraminic acid on transplant rejection risk. Xenotransplantation 18, 1–5.

Phelps, C.J., Koike, C., Vaught, T.D., Boone, J., Wells, K.D., Chen, S.H., et al., 2003. Production of α1,3-galactosyltransferase-deficient pigs. Science 299, 411–414.

Posekany, K.J., Pittman, H.K., Bradfield, J.F., Haisch, C.E., Verbanac, K.M., 2002. Induction of cytolytic anti-Gal antibodies in α-1,3-galactosyltransferase gene knockout mice by oral inoculation with *Escherichia coli* O86:B7 bacteria. Infect. Immun. 70, 6215–6222.

Radic, M.Z., Zouali, M., 1996. Receptor editing, immune diversification, and self-tolerance. Immunity 5, 505–511.

Reemtsma, K., McCracken, B.H., Schlegel, J.U., Pearl, M.A., Pearce, C.W., Dewitt, C.W., et al., 1964. Renal heterotransplantation in man. Ann. Surg. 160, 384–410.

Reuven, E.M., Leviatan Ben-Arye, S., Marshanski, T., Breimer, M.E., Yu, H., Fellah-Hebia, I., et al., 2016. Characterization of immunogenic Neu5Gc in bioprosthetic heart valves. Xenotransplantation 23, 381–392.

Rose, A.G., Cooper, D.K.C., Human, P.A., Reichenspurner, H., Reichart, B., 1991. Histopathology of hyperacute rejection of the heart-experimental and clinical observations in allografts and xenografts. J. Heart Transplant. 10, 223–234.

Salama, A., Evanno, G., Harb, J., Soulillou, J.P., 2015. Potential deleterious role of anti-Neu5Gc antibodies in xenotransplantation. Xenotransplantation 22, 85–94.

Salama, A., Evanno, G., Lim, N., Rousse, J., Le Berre, L., Nicot, A., et al., February 14, 2017. Anti-Gal and anti-Neu5Gc responses in nonimmunosuppressed patients following treatment with rabbit anti-thymocyte polyclonal IgGs. Transplantation [Epub ahead of print].

Sandel, P.C., Monroe, J.G., 1999. Negative selection of immature B cells by receptor editing or deletion is determined by site of antigen encounter. Immunity 10, 289–299.

Sandrin, M.S., Vaughan, H.A., Dabkowski, P.L., McKenzie, I.F.C., 1993. Anti-pig IgM antibodies in human serum react predominantly with Gal (αl-3)Gal epitopes. Proc. Natl. Acad. Sci. U.S.A. 90, 11391–11395.

Sandrin, M.S., Fodor, W.L., Mouhtouris, E., Osman, N., Cohney, S., Rollins, S.A., et al., 1995. Enzymatic remodeling of the carbohydrate surface of a xenogenic cell substantially reduces human antibody binding and complement-mediated cytolysis. Nat. Med. 1, 1261–1267.

Scobie, L., Padler-Karavani, V., Le Bas-Bernardet, S., Crossan, C., Blaha, J., Matouskova, M., 2013. Long-term IgG response to porcine Neu5Gc antigens without transmission of PERV in burn patients treated with porcine skin xenografts. J. Immunol. 191, 2907–2915.

Shafi, R., Iyer, S.P., Ellies, L.G., O'Donnell, N., Marek, K.W., Chui, D., et al., 2000. The O-GlcNAc transferase gene resides on the X chromosome and is essential for embryonic stem cell viability and mouse ontogeny. Proc. Natl. Acad. Sci. U.S.A. 97, 5735–5739.

Sharabi, Y., Sachs, D.H., 1989. Mixed chimerism and permanent specific transplantation tolerance induced by a non-lethal preparative regimen. J. Exp. Med. 169, 493–502.

Simon, P.M., Neethling, F.A., Taniguchi, S., Goode, P.L., Zopf, D., Hancock, W.W., et al., 1998. Intravenous infusion of Galα1-3Gal oligosaccharides in baboon delays hyperacute rejection of porcine heart xenografts. Transplantation 56, 346–353.

Smith, D.F., Larsen, R.D., Mattox, S., Lowe, J.B., Cummings, R.D., 1990. Transfer and expression of a murine UDP-Gal:β-D-Gal- α1,3-galactosyltransferase gene in transfected Chinese hamster ovary cells. Competition reactions between the α1,3-galactosyltransferase and the endogenous α2,3-sialyltransferase. J. Biol. Chem. 265, 6225–6234.

Springer, S.A., Diaz, S.L., Gagneux, P., 2014. Parallel evolution of a self-signal: humans and New World monkeys independently lost the cell surface sugar Neu5Gc. Immunogenetics 66, 671–674.

Starzl, T.E., 1964. Renal homografts in patients with major donor recipient blood group incompatibilities. Surgery 55, 195–200.

Starzl, T.E., Marchioro, T.L., Peters, G.N., Kirkpatrick, C.H., Wilson, W.E., Porter, K.A., et al., 1964. Renal heterotransplantation from baboon to man: experience with 6 cases. Transplantation 2, 752–776.

Stone, K.R., Ayala, G., Goldstein, J., Hurst, R., Walgenbach, A., Galili, U., 1998. Porcine cartilage transplants in the cynomolgus monkey. III. Transplantation of α-galactosidase-treated porcine cartilage. Transplantation 65, 1577–1583.

Stone, K.R., Abdel-Motal, U.M., Walgenbach, A.W., Turek, T.J., Galili, U., 2007a. Replacement of human anterior cruciate ligaments with pig ligaments: a model for anti-non-gal antibody response in long-term xenotransplantation. Transplantation 83, 211–219.

Stone, K.R., Walgenbach, A.W., Turek, T.J., Somers, D.L., Wicomb, W., Galili, U., 2007b. Anterior cruciate ligament reconstruction with a porcine xenograft: a serologic, histologic, and biomechanical study in primates. Arthroscopy 23, 411–419.

Tahara, H., Ide, K., Basnet, N.B., Tanaka, Y., Matsuda, H., Takematsu, H., et al., 2010. Immunological property of antibodies against *N*-glycolylneuraminic acid epitopes in cytidine monophospho-*N*-acetylneuraminic acid hydroxylase-deficient mice. J. Immunol. 184, 3269–3275.

Takahagi, Y., Fujimura, T., Miyagawa, S., Nagashima, H., Shigehisa, T., Shirakura, R., et al., 2005. Production of α1,3-galactosyltransferase gene knockout pigs expressing both human decay-accelerating factor and *N*-acetylglucosaminyltransferase III. Mol. Reprod. Dev. 71, 331–338.

Takemae, N., Ruttanapumma, R., Parchariyanon, S., Yoneyama, S., Hayashi, T., Hiramatsu, H., et al., 2010. Alterations in receptor-binding properties of swine influenza viruses of the H1 subtype after isolation in embryonated chicken eggs. J. Gen. Virol. 91, 938–948.

Tanemura, M., Miyagawa, S., Ihara, Y., Matsuda, H., Shirakura, R., Taniguchi, N., 1997. Significant downregulation of the major swine xenoantigen by *N*-acetylglucosaminyltransferase III gene transfection. Biochem. Biophys. Res. Chem. 235, 359–364.

Tanemura, M., Miyagawa, S., Koyota, S., Koma, M., Matsuda, H., Tsuji, S., et al., 1998. Reduction of the major swine xenoantigen, the α-galactosyl epitope by transfection of the α2,3 sialyltransferase gene. J. Biol. Chem. 273, 16421–16425.

Tanemura, M., Yin, D., Chong, A.S., Galili, U., 2000a. Differential immune response to α-gal epitopes on xenografts and allografts: implications for accommodation in xenotransplantation. J. Clin. Investig. 105, 301–310.

Tanemura, M., Maruyama, S., Galili, U., 2000b. Differential expression of α-gal epitopes (Galα1-3Galβ1-4GlcNAc-R) on pig and mouse organs. Transplantation 69, 187–190.

Taniguchi, S., Cooper, D.K., 1997. Clinical xenotransplantation: past, present and future. Ann. R. Coll. Surg. Engl. 79, 13–19.

Tasaki, M., Wamala, I., Tena, A., Villani, V., Sekijima, M., Pathiraja, V., et al., 2015. High incidence of xenogenic bone marrow engraftment in pig-to-baboon intra-bone bone marrow transplantation. Am. J. Transplant. 15, 974–983.

Tearle, R.G., Tange, M.J., Zannettino, Z.L., Katerelos, M., Shinkel, T.A., Van Denderen, B.J., et al., 1996. The α-1,3-galactosyltransferase knockout mouse. Implications for xenotransplantation. Transplantation 61, 13–19.

Tena, A., Kurtz, J., Leonard, D.A., Dobrinsky, J.R., Terlouw, S.L., Mtango, N., et al., 2014. Transgenic expression of human CD47 markedly increases engraftment in a murine model of pig-to-human hematopoietic cell transplantation. Am. J. Transplant. 14, 2713–2722.

Teranishi, K., Mañez, R., Awwad, M., Cooper, D.K., 2002. Anti-Galα1-3Gal IgM and IgG antibody levels in sera of humans and Old World non-human primates. Xenotransplantation 9, 148–154.

Thall, A.D., 1999. Generation of α1,3galactosyltransferase deficient mice. Subcell. Biochem. 32, 259–279.

Thall, A.D., Maly, P., Lowe, J.B., 1995. Oocyte Galα1,3Gal epitopes implicated in sperm adhesion to the zona pellucida glycoprotein ZP3 are not required for fertilization in the mouse. J. Biol. Chem. 270, 21437–21440.

Thompson, P., Badell, I.R., Lowe, M., Cano, J., Song, M., Leopardi, F., et al., 2011. Islet xenotransplantation using gal-deficient neonatal donors improves engraftment and function. Am. J. Transplant. 11, 2593–2602.

Tseng, Y.L., Kuwaki, K., Dor, F.J., Shimizu, A., Houser, S., Hisashi, Y., et al., 2005. α1,3Galactosyltransferase gene-knockout pig heart transplantation in baboons with survival approaching six months. Transplantation 80, 1493–1500.

Tseng, Y.L., Moran, K., Dor, F.J., Sanderson, T.M., Li, W., Lancos, C.J., et al., 2006. Elicited antibodies in baboons exposed to tissues from α1,3-galactosyltransferase gene-knockout pigs. Transplantation 81, 1058–1062.

van der Windt, D.J., Bottino, R., Casu, A., Campanile, N., Smetanka, C., He, J., et al., 2009. Long-term controlled normoglycemia in diabetic non-human primates after transplantation with hCD46 transgenic porcine islets. Am. J. Transplant. 9, 2716–2726.

Varki, A., 2010. Colloquium paper: uniquely human evolution of sialic acid genetics and biology. Proc. Natl. Acad. Sci. U.S.A. 107, 8939–8946.

Watier, H., Guillaumin, J.M., Vallee, I., Thibault, G., Gruel, Y., Lebranchu, Y., et al., 1996. Human NK cell-mediated direct and IgG-dependent cytotoxicity against xenogeneic porcine endothelial cells. Transpl. Immunol. 4, 293–299.

White, D.J., Yannoutsos, N., 1996. Production of pigs transgenic for human DAF to overcome complement-mediated hyperacute xenograft rejection in man. Res. Immunol. 147, 88–94.

Wilbrandt, R., Tung, K.S., Deodhar, S.D., Nakamoto, S., Kolff, W.J., 1969. ABO blood group incompatibility in human renal homotransplantation. Am. J. Clin. Pathol. 51, 15–23.

Xu, Y., Lorf, T., Sablinski, T., Gianello, P., Bailin, M., Monroy, R., et al., 1998. Removal of anti-porcine natural antibodies from human and nonhuman primate plasma in vitro and in vivo by a Galα1-3Galβ1-4Glc-R immunoaffinity column. Transplantation 65, 172–179.

Yamada, K., Yazawa, K., Shimizu, A., Iwanaga, T., Hisashi, Y., Nuhn, M., et al., 2005. Marked prolongation of porcine renal xenograft survival in baboons through the use of α1,3-galactosyltransferase gene-knockout donors and the co-transplantation of vascularized thymic tissue. Nat. Med. 11, 32–34.

Yang, Y.G., deGoma, E., Ohdan, H., Bracy, J.L., Xu, Y., Iacomini, J., et al., 1998. Tolerization of anti-Galα1-3Gal natural antibody-forming B cells by induction of mixed chimerism. J. Exp. Med. 187, 1335–1342.

Yeh, P., Ezzelarab, M., Bovin, N., Hara, H., Long, C., Tomiyama, K., et al., 2010. Investigation of potential carbohydrate antigen targets for human and baboon antibodies. Xenotransplantation 17, 197–206.

Zhao, Z., Termignon, J.L., Cardoso, J., Chéreau, C., Gautreau, C., Calmus, Y., et al., 1994. Hyperacute xenograft rejection in the swine-to-human donor-recipient combination. In vitro analysis of complement activation. Transplantation 57, 245–249.

Zhu, A., Hurst, R., 2002. Anti-*N*-glycolylneuraminic acid antibodies identified in healthy human serum. Xenotransplantation 9, 376–381.

CHAPTER

7

Anti-Gal IgE Mediates Allergies to Red Meat

OUTLINE

EARLY WARNINGS

The 1980s were the first decade in which many studies were initiated on the production of therapeutic recombinant proteins. A proportion of these proteins is naturally produced as glycoproteins carrying carbohydrate chains that are required for the biological activity of the recombinant proteins (e.g., folding to form an appropriate tridimensional structure), as well as maintaining the *in vivo* half-life of the molecule. Recombinant glycoproteins had to be produced in mammalian cell lines because bacteria and yeast expression systems lack the glycosylation machinery for synthesis of O-linked or of N-linked carbohydrate chains of the complex type. Because nonprimate mammalian cells have catalytically active α1,3galactosyltransferase (α1,3GT) that synthesizes α-gal epitopes (see Chapter 1), N-linked carbohydrate chains on the recombinant glycoprotein produced in such cells may carry these epitopes. In the mid-1980s, the findings of anti-Gal as an abundant natural antibody in humans (Galili et al., 1984) and the interaction of this antibody with mammalian α-gal epitopes (Galili et al., 1985, 1987, 1988) were reported. Based on this information, researchers in

The Natural Anti-Gal Antibody as Foe Turned Friend in Medicine
http://dx.doi.org/10.1016/B978-0-12-813362-0.00007-5

glycobiology warned that infusion of therapeutic recombinant glycoproteins with α-gal epitopes into humans may result in some individuals, in allergic reactions and even anaphylactic shock, if the patient is allergic to the α-gal epitope, i.e., if the patient produces anti-Gal IgE antibodies (Rademacher et al., 1988). Indeed, α-gal epitopes were identified on recombinant human interferon-β produced in mouse C127 cells (Kagawa et al., 1988) and on recombinant human Factor VIII produced in baby hamster kidney (BHK) cells (Hironaka et al., 1992). Similarly, a proportion of therapeutic mouse monoclonal antibodies and monoclonal antibodies produced in human/mouse heterohybridomas was found to carry α-gal epitopes synthesized by the α1,3GT of the mouse lymphocytes and of the mouse myeloma cells serving as the myeloma fusion partner (Borrebaeck et al., 1993; Montaño and Romano, 1994). The α-gal epitopes of these monoclonal antibodies readily bound anti-Gal. Moreover, *in vivo* binding of anti-Gal to such monoclonal antibodies decreased their half-life in the circulation (Borrebaeck et al., 1993). Based on the warning of possible allergic reaction and anaphylactic shock mediated by anti-Gal IgE, the FDA requested in the late 1980s and in the 1990s information on α-gal epitopes linked to recombinant therapeutic glycoproteins and on the possible induction of allergic reactions by such glycoproteins. No reports on such allergic reactions appeared during the two decades since that warning of Rademacher et al. (1988) was published. The first indication that humans can produce anti-Gal IgE, which mediates allergic reactions to the α-gal epitope was found in cancer patients treated with the monoclonal antibody cetuximab (Chung et al., 2008).

ANTI-GAL IgE MEDIATING ALLERGIES IN CETUXIMAB-TREATED PATIENTS

Cetuximab is a humanized monoclonal antibody produced by a hybridoma in which the myeloma fusion partner was SP2/0. This cell line was previously found to present ~1.2×10^6 α-gal epitopes/cell (Galili et al., 1988), implying that similar to other mouse cells, SP2/0 cells contain an active α1,3GT. Cetuximab binds specifically to epidermal growth factor receptor (EGFR) and is used for treatment of metastatic colorectal cancer and squamous cell carcinoma of the head and neck (Cunningham et al., 2004; Chung et al., 2005; Bonner et al., 2006). In general, 2%–5% of the cancer patients infused with cetuximab display an allergic reaction to the antibody within periods up to 1 h. This suggests that the IgE antibody mediating such allergic reactions has been present in the treated patients prior to the treatment. Patients in the South of the United States have much higher incidence of allergies, reaching up to 22% (O'Neil et al., 2007; Chung et al., 2008; Maier et al., 2015). In a small proportion of the patients, the allergic reactions are very intense, resulting in an anaphylactic shock. Approximately half of the patients that display anaphylactic reaction had in their serum IgE antibodies to cetuximab (Chung et al., 2008), as measured by an IgE fluorometric enzyme immunoassay called CAP assay (Erwin et al., 2005).

The presence of α-gal epitopes on cetuximab was determined by a study using physicochemical methods for characterizing the carbohydrate chains on this monoclonal antibody, which is an IgG_1 antibody. This study found that among the 21 different oligosaccharide structures on cetuximab, ~30% have the α-gal epitope (as that illustrated in Fig. 2A of Chapter 1), both on the Fab and Fc regions of the antibody (Qian et al., 2007). Thus, similar to the hybridomas studied in the 1990s (Borrebaeck et al., 1993; Montaño and Romano, 1994), the mouse α1,3GT in the cetuximab hybridoma synthesized α-gal epitopes on the carbohydrate chains of the antibody, including on a carbohydrate chain of the Fab region. In a large proportion of the patients producing anti-cetuximab IgE antibodies and displaying allergic response to cetuximab, these antibodies were found to bind to the α-gal epitope on the cetuximab monoclonal antibody, whereas no binding of anti-Gal IgE was detected with cetuximab produced in Chinese hamster ovary (CHO) cells that lack α-gal epitopes (Chung et al., 2008). Among healthy individuals, 0.6% of the population in Boston area was found to produce these anti-Gal IgE antibodies. However, in Tennessee, as many as 20.8% of the population were reported to produce such IgE antibodies (Chung et al., 2008).

ANTI-GAL IgE AND RED MEAT ALLERGIES

As detailed in Chapter 1, the α-gal epitope is produced in large amounts in nonprimate mammals (Galili et al., 1987, 1988) and is abundant in red meat including beef, pork, and lamb. It is probable that an average hamburger portion contains hundreds of billions to many trillions of α-gal epitopes. Thus, following the identification of α-gal epitopes as the target of anti-cetuximab IgE antibody in patients allergic to cetuximab (Chung et al., 2008), it was of interest to determine whether similar antibodies are produced in individuals that are allergic to red meat. By performing assays for detecting anti-Gal IgE, as in patients treated with cetuximab, Commins et al. (2009) indicated that patients with a history of allergic reactions in the form of anaphylaxis, angioedema, or urticaria, 3–6h after the ingestion of red meat have high activity of anti-Gal IgE antibody. It is of interest to note that infusion of cetuximab resulted in allergic reaction within only 20–60min, suggesting that the longer time it takes for red meat α-gal epitopes to reach anti-Gal IgE on basophils and on mast cells reflects the required period for passage of the α-gal epitope on glycoconjugates through the gastrointestinal wall. Patients allergic to red meat displayed no respiratory allergies in the form of asthma (Commins et al., 2012).

The identity of anti-Gal IgE as a cause for red meat allergy was confirmed in patients reporting such allergies in other clinical centers (Nunez et al., 2011; cf., Platts-Mills et al., 2015), including allergies to pig kidney (Morisset et al., 2012; Fischer et al., 2014). It should be noted that pig kidney is one of the richest organs for α-gal epitopes in pigs. A very high concentration of these epitopes is found in the large amounts of laminin within the glomeruli and in the brush border of the epithelium lining the proximal tubules of the kidneys (Tanemura et al., 2000). Accordingly, analysis of proteins that readily bind anti-Gal IgE in bovine meat extract demonstrated high binding to the multiple α-gal epitopes on laminin (Takahashi et al., 2014). These findings also correlate with the very high number of α-gal epitopes on mouse (~50 epitopes/molecule) (Arumugham et al., 1986) and bovine laminin (Mohan and Spiro, 1986). The overall concentration of α-gal epitopes in pig kidney was further found to be ~500-fold higher than that in mouse kidney (Tanemura et al., 2000).

Allergy to food containing bovine-derived gelatin also was found to be associated with production of anti-Gal IgE (Caponetto et al., 2013; Uyttebroek et al., 2014). This finding is in accord with the report that α-gal epitopes are abundant in bovine collagen α-1 (VI) chain (Takahashi et al., 2014). It should be noted, however, that cells of bovine, porcine, and ovine origin present 10^6—3×10^7 α-gal epitopes per cell (Galili et al., 1988), which also are likely to contribute to allergic reactions in individuals producing anti-Gal IgE. Patients producing anti-Gal IgE and who are allergic to red meat are not allergic to poultry and fish (Commins et al., 2009). This observation is in agreement with the fact that birds and fish, like other non-mammalian vertebrates, do not synthesize α-gal epitopes (Galili et al., 1988). It is of interest to note that bovine, ovine, and horse milk also contain α-gal epitopes (Urashima et al., 2001, 2013), a finding that explains the positive results of skin tests with cow's milk substances in producers of anti-Gal IgE (Commins et al., 2009). Recently, this type of allergy has been designated "α-gal syndrome" (Fischer et al., 2016). In this syndrome, glycoconjugates carrying α-gal epitopes can interact with anti-Gal IgE on the cell membranes of mast cells and basophils in individuals producing this class of anti-Gal. This interaction induces degranulation of mast cells and basophils. The histamine, serotonin and other substances in the released granules initiate the allergic reactions (Fig. 1).

PRODUCTION OF ANTI-GAL IgE FOLLOWING TICK BITES

Patients with red meat allergies, included in the study of Commins et al. (2009), reported being bitten by ticks before having allergy symptoms. Based on this information, Commins et al. (2011) identified three individuals who initially lacked anti-Gal IgE, however, after they were bitten by the tick *Amblyomma americanum* (Lone Star tick), common in the South of the United States, they seroconverted and produced anti-Gal IgE. This seroconversion also was associated with becoming allergic to red meat in other continents following bites by ticks of various species. These included bites of the tick *Ixodes ricinus* in Europe (Hamsten et al., 2013), *Haemaphysalis longicornis* in Asia (Chinuki et al., 2016), and *Ixodes holocyclus* in Australia (Van Nunen et al., 2009; van Nunen, 2015).

The cause for production of anti-Gal IgE following tick bites is unclear at present. Class (isotype) switch of B cells from IgM or IgG_1 producing B cells to IgE producing cells is a complex process that is induced by secretion of IL4 from Th2 cells, while the level of interferon-γ secreted from Th1 cells is very low (Tong and Wesemann, 2015). As discussed in Chapter 1, there are no ionic bonds between anti-Gal and the α-gal epitope, therefore, the affinity of this antibody to the carbohydrate ligand is relatively low (Galili and Matta, 1996). This suggests that production of anti-Gal IgE which effectively binds to α-gal epitopes, may require conversion of IgG_1 anti-Gal B cells (i.e., B cells that increased their antibody affinity by affinity maturation) into IgE producing B cells. It is further probable that the appearance of anti-Gal IgE B cells is not the result only of elicited anti-Gal response due to increased stimulation of the immune system by α-gal or α-gal-like epitopes on various glycoconjugates. This is suggested from the experience with patients receiving porcine heart valve for replacement of impaired valves in their heart. Porcine heart valve present multiple α-gal epitopes, which stimulate the immune system to produce anti-Gal at elevated titers (Konakci et al., 2005; Bloch et al., 2011). Nevertheless, except for three patients with porcine heart valve implant who displayed elevated anti-Gal IgE and red meat allergy prior to implantation (Mozzicato et al., 2014), no such allergies have been reported in patients with porcine heart valve implants. These three patients are tolerating the valve replacement.

As suggested by several investigators (Steinke et al., 2015; van Nunen, 2015; Platts-Mills et al., 2015), the induction of anti-Gal B cells to undergo class switch into IgE producing cells is probably stimulated by some substances in the saliva of the tick, which is introduced into the skin during biting. It would be of interest to determine whether such substances stimulate Th2 T cells to secrete IL4, activate anti-Gal B cells to undergo class switch to IgE production, or both. These assumptions are supported by the observation that tick saliva injection results in extensive activation of Th2 cells secreting IL4 and inhibition of Th1 cell activity in mice following successive infestations with the tick *Rhipicephalus sanguineus* (Ferreira and Silva, 1999). Moreover, Araujo et al. (2016) have demonstrated with α1,3galactosyltransferase knockout (GT-KO) mice that injection of *Amblyomma sculptum* tick saliva or exposure to feeding ticks induces both IgG and IgE anti-Gal antibodies. Identification of substances in tick saliva inducing anti-Gal IgE production will enable unraveling the mechanism that enables external substances to induce class switch to IgE producing B cells and may further explain induction of allergies by other arthropods. Identification of such substances may also have practical applications because they may be converted into vaccines that induce production of antibodies neutralizing these substances in individuals that are likely to be bitten by ticks.

AVOIDING ANTI-GAL IgE-MEDIATED ALLERGIC REACTIONS

Prevention of allergies and anaphylaxis by α-gal epitopes may be achieved by two approaches: (1) Identification of patients that are allergic to the α-gal epitope and (2) Prevention of the use of therapeutic glycoproteins and monoclonal antibodies carrying α-gal epitopes.

Identification of patients allergic to α-gal epitopes

Much of the research associated with demonstration of anti-Gal IgE production has been performed using the CAP assay. This is an ELISA-like assay (Erwin et al., 2005), performed in a clinical lab, using as solid-phase antigen bovine thyroglobulin (670 kDa) that carries 11 α-gal epitopes per molecule (Spiro and Bhoyroo, 1984). Rabbit anti-human IgE antibody with coupled β-galactosidase is used as secondary antibody for measuring the extent of IgE binding by a fluorimeter (Chung et al., 2008). An alternative test, which can be performed in the clinic, is a skin test. Skin tests performed with beef, pork, or lamb extracts in patients producing anti-Gal IgE provide positive results within 15 min (Commins et al., 2009). A similar assay can be performed with glycolipids carrying α-gal epitopes (referred to as α-gal glycolipid) to directly indicate allergic reaction to α-gal epitopes. A convenient source of α-gal glycolipids is rabbit red blood cells (RBC). All neutral glycolipids in rabbit are α-gal glycolipids except for ceramide trihexoside with the structure Galα1-4Galβ1-4Glc-Cer (which is present also in human RBC membranes) (Ogawa and Galili, 2006; Galili et al., 2007). As detailed in Chapter 10, α-gal glycolipids can be readily extracted from rabbit RBC and dissolved in water as micelles. In a clinical trial evaluating the conversion of tumors into vaccines by intratumoral injection of α-gal glycolipids, skin test with these glycolipids was used to determine whether the treated patients became allergic to α-gal epitope (Fig. 1) (Albertini et al., 2016). No seroconversion was observed in the treated patients. Similarly, α-gal nanoparticles made of natural or synthetic α-gal glycolipids and phospholipids, and which present ~10^{15} α-gal epitopes/mg (described in Chapter 12) (Wigglesworth et al., 2011), may serve as a useful substance for determining allergic response to α-gal epitopes by a skin test (Fig. 1).

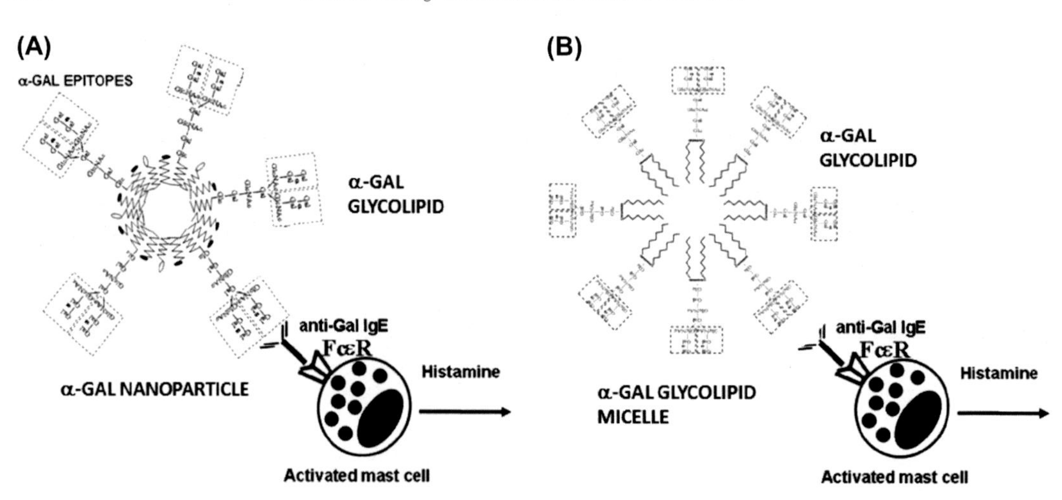

FIGURE 1 Suggested allergens for performing skin test determining allergic response to α-gal epitopes. (A) Use of α-gal nanoparticles, (B) Use of α-gal glycolipids in the form of micelles. α-Gal epitopes on both α-gal nanoparticles and α-gal glycolipid micelles interact with anti-Gal IgE that is bound to Fcε receptors on mast cells and basophils. This interaction results in activation of mast cells and basophils, which degranulate and release histamine and other substances that mediated allergic reactions. The α-gal glycolipids for constructing α-gal nanoparticles or forming α-gal micelles can be obtained from natural sources (e.g., rabbit RBC membranes) or as synthetic glycolipids.

Use of therapeutic recombinant glycoproteins and monoclonal antibodies devoid of α-gal epitopes

If the α-gal epitope synthesis is unavoidable in the production process, passing of glycoproteins through a column (e.g., agarose column) with bound recombinant α-galactosidase results in the cleavage of the terminal α-galactosyl unit and destruction of α-gal epitopes (Fig. 2). This enzyme was found to be very effective in removal of α-gal epitopes from porcine tendon that is converted into a bio-implant for reconstruction of ruptured anterior cruciate ligaments (Stone et al., 2007). Because recombinant α-galactosidase is a highly stable enzyme, an agarose column with coupled recombinant α-galactosidase may be used multiple times without loss of catalytic activity. Alternatively, production of recombinant glycoproteins or monoclonal antibodies may be performed in CHO cells, which usually lack α-gal epitopes and are effective cells in producing recombinant glycoproteins. Accordingly, cetuximab produced in CHO cells does not bind anti-Gal IgE, whereas cetuximab produced in hybridoma cells carries α-gal epitopes (Qian et al., 2007; Chung et al., 2008). However, as indicated below, CHO clones used for production of recombinant glycoproteins should be confirmed for the lack of α-gal epitopes prior to production of the recombinant glycoprotein.

Absence versus synthesis of α-gal epitopes in CHO cell clones—Hamsters have active α1,3GT in their cells, which synthesize α-gal epitopes and express them or their cell membranes and on secreted glycoproteins (Thall and Galili, 1990; Thall et al., 1991). Accordingly, recombinant human Factor VIII glycoprotein produced in transfected BHK cells carry α-gal epitopes on some of the N-linked carbohydrate chains (Hironaka et al., 1992). However, CHO cells that are widely used for production of recombinant glycoproteins

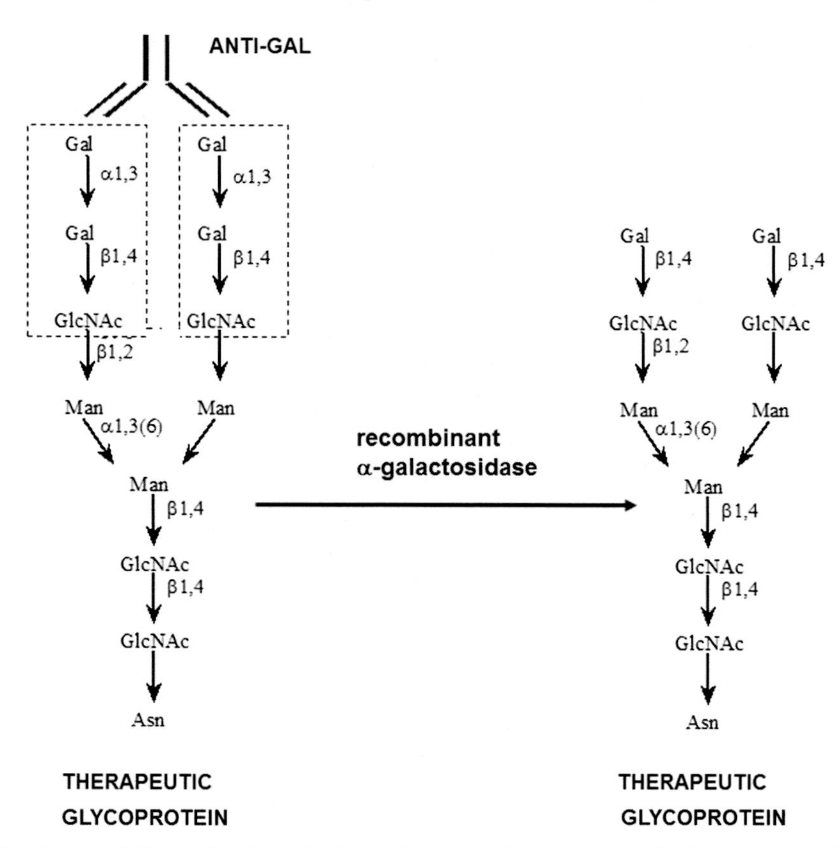

FIGURE 2 Suggested elimination of α-gal epitopes from recombinant therapeutic glycoproteins and monoclonal antibodies to avoid allergic reactions by anti-Gal IgE. The glycoproteins and antibodies are passed through an agarose column with coupled recombinant α-galactosidase. This solid-phase enzyme cleaves the terminal α-galactosyl of α-gal epitopes, exposing *N*-acetyllactosamine (Galβ1-4GlcNAc-R) epitopes, which are naturally present on human cells and do not bind the anti-Gal antibody.

and monoclonal antibodies do not synthesize α-gal epitopes on glycoproteins (Kagawa et al., 1988; Takeuchi et al., 1988; Smith et al., 1990) and do not bind anti-Gal (Winand et al., 1993), although they also originate from hamster, as BHK cells. Since the CHO cell line is an old cell line that was cultured already in the 1950s in a lab that used human serum for some of the tissue culture media (Tjio and Puck, 1958), it may be possible that inadvertently, the cells were cultured in a medium containing human serum. If such culturing occurred, the natural anti-Gal antibody in human serum used to supplement the culture medium could select for the growth of CHO cells in which the α1,3GT gene expression was accidentally suppressed (e.g., impaired promoter activity). This could result in the growth of CHO cells lacking α1,3GT activity and the establishment of the CHO cell line devoid of α-gal epitopes, which is presently used in most labs. However, there have been several revertant clones of CHO cells that were found to produce glycoproteins carrying α-gal epitopes (Ashford et al., 1993; Bosques et al., 2010).

This suggests that whatever has been the reason for the suppression of α1,3GT activity in CHO cells for >30 years, some clones "succeeded" in overcoming that suppressive mechanism, resulting in production of active α1,3GT that synthesizes α-gal epitopes. These observations suggest that prior to production of a recombinant glycoprotein in CHO cells, the cells should be assayed for synthesis of α-gal epitopes, to avoid the use of a therapeutic glycoprotein that may bind anti-Gal IgE.

α-Gal epitopes on milk recombinant glycoproteins—A technology that is being studied as an alternative to production of therapeutic recombinant glycoproteins in cells, such as CHO cells, is production of such glycoproteins in mammary glands and their secretion in milk of transgenic farm animals such as goats, pigs, or cows (Velander et al., 1997; Colman, 1998; Sánchez et al., 2014). The large amounts of milk obtained from these animals are expected to provide corresponding large amounts of recombinant glycoproteins. Studies on oligosaccharides in milk produced by these mammals demonstrated the presence of oligosaccharides with the α-gal epitope, implying that mammary glands of these mammals contain active α1,3GT (Urashima et al., 1991, 2001, 2013). Thus, it is possible that recombinant glycoproteins produced in these animals will have α-gal epitopes on some of their carbohydrate chains. If these epitopes are found on recombinant glycoproteins in the milk, the α-gal epitope may be destroyed by α-galactosidase. Alternatively, the use of animals with disrupted α1,3GT gene (i.e., knockout animals for this gene), which also are transgenic for the recombinant glycoprotein, may provide milk containing recombinant glycoproteins that are devoid of α-gal epitopes.

CONCLUSIONS

Immunoglobulin class (isotype) switch to anti-Gal IgE was first observed in cancer patients receiving an immunotherapy treatment by infusion of the monoclonal antibody cetuximab (anti-EGFR antibody). When produced in a hybridoma, this antibody carries the α-gal epitope, which binds anti-Gal IgE antibody in these patients, inducing allergic and anaphylactic reactions. Studies on the production of anti-Gal IgE indicated that it is rare in Northern regions of the United States (<1%) but is frequently found Southern regions (>20%). The production of anti-Gal IgE was further found to correlate with allergies to red meat (beef, pork, and lamb). One of the reasons for class switch from anti-Gal IgM or IgG_1 to IgE was found to be "lone star" (*A. americanum*) tick bites. Bites of the ticks *I. ricinus* in Europe, *H. longicornis* in Asia and *I. holocyclus* in Australia were found to cause similar seroconversion to anti-Gal IgE and appearance of allergic reactions to red meat. Such allergies also are found with pork kidneys because they contain very large amounts of α-gal epitopes. The substances in tick saliva which stimulate the class switch in anti-Gal B cells into IgE producing cells have not been identified yet. Prevention of allergic reactions to α-gal epitopes may be achieved by identifying patients producing anti-Gal IgE, either by the use of a lab test or by a skin test with natural or synthetic α-gal epitopes linked to lipids or to other molecules. Use of therapeutic natural or recombinant glycoproteins and monoclonal antibodies lacking α-gal epitopes will also prevent anti-Gal IgE-mediated allergic reactions. This may be achieved by enzymatic destruction of α-gal epitopes with recombinant α-galactosidase or by using eukaryotic expression systems confirmed not to include the biosynthetic pathway for synthesis of α-gal epitopes. Because such biosynthetic pathways are also present in mammary glands of nonprimate mammals, recombinant therapeutic glycoproteins produced in mammary glands and secreted in milk of transgenic farm animals should be evaluated for presence of α-gal epitopes. If present, the α-gal epitopes may be destroyed enzymatically with α-galactosidase. Alternatively, the milk may be produced in knockout mammals for the α1,3GT gene, i.e., mammals that lack the ability to synthesize the α-gal epitope. Lastly, it would be of interest to determine whether red meat from cows, pigs, or lambs engineered to lack α-gal epitopes by disruption of the α1,3GT (*GGTA1*) gene, can be consumed by individuals producing anti-Gal IgE without having allergic reactions to this food.

References

Albertini, M.R., Ranheim, E.A., Zuleger, C.L., Sondel, P.M., Hank, J.A., Bridges, A., et al., 2016. Phase I study to evaluate toxicity and feasibility of intratumoral injection of α-gal glycolipids in patients with advanced melanoma. Cancer Immunol. Immunother. 65, 897–907.

Araujo, R.N., Franco, P.F., Rodrigues, H., Santos, L.C., McKay, C.S., Sanhueza, C.A., et al., 2016. *Amblyomma sculptum* tick saliva: α-Gal identification, antibody response and possible association with red meat allergy in Brazil. Int. J. Parasitol. 46, 213–220.

Arumugham, R.G., Hsieh, T.C., Tanzer, M.L., Laine, R.A., 1986. Structure of the asparagine-linked sugar chains of laminin. Biochim. Biophys. Acta 883, 112–126.

Ashford, D.A., Alafi, C.D., Gamble, V.M., Mackay, D.J., Rademacher, T.W., Williams, P.J., et al., 1993. Site-specific glycosylation of recombinant rat and human soluble CD4 variants expressed in Chinese hamster ovary cells. J. Biol. Chem. 268, 3260–3267.

Bloch, O., Golde, P., Dohmen, P.M., Posner, S., Konertz, W., Erdbrügger, W., 2011. Immune response in patients receiving a bioprosthetic heart valve: lack of response with decellularized valves. Tissue Eng. Part A 17, 19–20.

Bonner, J.A., Harari, P.M., Giralt, J., Azarnia, N., Shin, D.M., Cohen, R.B., et al., 2006. Radiotherapy plus cetuximab for squamous cell carcinoma of the head and neck. N. Engl. J. Med. 354, 567–578.

Borrebaeck, C.K., Malmborg, A.C., Ohlin, M., 1993. Does endogenous glycosylation prevent the use of mouse monoclonal antibodies as cancer therapeutics? Immunol. Today 14, 477–479.

Bosques, C.J., Collins, B.E., Meador 3rd, J.W., Sarvaiya, H., Murphy, J.L., Dellorusso, G., et al., 2010. Chinese hamster ovary cells can produce galactose-α-1,3-galactose antigens on proteins. Nat. Biotechnol. 28, 1153–1156.

Caponetto, P., Fischer, J., Biedermann, T., 2013. Gelatin-containing sweets can elicit anaphylaxis in a patient with sensitization to galactose-α-1,3-galactose. J. Allergy Clin. Immunol. Pract. 1, 302–303.

Chinuki, Y., Ishiwata, K., Yamaji, K., Takahashi, H., Morita, E., 2016. *Haemaphysalis longicornis* tick bites are a possible cause of red meat allergy in Japan. Allergy 71, 421–425.

Chung, K.Y., Shia, J., Kemeny, N.E., Shah, M., Schwartz, G.K., Tse, A., et al., 2005. Cetuximab shows activity in colorectal cancer patients with tumors that do not express the epidermal growth factor receptor by immunohistochemistry. J. Clin. Oncol. 23, 1803–1810.

Chung, C.H., Mirakhur, B., Chan, E., Le, Q.T., Berlin, J., Morse, M., et al., 2008. Cetuximab-induced anaphylaxis and IgE specific for galactose-α-1,3-galactose. N. Engl. J. Med. 358, 1109–1117.

Colman, A., 1998. Production of therapeutic proteins in the milk of transgenic livestock. Biochem. Soc. Symp. 63, 141–147.

Commins, S.P., Satinover, S.M., Hosen, J., Mozena, J., Borish, L., Lewis, B.D., et al., 2009. Delayed anaphylaxis, angioedema, or urticaria after consumption of red meat in patients with IgE antibodies specific for galactose-α-1,3-galactose. J. Allergy Clin. Immunol. 123, 426–433.

Commins, S.P., James, H.R., Kelly, L.A., Pochan, S.L., Workman, L.J., Perzanowski, M.S., et al., 2011. The relevance of tick bites to the production of IgE antibodies to the mammalian oligosaccharide galactose-α-1,3-galactose. J. Allergy Clin. Immunol. 127, 1286–1293.

Commins, S.P., Kelly, L.A., Rönmark, E., James, H.R., Pochan, S.L., Peters, E.J., et al., 2012. Galactose-α-1,3-galactose-specific IgE is associated with anaphylaxis but not asthma. Am. J. Respir. Crit. Care. Med. 185, 723–730.

Cunningham, D., Humblet, Y., Siena, S., Khayat, D., Bleiberg, H., Santoro, A., et al., 2004. Cetuximab monotherapy and cetuximab plus irinotecan in irinotecan-refractory metastatic colorectal cancer. N. Engl. J. Med. 351, 337–345.

Erwin, E.A., Custis, N.J., Satinover, S.M., Perzanowski, M.S., Woodfolk, J.A., Crane, J., et al., 2005. Quantitative measurement of IgE antibodies to purified allergens using streptavidin linked to a high-capacity solid phase. J. Allergy Clin. Immunol. 115, 1029–1035.

Ferreira, B.R., Silva, J.S., 1999. Successive tick infestations selectively promote a T-helper 2 cytokine profile in mice. Immunology 96, 434–439.

Fischer, J., Hebsaker, J., Caponetto, P., Platts-Mills, T.A., Biedermann, T., 2014. Galactose-α-1,3-galactose sensitization is a prerequisite for pork-kidney allergy and cofactor-related mammalian meat anaphylaxis. J. Allergy Clin. Immunol. 134, 755–775.

Fischer, J., Yazdi, A.S., Biedermann, T., 2016. Clinical spectrum of α-Gal syndrome: from immediate-type to delayed immediate-type reactions to mammalian innards and meat. Allergo J. Int. 25, 55–62.

Galili, U., Rachmilewitz, E.A., Peleg, A., Flechner, I., 1984. A unique natural human IgG antibody with anti-α-galactosyl specificity. J. Exp. Med. 160, 1519–1531.

Galili, U., Macher, B.A., Buehler, J., Shohet, S.B., 1985. Human natural anti-α-galactosyl IgG. II. The specific recognition of α(1→3)-linked galactose residues. J. Exp. Med. 162, 573–582.

Galili, U., Clark, M.R., Shohet, S.B., Buehler, J., Macher, B.A., 1987. Evolutionary relationship between the anti-Gal antibody and the Galα1-3Gal epitope in primates. Proc. Natl. Acad. Sci. U.S.A. 84, 1369–1373.

Galili, U., Shohet, S.B., Kobrin, E., Stults, C.L.M., Macher, B.A., 1988. Man, apes, and Old World monkeys differ from other mammals in the expression of α-galactosyl epitopes on nucleated cells. J. Biol. Chem. 263, 17755–17762.

Galili, U., Matta, K.L., 1996. Inhibition of anti-Gal IgG binding to porcine endothelial cells by synthetic oligosaccharides. Transplantation 62, 256–262.

Galili, U., Wigglesworth, K., Abdel-Motal, U.M., 2007. Intratumoral injection of α-gal glycolipids induces xenograft-like destruction and conversion of lesions into endogenous vaccines. J. Immunol. 178, 4676–4687.

Hamsten, C., Starkhammar, M., Tran, T.A., Johansson, M., Bengtsson, U., Ahlen, G., et al., 2013. Identification of galactose-α-1,3-galactose in the gastrointestinal tract of the tick *Ixodes ricinus*; possible relationship with red meat allergy. Allergy 68, 549–552.

Hironaka, T., Furukawa, K., Esmon, P.C., Fournel, M.A., Sawada, S., Kato, M., 1992. et al., Comparative study of the sugar chains of factor VIII purified from human plasma and from the culture media of recombinant baby hamster kidney cells. J. Biol. Chem. 263, 17508–17515.

Kagawa, Y., Takasaki, S., Utsumi, J., Hosoi, K., Shimizu, H., Kochibe, N., et al., 1988. Comparative study of the asparagine-linked sugar chains of natural human interferon-β1 and recombinant human interferon-β1 produced by three different mammalian cells. J. Biol. Chem. 267, 8012–8020.

Konakci, K.Z., Bohle, B., Blumer, R., Hoetzenecker, W., Roth, G., Moser, B., et al., 2005. α-Gal on bioprosthesis: xenograft immune response in cardiac surgery. Eur. J. Clin. Investig. 35, 17–23.

Maier, S., Chung, C., Morse, M., Platts-Mills, T.A., Townes, L., Mukhopadhyay, P., et al., 2015. A retrospective analysis of cross-reacting cetuximab IgE antibody and its association with severe infusion reactions. Cancer Med. 4, 36–42.

Mohan, P.S., Spiro, R.G., 1986. Macromolecular organization of basement membranes. Characterization and comparison of glomerular basement membrane and lens capsule components by immunochemical and lectin affinity procedures. J. Biol. Chem. 261, 4328–4336.

Montaño, R.F., Romano, E.L., 1994. Human monoclonal anti-Rh antibodies produced by human-mouse heterohybridomas express the Galα1-3Gal epitope. Hum. Antibodies Hybrid. 5, 152–156.

Morisset, M., Richard, C., Astier, C., Jacquenet, S., Croizier, A., Beaudouin, E., et al., 2012. Anaphylaxis to pork kidney is related to IgE antibodies specific for galactose-α-1,3-galactose. Allergy 67, 699–704.

Mozzicato, S.M., Tripathi, A., Posthumus, J.B., Platts-Mills, T.A., Commins, S.P., 2014. Porcine or bovine valve replacement in 3 patients with IgE antibodies to the mammalian oligosaccharide galactose-α-1,3-galactose. J. Allergy Clin. Immunol. Pract. 2, 637–638.

Nunez, R., Carballada, F., Gonzalez-Quintela, A., Gomez-Rial, J., Boquete, M., Vidal, C., 2011. Delayed mammalian meat-induced anaphylaxis due to galactose-α-1,3-galactose in 5 European patients. J. Allergy Clin. Immunol. 128, 1122–1124.

Ogawa, H., Galili, U., 2006. Profiling terminal *N*-acetyllactosamines of glycans on mammalian cells by an immuno-enzymatic assay. Glycoconj. J. 23, 663–674.

O'Neil, B.H., Allen, R., Spigel, D.R., Stinchcombe, T.E., Moore, D.T., Berlin, J.D., et al., 2007. High incidence of cetuximab-related infusion reactions in Tennessee and North Carolina and the association with atopic history. J. Clin. Oncol. 25, 3644–3648.

Platts-Mills, A.E., Schuyler, A.J., Hoyt, A.E., Commins, S.P., 2015. Delayed anaphylaxis involving IgE to galactose-α-1,3galactose. Curr. Allergy Asthma Rep. 15, 512.

Qian, J., Liu, T., Yang, L., Daus, A., Crowley, R., Zhou, Q., 2007. Structural characterization of N-linked oligosaccharides on monoclonal antibody cetuximab by the combination of orthogonal matrix assisted laser desorption/ionization hybrid quadrupole-quadrupole time-of-flight tandem mass spectrometry and sequential enzymatic digestion. Anal. Biochem. 364, 8–18.

Rademacher, T.W., Parekh, R.B., Dwek, R.A., 1988. Glycobiology. Annu. Rev. Biochem. 57, 785–838.

Sánchez, O., Barrera, M., Farnós, O., Parra, N.C., Salgado, E.R., Saavedra, P.A., et al., 2014. Effectiveness of the E2-classical swine fever virus recombinant vaccine produced and formulated within whey from genetically transformed goats. Clin. Vaccine Immunol. 21, 1628–1634.

Smith, D.F., Larsen, R.D., Mattox, S., Lowe, J.B., Cummings, R.D., 1990. Transfer and expression of a murine UDP-Gal:β-D-Gal-α1,3-galactosyltransferase gene in transfected Chinese hamster ovary cells. Competition reactions between the α1,3-galactosyltransferase and the endogenous α2,3-sialyltransferase. J. Biol. Chem. 265, 6225–6234.

Spiro, R.G., Bhoyroo, V.D., 1984. Occurrence of α-D-galactosyl residues in the thyroglobulin from several species. Localization in the saccharide chains of the complex carbohydrate units. J. Biol. Chem. 259, 9858–9866.

Steinke, J.W., Platts-Mills, T.A., Commins, S.P., 2015. The α-gal story: lessons learned from connecting the dots. J. Allergy Clin. Immunol. 135, 589–596.

Stone, K.R., Abdel-Motal, U.M., Walgenbach, A.W., Turek, T.J., Galili, U., 2007. Replacement of human anterior cruciate ligaments with pig ligaments: a model for anti-non-gal antibody response in long-term xenotransplantation. Transplantation 83, 211–219.

Takahashi, H., Chinuki, Y., Tanaka, A., Morita, E., 2014. Laminin γ-1 and collagen α-1 (VI) chain are galactose-α-1,3-galactose-bound allergens in beef. Allergy 69, 199–207.

Takeuchi, M., Takasaki, S., Miyazaki, H., Kato, T., Hoshi, S., Kochibe, N., et al., 1988. Comparative study of the asparagine-linked sugar chains of human erythropoietins purified from urine and the culture medium of recombinant Chinese hamster ovary cells. J. Biol. Chem. 263, 3657–3663.

Tanemura, M., Maruyama, S., Galili, U., 2000. Differential expression of α-gal epitopes (Galα1-3Galβ1-4GlcNAc-R) on pig and mouse organs. Transplantation 69, 187–190.

Thall, A.Galili, U., 1990. The differential expression of Galα1-3Galß1-4GlcNAc-R residues on mammalian secreted *N*-glycosylated glycoproteins. Biochemistry 29, 3959–3965.

Thall, A., Etienne-Decerf, J., Winand, R., Galili, U., 1991. The α-galactosyl epitope on mammalian thyroid cells. Acta Endocrinol. 124, 692–699.

Tjio, J.H., Puck, T.T., 1958. Genetics of somatic mammalian cells. II. Chromosomal constitution of cells in tissue culture. J. Exp. Med. 108, 259–268.

Tong, P., Wesemann, D.R., 2015. Molecular mechanisms of IgE class switch recombination. Curr. Top. Microbiol. Immunol. 388, 21–37.

Urashima, T., Saito, T., Ohmisya, K., Shimazaki, K., 1991. Structural determination of three neutral oligosaccharides in bovine (Holstein-Friesian) colostrum, including the novel trisaccharide; GalNAcα1-3Galβ1-4Glc. Biochim. Biophys. Acta 1073, 225–229.

Urashima, T., Saito, T., Nakamura, T., Messer, M., 2001. Oligosaccharides of milk and colostrum in non-human mammals. Glycoconj. J. 18, 357–371.

Urashima, T., Taufik, E., Fukuda, K., Asakuma, S., 2013. Recent advances in studies on milk oligosaccharides of cows and other domestic farm animals. Biosci. Biotechnol. Biochem. 77, 455–466.

Uyttebroek, A., Sabato, V., Bridts, C.H., De Clerck, L.S., Ebo, D.G., 2014. Anaphylaxis to succinylated gelatin in a patient with a meat allergy: galactose-α(1, 3)-galactose (α-gal) as antigenic determinant. J. Clin. Anesth. 26, 574–576.

Van Nunen, S.A., O'Connor, K.S., Clarke, L.R., Boyle, R.X., Fernando, S.L., 2009. An association between tick bite reactions and red meat allergy in humans. Med. J. Aust. 190, 510–511.

van Nunen, S., 2015. Tick-induced allergies: mammalian meat allergy, tick anaphylaxis and their significance. Asia Pac. Allergy 5, 3–16.

Velander, W.H., Lubon, H., Drohan, W.N., 1997. Transgenic livestock as drug factories. Sci. Am. 276, 70–74.

Wigglesworth, K.M., Racki, W.J., Mishra, R., Szomolanyi-Tsuda, E., Greiner, D.L., Galili, U., 2011. Rapid recruitment and activation of macrophages by anti-Gal/α-Gal liposome interaction accelerates wound healing. J. Immunol. 186, 4422–4432.

Winand, R.J., Anaraki, F., Etienne-Decerf, J., Galili, U., 1993. Xenogeneic thyroid-stimulating hormone-like activity of the human natural anti-Gal antibody. Interaction of anti-Gal with porcine thyrocytes and with recombinant human thyroid-stimulating hormone receptors expressed on mouse cells. J. Immunol. 151, 3923–3934.

CHAPTER

8

Anti-Gal and Autoimmunity

OUTLINE

INTRODUCTION

The objective of this chapter is to draw attention to the possible contribution of the natural anti-Gal antibody to autoimmune disorders in humans. A major cause for autoimmune disorders is the *de novo* production of antibodies against self-antigens. These antibodies may appear because of impaired immune tolerance mechanism and because of production of cross-reactive antibodies against pathogens, which also interact with normal self-antigens. This chapter discusses an additional possible cause for autoimmunity: Association of anti-Gal (and/or other natural antibodies) with autoimmune reactivity against *de novo* appearing antigens that can bind the natural antibody. In a study published few years following the discovery of the natural anti-Gal antibody, it was suggested that *in situ* aberrant expression of α-gal epitopes on human cells might result in binding of anti-Gal to these epitopes and induction of an autoimmune process causing the destruction of cells expressing these epitopes (Galili, 1989). As described in Chapters 1 and 2, human cells lack α-gal epitopes because the α1,3galactosyltransferase (α1,3GT) synthesizing these epitopes is absent in humans (Galili et al., 1987, 1988a). Nevertheless, low binding of the natural anti-Gal antibody or of the lectin *Bandeiraea* (*Griffonia*) *simplicifolia* IB4 (BS lectin that binds to α-gal epitopes) was observed with some human normal and malignant cells including: human placenta cells (Christiane et al., 1992), normal senescent red blood cells (RBC)

The Natural Anti-Gal Antibody as Foe Turned Friend in Medicine
http://dx.doi.org/10.1016/B978-0-12-813362-0.00008-7

(Galili, 1986a), RBC in hemoglobinopathies such as β-thalassemia (Galili et al., 1983, 1984) and sickle cell anemia (Galili et al., 1986b), freshly obtained human mammary carcinoma cells, and the human mammary carcinoma cell line MCF7 (Castronovo et al., 1989; Galili, 1989; Petryniak et al., 1991). However, no biochemical studies have demonstrated spontaneous synthesis and presentation of the carbohydrate structure Galα1-3Galβ1-4GlcNAc-R (i.e., the α-gal epitope) in human cells. The observations described in this chapter raise the possibility that anti-Gal may contribute to the pathogenesis of some autoimmune disorders by binding to α-gal-like epitopes, which have a structure resembling the α-gal epitope and which are synthesized by enzymes different from the mammalian α1,3GT (Chapter 1). An analogous example for such autoimmunity is the extensive inflammatory reactions observed in the heart, liver, or intestine of patients with Chagas' disease, which are discussed in Chapter 4. Hypothetical mechanisms mediating presentation of α-gal or α-gal-like epitopes on human cells are discussed at the end of the present chapter. Identification of α-gal or α-gal-like epitopes in affected human tissues may be a difficult task because the concentration of these epitopes in pathologic tissue specimens is likely to be low. The reason is that cells with high concentration of such aberrantly expressed epitopes will be destroyed *in situ* by anti-Gal in a manner similar to hyperacute rejection of xenograft cells (see Chapter 6). Therefore, development of appropriate autoimmune experimental models in anti-Gal producing animals will be important for the exploration of the suggested contribution of this antibody to autoimmunity.

ELEVATION IN ANTI-GAL TITERS IN SOME AUTOIMMUNE DISORDERS

The putative association between anti-Gal and autoimmunity emerged from observations on elevated activity of anti-Gal in the blood of patients with some autoimmune disorders. The first report of such an increase in anti-Gal activity was in patients with disorders of the thyroid gland. When compared with healthy population, anti-Gal was found to be several-fold higher in patients with nontreated active Graves' disease (GD), including those with exophthalmos (protrusion of the eyeball often observed in patients with thyroid disorders) and patients with nontoxic goiter (enlargement of the thyroid gland without abnormal activity of the gland) (Etienne-Decerf et al., 1987; Knobel et al., 1999). In contrast, patients cured of GD or those with nonprogressive goiter were found to have normal titers of anti-Gal. Studies of Fullmer et al. (2005) reported on anti-Gal elevation only in GD patients with ophthalmopathy. Elevation of anti-Gal, in particular of the IgA class, was also reported in patients with Henoch–Schönlein purpura (Davin et al., 1987), which is an acute IgA-mediated disorder characterized by inflammation of small blood vessels in the skin, gastrointestinal tract, kidneys, and joints. Elevation in anti-Gal activity was also reported in Crohn's disease and ulcerative colitis (D'Alessandro et al., 2002), in patients with scleroderma (Gabrielli et al., 1992), myelofibrosis (Leoni et al., 1993), renal injury due to treatment in patients with rheumatoid arthritis (Malaise et al., 1986), and in inner ear diseases associated with hearing loss (Klein et al., 1989). One interpretation of such observations may be an abnormal expression of α-gal epitopes or α-gal-like epitopes on cells of the patient, which may stimulate the immune system to increase production of anti-Gal. Anti-Gal B cells in healthy individuals comprise ~1% of B cells, most of which are quiescent (Galili et al., 1993). Thus, it is possible that such epitopes may activate these quiescent anti-Gal B cells to increase anti-Gal production. Anti-Gal B cells also can be activated by various pathogens such as *Escherichia coli* (Springer and Horton, 1969; Davin et al., 1987; Posekany et al., 2002) or by protozoa (see Chapter 4). Therefore, one reason for elevation in anti-Gal activity may be damage to the intactness of barriers to internal or external environments. Such barriers may include gastrointestinal wall and barrier of the urinary track or skin. Bacteria presenting α-gal-like epitopes, which cross such barriers, may stimulate the immune system to produce anti-Gal. In addition, infections inducing polyclonal activation of B cells also may result in elevation of anti-Gal activity, as well as elevation in activity of other antibodies. Thus, identification of cells or tissues presenting anti-Gal-binding epitopes in an autoimmune disorder and identification of molecules on the cell membranes binding this antibody are required to establish the role of anti-Gal in an autoimmune disorder. Anti-Gal-mediated autoimmunity has not been studied in experimental animal models because mice, rats, rabbits, guinea pigs, and pigs synthesize α-gal epitopes, like other nonprimate mammals, and thus they cannot produce the anti-Gal antibody. The studies described below on anti-Gal in GD provided a partial support to the assumption that anti-Gal contributes to the pathogenesis of this disease.

ANTI-GAL INTERACTION WITH GRAVES' DISEASE THYROCYTES

GD is a disease of the thyroid. In healthy individuals, the production of the thyroid hormones triiodothyronine (T_3) and thyroxine (T_4) by thyrocytes is regulated by the thyroid-stimulating hormone (TSH—thyrotropin), which is produced by the pituitary. TSH reaches the thyroid via the blood circulation and binds to TSH receptors (TSHR) on the follicular thyroid cells (thyrocytes). As a result of this interaction, the cytoplasmic tail of TSHR activates its attached G protein signal cascade, resulting in elevated production of cAMP by adenylyl cyclase. The *de novo* produced cAMP initiates the signal transduction that induces a range of metabolic activities characteristic to thyrocytes. These include pumping of iodine into the thyrocytes, thyroglobulin synthesis, iodination of tyrosines in thyroglobulin (i.e., organification of iodine), production and secretion of the thyroid hormones T_3 and T_4. At high concentrations of TSH, thyrocytes are induced to proliferate. In GD, the immune system produces autoantibodies that bind the TSHR. This interaction mimics the stimulatory effect of TSH on the TSHR; however, this stimulatory effect is unregulated and continuously activates the thyrocytes. This results in proliferation of thyrocytes, elevated production of thyroglobulin and of T_3 and T_4. The TSHR is a membrane-anchored glycoprotein that carries 5–7 *N*-linked carbohydrate chains (Nagayama et al., 1989; Parmentier et al., 1989; Akamizu et al., 1990; Misrahi et al., 1990). The observation on the elevated anti-Gal activity in GD patients (Etienne-Decerf et al., 1987) raised the question whether anti-Gal is an antibody that aberrantly binds to TSHR on GD thyrocytes and activates these thyrocytes similar to TSH/TSHR activation. As a first step to addressing this question, the ability of anti-Gal to bind and activate TSHR known to carry α-gal epitopes was evaluated.

Anti-Gal-mediated activation of thyroid-stimualting hormone receptors on porcine thyrocytes

Immunostaining by anti-Gal of porcine thyrocyte proteins separated by sodium dodecyl sulfate polyacrylamide gel electrophoresis demonstrated in Western blots the

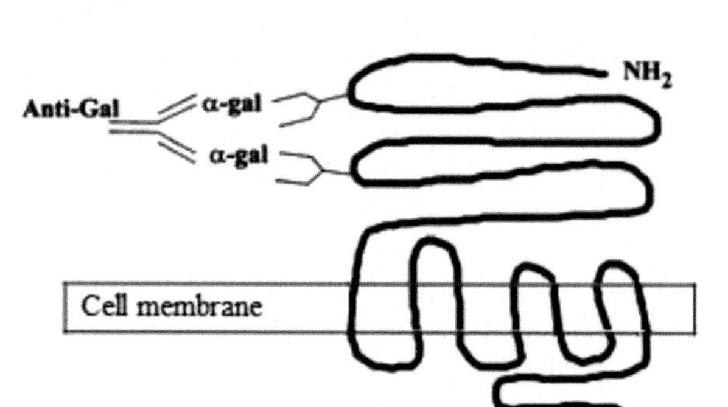

FIGURE 1 Schematic illustration of thyroid-stimulating hormone receptor (TSHR) with two out of five to seven *N*-linked carbohydrate chains. Some of these carbohydrate chains on porcine TSHR are likely to carry α-gal epitopes. It was hypothesized by Winand et al. (1993) that binding of the natural anti-Gal antibody to these α-gal epitopes may result in activation of the TSHR a manner similar to the activation observed following binding of thyroid-stimulating hormone (TSH—thyrotropin) to TSHR. *Reprinted from Galili, U., 1999. Graves' disease as a model for anti-Gal involvement in autoimmune diseases pathogenesis. Subcell. Biochem. 32, 339–360, with permission.*

presence multiple α-gal epitopes on porcine thyroid glycoproteins (Thall et al., 1991). Because TSHR has 5–7 *N*-linked carbohydrate chains, it was reasonable to assume that in pigs some of these carbohydrate chains may be capped by α-gal epitopes (Fig. 1). Thus, it was of interest to determine whether binding of anti-Gal to porcine thyrocytes has metabolic effects on these cells. Thyrocytes were obtained from porcine thyroid gland by trypsinization and were cultured as monolayers. These thyrocytes were readily induced by bovine TSH to pump in iodine, as measured by uptake of ^{125}I (Fig. 2A). Similar to incubation with TSH, incubation of porcine thyrocytes with human anti-Gal, purified from normal AB sera, increased ^{125}I uptake by almost 200% above the basal level (Fig. 2B) (Winand et al., 1993). Anti-Gal added to porcine thyrocyte cultures at 50 µg/mL further stimulated the cells to proliferate, as measured by ^{3}H-thymidine incorporation (Fig 2C). Incubation of these thyrocytes with the serum of GD patients activated the cells as indicated by the increase in cAMP production above the basal level (Fig. 2D). This stimulation was expected because of the presence of anti-TSHR antibodies in these sera. However, if the sera were depleted of anti-Gal by passage through a column of synthetic α-gal epitopes (Galα1-3Galβ1-4GlcNAc), >70% of the stimulatory activity of serum antibodies was eliminated in most sera (Fig. 2D) (Winand et al., 1993). A similar passage of the serum through a control column with synthetic *N*-acetyllactosamine epitopes (Galβ1-4GlcNAc), resulted in no decrease in the stimulatory activity of the GD sera (Fig. 2D). Overall, these observations on anti-Gal isolated from normal human serum and anti-Gal in sera of GD patients suggested that binding of this antibody to α-gal epitopes on porcine thyrocytes indeed can mimic the effects of TSH binding to TSHR on the thyrocytes.

Anti-Gal binding to human recombinant thyroid–stimualting hormone receptors on mouse cells

The α-gal epitope is expressed on multiple glycoproteins of porcine thyrocytes, in addition to TSHR (Thall et al., 1991). To validate that the effects of human anti-Gal on porcine thyrocytes are the result of antibody binding to α-gal epitopes on TSHR, the human TSHR

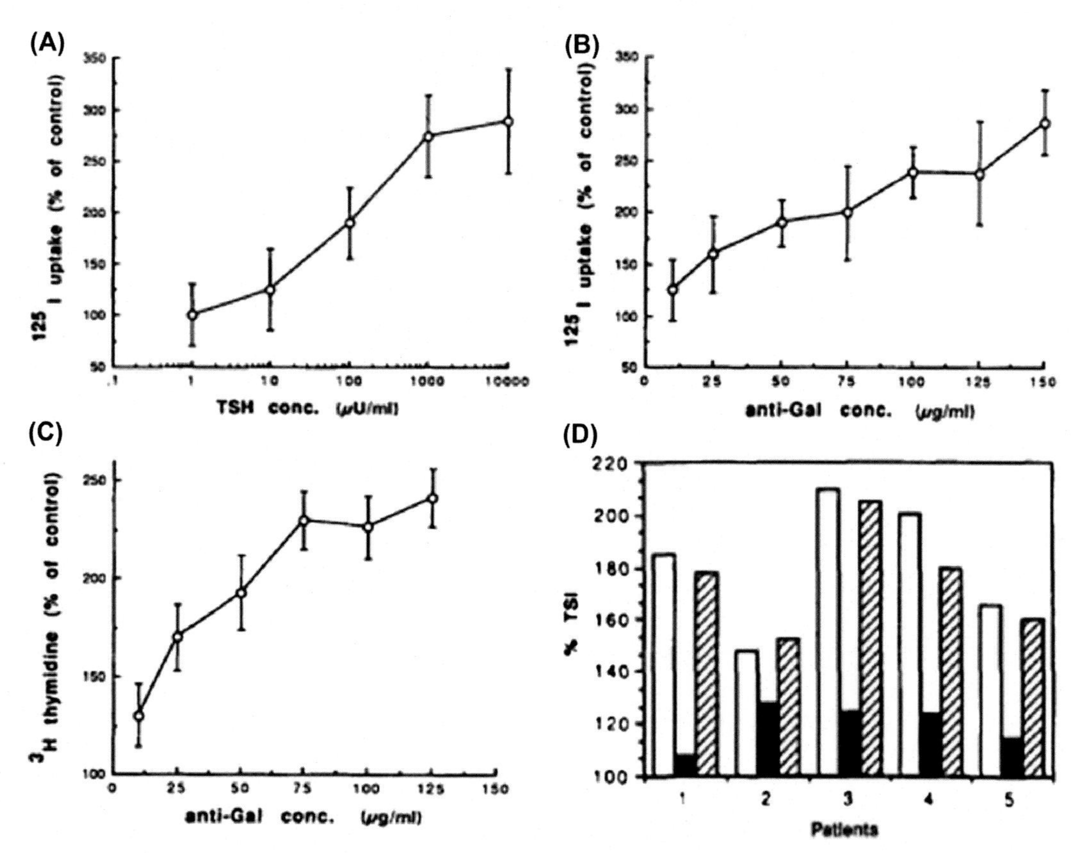

FIGURE 2 Activating effects of human anti-Gal on porcine thyrocytes. (A) Activation of porcine thyrocytes by thyroid-stimulating hormone for ^{125}I uptake during 5h incubation. The basal ^{125}I uptake into the thyrocytes is presented as 100%. (B) Increase in ^{125}I uptake in porcine thyrocytes incubated with the natural anti-Gal antibody. (C) Proliferation of porcine thyrocytes incubated for 24h with the natural anti-Gal antibody, as measured by ^{3}H-thymidine incorporation. (D) Effect of anti-Gal depletion on the ability of sera from five GD patients to stimulate porcine thyrocytes to produce cAMP. Open bars, stimulatory activity of serum; closed bars, stimulatory activity of sera depleted of anti-Gal on affinity columns with synthetic Galαl-3Galβl-4GlcNAc-R (i.e., α-gal epitopes); hatched bars, stimulatory activity of sera passed through control Galβl-4GlcNAc-R columns. Basal 100% level is that of porcine thyrocytes incubated with pooled normal sera. Sera were diluted 1:10 in the culture medium. Data in (A–C) presented as mean ± SD of six experiments. *Reprinted from Winand, R.J., Anaraki, F., Etienne-Decerf, J., Galili, U., 1993. Xenogeneic thyroid-stimulating hormone-like activity of the human natural anti-Gal antibody. Interaction of anti-Gal with porcine thyrocytes and with recombinant human thyroid-stimulating hormone receptors expressed on mouse cells. J. Immunol. 151, 3923–3934, with permission.*

gene (Nagayama et al., 1989) was transfected into mouse 3T3 fibroblast and into Chinese hamster ovary (CHO) cell lines (Winand et al., 1993). Whereas α-gal epitopes are the most abundant carbohydrate epitopes on 3T3 fibroblasts (Santer et al., 1989), they are completely absent on CHO cells (Kagawa et al., 1988; Takeuchi et al., 1988; Smith et al., 1990) (see discussion of α-gal and CHO cells in Chapter 7). Binding of ^{125}I-labeled TSH to the *de novo* expressed TSHR on both cell lines indicated that the human TSHR is presented on the transfected 3T3 and on transfected CHO cells (Fig. 3A). As expected, human anti-Gal (purified

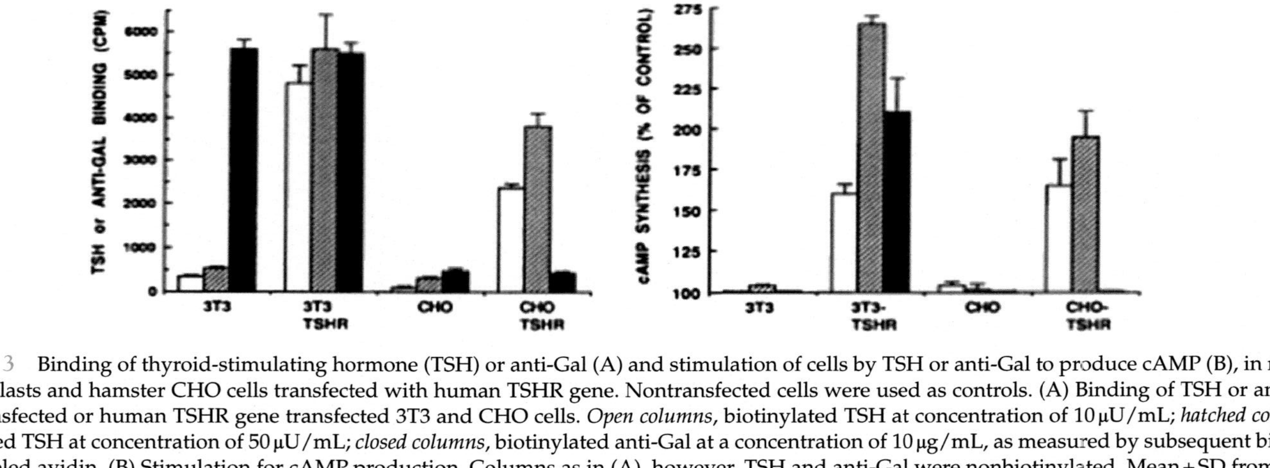

FIGURE 3 Binding of thyroid-stimulating hormone (TSH) or anti-Gal (A) and stimulation of cells by TSH or anti-Gal to produce cAMP (B), in mouse 3T3 fibroblasts and hamster CHO cells transfected with human TSHR gene. Nontransfected cells were used as controls. (A) Binding of TSH or anti-Gal to nontransfected or human TSHR gene transfected 3T3 and CHO cells. *Open columns*, biotinylated TSH at concentration of 10 µU/mL; *hatched columns*, biotinylated TSH at concentration of 50 µU/mL; *closed columns*, biotinylated anti-Gal at a concentration of 10 µg/mL, as measured by subsequent binding of ^{125}I-labeled avidin. (B) Stimulation for cAMP production. Columns as in (A), however, TSH and anti-Gal were nonbiotinylated. Mean + SD from three experiments. *Reprinted from Winand, R.J., Anaraki, F., Etienne-Decerf, J., Galili, U., 1993. Xenogeneic thyroid-stimulating hormone-like activity of the human natural anti-Gal antibody. Interaction of anti-Gal with porcine thyrocytes and with recombinant human thyroid-stimulating hormone receptors expressed on mouse cells. J. Immunol. 151, 3923–3934, with permission.*

from human AB serum) readily bound to nontransfected and transfected 3T3 fibroblasts and did not bind to nontransfected or transfected CHO cells, confirming that these hamster cells lack α-gal epitopes. The activation of both cells by TSH to produce cAMP further indicated that the recombinant human TSHR was functional in both transfected cell lines and activated the G protein cascade, resulting in cAMP production (Fig. 3B). A similar activation of the human recombinant TSHR, for induction of cAMP production, was observed with anti-Gal in 3T3 fibroblasts that expressed the human TSHR but not in CHO cells expressing this receptor. These findings implied that anti-Gal binding to α-gal epitopes on recombinant TSHR presented on 3T3 fibroblasts activated the receptor, whereas in the absence of α-gal epitopes, as on the recombinant human TSHR on CHO cells, anti-Gal had no activating effect (Fig. 3B) (Winand et al., 1993).

Anti-Gal activation of Graves' disease thyrocytes

The observations on human anti-Gal binding and activating TSHR with α-gal epitopes on porcine thyrocytes and on recombinant human TSHR expressed on mouse 3T3 fibroblasts led to the evaluation of similar binding and activation in thyrocytes of GD patients. GD thyrocyte cell cultures were prepared from GD thyroid tissues resected at the University of Liege hospital. The thyroid tissues underwent proteolytic digestion and the resulting cell suspensions were washed 10 times to remove GD antibodies that bind to the thyrocytes and prevent their attachment to tissue culture plates (Winand and Kohn, 1975). Normal thyrocytes were isolated from normal portion of thyroid tissue specimens resected from patients with "cold" nodules (i.e., thyroid tissues displaying no increased iodine uptake) as part of their treatment. Normal and GD thyrocytes were plated in 60 mm tissue culture dishes in 199 medium supplemented with 10% fetal calf serum as detailed in Winand et al. (1994).

The cultured thyrocytes were tested for binding of anti-Gal by incubation for 3 h at 37°C with 10 μg/mL biotinylated anti-Gal isolated from normal AB serum, followed by washes of the cells and measurement of ^{125}I-labeled avidin binding (Winand et al., 1994). Normal human thyrocytes displayed low binding of anti-Gal (on average 450 cpm per 10^6 cells), whereas GD thyrocytes bound natural anti-Gal at a level that was on average sixfold higher (Fig. 4A). The extent of anti-Gal binding to GD thyrocytes was ~10-fold lower than that observed with porcine thyrocytes, which displayed binding of ~25,000 cpm per 10^6 cells (not shown). The molecules binding anti-Gal on cultured GD thyrocytes have not been identified; however, this antibody binding readily activated the GD thyrocytes. Such activation was demonstrated in measurements of cAMP production (Fig. 4B) in presence of anti-Gal. Normal thyrocytes incubated for 2 h with 50 μg/mL of this natural antibody did not display any increase in cAMP synthesis in comparison with the basal level of cAMP production in the absence of the antibody. In contrast, cAMP synthesis in GD thyrocytes in presence of anti-Gal was 2- to 4.5-fold higher than the basal level (Fig. 4B). Another indication for anti-Gal-mediated activation of GD thyrocytes by the natural anti-Gal antibody was ^{125}I uptake into the cells. Thyrocytes were incubated for 24 h with anti-Gal then Na^{125}I (10 μCi per dish) was added for 5 h. Whereas normal thyrocytes displayed no uptake of ^{125}I when incubated with 50 μg/mL anti-Gal, the GD thyrocytes incubated with this antibody internalized iodine at a level of 2000–10,000 cpm per 10^6 cells (Fig. 4C). This uptake was followed by organification of ^{125}I

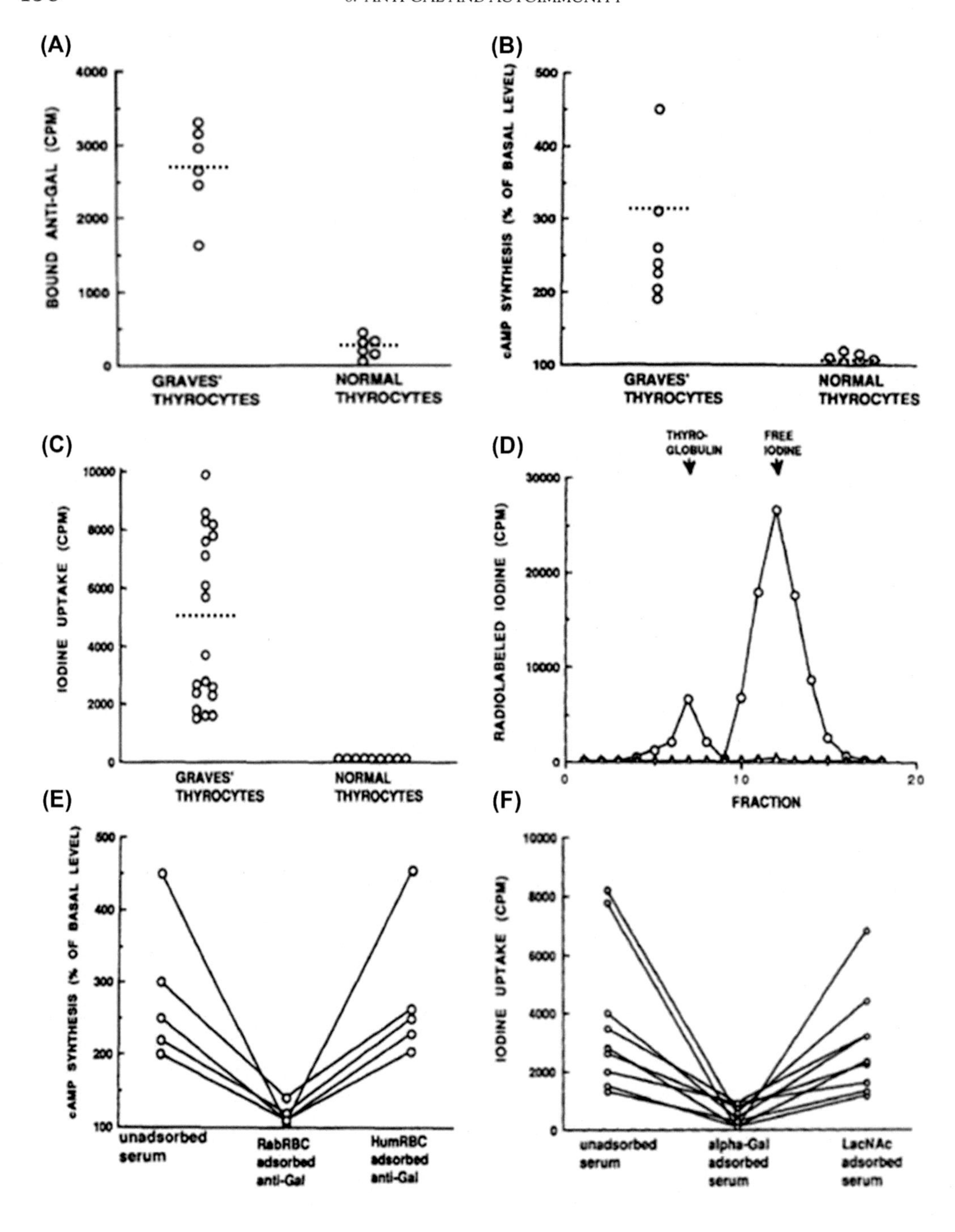
(A)
BOUND ANTI-GAL (CPM)
4000
3000
2000
1000
0
GRAVES' THYROCYTES
NORMAL THYROCYTES
(B)
cAMP SYNTHESIS (% OF BASAL LEVEL)
500
400
300
200
100
GRAVES' THYROCYTES
NORMAL THYROCYTES
(C)
IODINE UPTAKE (CPM)
10000
8000
6000
4000
2000
0
GRAVES' THYROCYTES
NORMAL THYROCYTES
(D)
THYRO-GLOBULIN
FREE IODINE
RADIOLABELED IODINE (CPM)
30000
20000
10000
0
0
10
20
FRACTION
(E)
cAMP SYNTHESIS (% OF BASAL LEVEL)
500
400
300
200
100
unadsorbed serum
RabRBC adsorbed anti-Gal
HumRBC adsorbed anti-Gal
(F)
IODINE UPTAKE (CPM)
10000
8000
6000
4000
2000
0
unadsorbed serum
alpha-Gal adsorbed serum
LacNAc adsorbed serum

within the GD thyrocytes, i.e., the iodine was linked to tyrosines on thyroglobulin molecules. This could be demonstrated by size fractionation, on Sephadex G-10 columns, of the cytosolic content of GD and normal thyrocytes incubated with anti-Gal and ^{125}I. No organification of iodine (i.e., ^{125}I linked to thyroglobulin) was detected in the thyroglobulin containing fractions from normal human thyrocytes, whereas a significant proportion of the radiolabeled iodine taken up by GD thyrocytes was linked to thyroglobulin in fractions of this protein (Fig. 4D) (Winand et al., 1994).

It is well established that anti-TSHR antibodies, *de novo* produced in GD patients, stimulate the activity of GD thyrocytes (cf., Akamizu et al., 1995). The studies described above on the selective ability of human anti-Gal to stimulate the activity of GD thyrocytes, but not of normal human thyrocytes, raised the question to what extent anti-Gal in the serum of GD patients can stimulate autologous GD thyrocytes. This was studied by incubation of cultured GD thyrocytes with autologous serum diluted 1:10 and measuring the activation of these thyrocytes to synthesize cAMP and for uptake of ^{125}I in comparison with the same serum depleted of anti-Gal. Depletion of anti-Gal was performed by adsorption of the serum on glutaraldehyde-fixed rabbit RBC, which are rich with α-gal epitopes (Galili et al., 1987), or on glutaraldehyde-fixed human blood type O RBC, used as control. Free reactive glutaraldehyde groups on the RBC were blocked by incubation with glycine prior to the adsorption process. A second method of anti-Gal depletion was by passing the serum through a column of Synsorb silica beads that carry on their surface multiple synthetic α-gal epitopes. A column with synthetic *N*-acetyllactosamine presenting silica beads (Galβ1-4GlcNAc-R) was used for control adsorption. Adsorption of GD serum on rabbit RBC resulted in >50% decrease in the ability of antibodies in the serum to stimulate autologous thyrocytes to produce cAMP, whereas adsorption on the control human RBC did not alter the stimulatory activity of antibodies in the serum (Fig. 4E). Similarly, depletion of anti-Gal from sera of GD patients by synthetic α-gal epitopes linked to silica beads decreased 30%–95% of ^{125}I uptake by the autologous GD thyrocytes, whereas adsorption on control *N*-acetyllactosamine column did not alter ^{125}I uptake by these thyrocytes (Fig. 4F) (Winand et al., 1994). These observations support the assumption that a significant proportion of antibodies stimulating autologous thyrocytes in GD patients is anti-Gal antibodies that bind to unidentified molecules on thyrocyte cell membranes in these patients.

FIGURE 4 Activation of GD thyrocytes by natural and autologous anti-Gal antibody. (A) Binding of biotinylated natural anti-Gal antibody followed by ^{125}I-avidin to cultured GD and normal thyrocytes incubated with 10 μg/mL of the biotinylated antibody. (B) Synthesis of cAMP in GD and normal thyrocytes incubated for 2 h with 50 μg/mL natural anti-Gal antibody. (C) Uptake of ^{125}I by GD and normal thyrocytes incubated for 24 h with 50 μg/mL natural anti-Gal antibody then Na^{125}I (10 μCi per dish) was added for 5 h. The *dotted line* in Figures (A–C) represents the mean activity in GD and normal thyrocytes. (D) Organification (linking iodine to thyroglobulin) of ^{125}I internalized by GD (*circles*) and normal thyrocytes (*triangles*) in thyroglobulin from the thyrocytes cultured with anti-Gal and with ^{125}I, then the cytosol fractionated on Sephadex-G10 columns. (E) Changes in stimulatory effect of GD sera on cAMP synthesis by autologous thyrocytes, following depletion of anti-Gal by adsorption on glutaraldehyde-fixed rabbit RBC. Adsorption of glutaraldehyde-fixed human blood group O RBC served as control lacking α-gal epitopes. (F) Changes in stimulation for ^{125}I uptake by GD sera depleted of anti-Gal by adsorption on synthetic α-gal epitopes linked to Synsorb silica beads, or on *N*-acetyllactosamine (LacNAc) linked to silica beads and serving as control that does not bind anti-Gal. The lines connect data obtained from the same GD serum, under the different adsorption procedures. *Reprinted from Winand, R.J., Winand-Devigne, J., Meurisse, M., Galili, U., 1994. Specific stimulation of Graves' disease thyrocytes by the natural anti-Gal antibody from normal and autologous serum. J. Immunol. 153, 1386–1395, with permission.*

Attempts to identify anti-Gal-binding molecules on Graves' disease thyrocytes

In an attempt to determine whether GD thyrocytes contain aberrant α1,3GT activity that may synthesize α-gal epitopes, microsomal fractions were isolated from homogenized GD thyroid tissue specimens, and separated from the membranes of the thyrocytes (Thall et al., 1991). The activity of α1,3GT was determined by linking ^{3}H-Gal to *N*-acetyllactosamine (Galβ1-4GlcNAc) on silica beads to form ^{3}H-labeled α-gal epitopes. No significant differences were observed in the α1,3GT activity between microsomal fractions of normal thyrocytes and GD thyrocytes (Thall et al., 1991). Measuring bovine thyroid α1,3GT activity by this assay resulted in ~100-fold higher synthesis of α-gal epitopes than that observed in human thyroid tissue.

A sensitive radioimmunoassay used in that study for measuring biotinylated natural anti-Gal binding to thyrocyte membranes demonstrated a low, but distinct binding to membranes of normal and GD thyrocytes, with no significant differences in binding between the two groups of specimens (Thall et al., 1991). This binding pattern differs from that observed when intact thyrocytes were studied for anti-Gal binding in Fig. 4. This observation raised the possibility that reorganization of anti-Gal-binding molecules on the cell membrane of GD thyrocytes results in the elevated binding of the antibody to the intact cells. Sequencing of the region of the α1,3GT gene (*GGTA1*) described in Fig. 3 of Chapter 2 revealed complete similarity between the human sequence described in that figure and that in DNA obtained from GD thyroid specimens (unpublished observations).

The identity of anti-Gal-binding molecules on GD thyrocytes remains unknown. The research on characterization of these molecules could not progress beyond these studies because of the untimely death of Dr. Roger J. Winand who headed this research and performed the major part of the studies. Dr. Winand was an endocrinologist at the University of Liege Hospital in Belgium and a pioneering researcher of *in vitro* studies on GD thyrocytes. He developed methods for culturing these cells, evaluating their functions and studying the effects of anti-Gal on the GD thyrocytes.

MECHANISMS THAT MAY INDUCE ANTI-GAL BINDING TO HUMANS CELLS

It is probable that the number of anti-Gal-binding molecules on the surface of GD thyrocytes is low. A high concentration of such molecules would result in anti-Gal-mediated destruction of the thyrocytes as in hyperacute rejection of xenograft, described in Chapter 6. There are several hypothetical possibilities for mechanisms which may result in expression of a low number of anti-Gal molecules on GD thyrocytes or which could contribute to other autoimmune disorders that may be associated with anti-Gal activity.

Aberrant presentation of x_2 glycolipids

As discussed in Chapter 1, studies on interaction of natural human anti-Gal and of monoclonal anti-Gal with various glycolipids demonstrated binding of these antibodies to the glycolipid x_2 with the structure GalNAcβ1-3Galβ1-4GlcNAcβ1-3Galβ1-4Glcβ1-ceramide, in addition to binding to α-gal glycolipids (Teneberg et al., 1996). The x_2 glycolipid is a minor glycolipid in human RBC (Kannagi et al., 1982a) and was also reported in human gastric tumor (Kannagi et al., 1982b). This glycolipid may be a cryptic molecule, which is exposed on senescent human normal, thalassemia and sickle cell anemia RBC. The exposed x_2 glycolipid may bind anti-Gal and label these RBC for removal from the circulation by macrophages of the reticuloendothelial system (Galili et al., 1983, 1984, 1986a,b). TSHR was found in thyroid cells to be associated with the ganglioside GD1 (glycolipid carrying sialic acid) (Kielczynski et al., 1994). Aberrantly expressed x_2 molecules may be associated with TSHR molecules and could be accessible to binding by anti-Gal. Binding of anti-Gal to x_2 glycolipids may cause perturbation of the TSHR, if this receptor is associated with the glycolipid, resulting in activation of the receptor. Another mammalian glycolipid which has terminal Galα1-3Gal epitope and which may be considered as candidate for anti-Gal binding is isoglobotrihexosylceramide (iGb3) with the structure Galα1-3Galβ1-4Glc-Cer, originally reported in dog intestine (Sung and Sweely, 1979). iGb3 is produced by iGb3 synthase on lactose type core structure (Taylor et al., 2003). However, neither the gene for iGb3 synthase nor iGb3 have been found in humans (Christiansen et al., 2008; Casals et al., 2009).

α-Gal mimetic peptides

Studies on phage-displayed peptide libraries led to the cloning of several α-gal mimetic peptides (Kooyman et al., 1996; Zhan et al., 2003; Lang et al., 2006), including peptides in mucin (Sandrin et al., 1997; Apostolopoulos et al., 1999). Similar to x_2, which may mediate binding of anti-Gal to senescent human RBC, α-gal mimetic peptides in mucin or in other cell surface proteins may explain the observed anti-Gal binding to mucin rich mammary carcinoma cells and cell lines (Castronovo et al., 1989) and to human thyrocytes cell membranes (Thall et al., 1991). If such mimetic peptides mediate anti-Gal binding to GD thyrocytes, then the lack of anti-Gal binding to normal human thyrocytes versus its binding to GD thyrocytes (Fig. 4) may be associated with unknown aberrant mechanisms of reorganization or changes in folding of proteins carrying these mimetic peptides, making them accessible to anti-Gal binding. α-Gal mimetic peptides may further stimulate anti-Gal B cells to increase production of this antibody.

Bacteria-mediated expression of α-gal-like epitopes on human cells

Anti-Gal is continuously produced in humans because of antigenic stimulation by bacteria of the gastrointestinal flora that present α-gal-like epitopes such as *E. coli*, *Serratia marcescens*, *Salmonella minnesota*, and *Klebsiella pneumonia* (Galili et al., 1988b). The lipopolysaccharides of these bacteria display extensive binding of human natural anti-Gal antibody. Indeed, lipopolysaccharides in these bacteria were reported to contain in their carbohydrate chains repeating Galα1-3Gal and Gal linked α1-3 to other carbohydrate units (Lüderitz et al., 1965; Bjorndal et al., 1971; Whitfield et al., 1991; Aucken et al., 1998). Accordingly, both gram-positive and gram-negative bacteria were found to have α1,3GT enzymes that synthesize α-gal-like epitopes (Endo and Rothfield, 1969; Han et al., 2012; Chen et al., 2016). Such enzymes may be able to synthesize α-gal epitopes using human carbohydrate chains as acceptors on human cells. As an example, human blood group O RBC were incubated with a sonicate of *K. pneumonia* and with UDP-Gal as the sugar donor. This resulted in synthesis of α-gal epitopes on a large proportion of the RBC, as indicated by the subsequent binding of human anti-Gal to the treated RBC (Hamadeh et al., 1996). This enzyme could be further purified from the lyzate in three size fractions. Theoretically, it is possible that a similar synthesis of α-gal epitopes by bacterial enzyme may occur in the body along the gastrointestinal tract or in tissues infected with bacteria producing enzymes that synthesize α-gal or α-gal-like epitopes. Appearance of such epitopes on normal cells will be followed by binding of the natural anti-Gal antibody that may serve as a trigger for the initiation of a local autoimmune-mediated inflammatory process that damages cells expressing these epitopes.

An additional mechanism for binding of anti-Gal to human cells presenting α-gal-like epitopes is the adhesion of bacterial wall fragments to human cells. Fragments of sonicated *E. coli O86* bacteria adhered to human HeLa cells and to normal human fibroblasts and mediated binding of anti-Gal to these cells (Galili et al., 1988b). This binding was observed even at a concentration of the antibody that is 2.5% of that found in normal human serum. As with synthesis of α-gal or α-gal-like epitopes by bacterial enzymes, adhesion of bacterial fragments that bind anti-Gal may result in an autoimmune process damaging normal cells that attach these fragments. Thus, it would be of interest to determine whether this mechanism of autoimmunity by anti-Gal binding to bacterial fragments, which adhere to cells of the gastrointestinal wall, contributes to chronic inflammatory processes in the gastrointestinal tract, such as ulcerative colitis and Crohn disease.

A second hypothetical mechanism for bacteria-mediated anti-Gal binding to human cells may be inferred from studies on the possible association between bacteria-mediated transfer of a sialic acid called *N*-glycolyl neuraminic acid (Neu5Gc) from red meat (beef, pork, and lamb) to human cell membranes. Humans produce natural anti-Neu5Gc antibodies against Neu5Gc, which is produced in other mammals, but not in humans (Zhu and Hurst, 2002; Tangvoranuntakul et al., 2003; Varki, 2010). These antibodies have been called also Hanganutziu-Deicher "heterophile" antibodies (Higashi et al., 1977; Merrick et al., 1987). Neu5Gc that is present in high concentration in meat was found to be transferred into bacterial wall (Taylor et al., 2010) and further transferred into membrane of human cells (Samraj et al., 2015). An *in vivo* bacterial-mediated transfer of meat Neu5Gc was observed in a mouse experimental model lacking Neu5Gc and producing anti-Neu5Gc antibodies. This transfer was found to result in inflammatory processes caused by binding of anti-Neu5Gc antibodies to the Neu5Gc carrying molecules inserted into the mouse cell membranes (Samraj et al., 2015). These findings have suggested that such transfer of glycoconjugates from red meat to human cells may mediate autoimmune reactions propagated by natural anti-Neu5Gc antibodies (Samraj et al., 2015). Because red meat, like other nonprimate mammalian tissues, contains high concentration of α-gal epitopes, it would be of interest to determine whether this bacterial mechanism also transfers glycoconjugates with α-gal epitopes from red meat to human cells. Insertion of these epitopes into human cell membranes may result in subsequent binding of anti-Gal to such human cells and destruction of the cells. Chapter 10 demonstrates a similar spontaneous insertion of α-gal glycolipids from micelles into human tumor cell membranes, resulting in binding of anti-Gal to the cells (Galili et al., 2007).

Infection with zoonotic viruses carrying the mammalian α1,3galactosyltransferase gene

The observations on the presence of oncogenes such as *src* and *ras* both in the genome of various vertebrates and in oncoviruses (Stehelin et al., 1976; Der et al., 1982; Santos et al., 1982; Parada et al., 1982) raise the possibility that in rare events, some viruses may be able to "pick up" genes from infected cells and subsequently introduce them by infection into human host cells. A hypothetical testable possibility is that some of the multiple α1,3GT transcripts in host cells, other than Old World monkeys, apes, or humans, may accidently become part of the genome of viruses infecting such cells. Infection of human cells by zoonotic viruses containing the mammalian α1,3GT gene *GGTA1* may result in production of α1,3GT in the host cells and subsequent synthesis of α-gal epitopes on glycolipids and glycoproteins of these cell. This scenario is analogous to the observed expression of α-gal epitopes on human HeLa cells within 12–24h after the infection of the cells with a replication defective adenovirus containing mouse α1,3GT cDNA (Deriy et al., 2002). As discussed in Chapter 2, if such viruses are enveloped viruses, many of them may be destroyed by anti-Gal following the infection of a human host. However, if the number of α-gal epitopes is low on such hypothetical viruses, or if the virus does not have envelope glycoproteins, infection of human cells may result in expression of α-gal epitopes on the infected cells and subsequent binding of anti-Gal. Cells expressing this epitope may be destroyed by the antibody. However, if the virus can proliferate prior to destruction of the infected cell and infect adjacent cells, such infection may become an autoimmune disorder. Alternatively, infected cells synthesizing α-gal epitopes on cell surface receptors may be activated because of anti-Gal binding to the receptor, as in the examples above of TSHR presenting α-gal epitopes on porcine thyrocytes or on murine 3T3 fibroblasts.

As emphasized above, the proposed options for the aberrant expression of anti-Gal binding molecules on human cells are hypothetical. However, they may be tested and provide information on the extent of anti-Gal contribution to the pathogenesis in various autoimmune disorders.

CONCLUSIONS

Autoimmune disorders usually occur when antibodies reactive with self-antigens, or autoreactive T cells, aberrantly appear. In view of the ubiquitous abundance of the natural anti-Gal antibody in humans, it is suggested that autoimmunity in some disorders may be associated with the *de novo* appearance of α-gal or α-gal-like epitopes on tissues that are affected by the autoimmune process. Binding of anti-Gal to these epitopes may mediate destruction of the target tissue, analogous to the anti-Gal destruction observed in xenotransplantation (Chapter 6), or in autoimmune-like phenomena in Chagas' disease (Chapter 4). Presentation of α-gal or α-gal-like epitopes on some cell receptors such as TSHR, or on molecules associated with these receptors may result in continuous activation and stimulation for proliferation of cells with aberrant expression of anti-Gal-binding molecules. Anti-Gal involvement in pathogenesis leading to autoimmunity is far from being understood; however, there are various observations that suggest possible contribution of this antibody to autoimmunity. One autoimmune disorder in which anti-Gal may contribute to the pathogenesis is GD. Many of the patients with GD display elevated titers of anti-Gal. Incubation of cultured GD thyrocytes with the natural anti-Gal antibody purified from normal AB sera, results in binding of anti-Gal to these cells, activation of the cells for cAMP synthesis, iodine incorporation, and cell proliferation, similar to the stimulatory effects of TSH (thyrotropin) on normal thyrocytes. Normal human cultured thyrocytes, however, do not bind anti-Gal and are not stimulated by this antibody.

Incubation of GD thyrocytes with autologous serum that was specifically depleted of anti-Gal resulted in 30%–95% decrease in the stimulatory activity of the antibodies in the serum of the GD patients.

The process that results in expression of anti-Gal-binding molecules on GD thyrocytes has not been identified. Several suggested hypothetical mechanisms that may induce anti-Gal binding to human cells are: aberrant expression of glycolipids such as x_2 that can bind anti-Gal; aberrant expression of α-gal mimetic peptides that bind this antibody; bacterial wall fragment adhesion or insertion of α-gal or α-gal-like carrying molecules into human cells; synthesis of such epitopes on cells by bacterial enzymes; and infection of cells by zoonotic viruses that carry the α1,3GT gene *GGTA1* and transport it from other mammals, into humans. Additional studies in elucidating anti-Gal role in autoimmune disorders may enable to determine which of these mechanisms or other unknown mechanisms result in aberrant expression of anti-Gal binding molecules on human cells.

References

Akamizu, T., Ibuyama, S., Saji, M., Kosugi, S., Kozak, C., McBride, O.W., et al., 1990. Cloning, chromosomal assignment, and regulation of the rat thyrotropin receptor: expression of the gene is regulated by thyrotropin, agents that increase cAMP levels, and thyroid autoantibodies. Proc. Natl. Acad. Sci. U.S.A. 87, 5677–5681.

Akamizu, T., Kohn, L.D., Mori, T., 1995. Molecular studies on thyrotropin (TSH) receptor and anti-TSH receptor antibodies. Endocr. J. 42, 617–627.

Apostolopoulos, V., Sandrin, M.S., McKenzie, I.F., 1999. Carbohydrate/peptide mimics: effect on MUC1 cancer immunotherapy. J. Mol. Med. (Berl.) 77, 427–436.

Aucken, H.M., Wilkinson, S.G., Pitt, T.L., 1998. Re-evaluation of the serotypes of *Serratia marcescens* and separation into two schemes based on lipopolysaccharide (O) and capsular polysaccharide (K) antigens. Microbiology 144, 639–653.

Bjorndal, H., Lindberg, B., Nimmich, W., 1971. Structural studies on *Klebsiella* 0 groups 1 and 6 lipopolysaccharides. Acta Chem. Scand. 25, 750.

Casals, F., Ferrer-Admetlla, A., Sikora, M., Ramírez-Soriano, A., Marquès-Bonet, T., Despiau, S., et al., 2009. Human pseudogenes of the ABO family show a complex evolutionary dynamics and loss of function. Glycobiology 19, 583–591.

Castronovo, V., Colin, C., Parent, B., Foidart, J.M., Lambotte, R., Mahieu, P., 1989. Possible role of human natural anti-Gal antibodies in the natural antitumor defense system. J. Natl. Cancer Inst. 81, 212–216.

Chen, C., Liu, B., Xu, Y., Utkina, N., Zhou, D., Danilov, L., et al., 2016. Biochemical characterization of the novel α-1,3-galactosyltransferase WclR from *Escherichia coli* O3. Carbohydr. Res. 430, 36–43.

Christiane, Y., Aghayan, M., Emonard, H., Lallemand, A., Mahieu, P., Foidart, J.M., 1992. Galactose α1-3 galactose and α-galactose antibody in normal and pathological pregnancies. Placenta 13, 475–487.

Christiansen, D., Milland, J., Mouhtouris, E., Vaughan, H., Pellicci, D.G., McConville, M.J., et al., 2008. Humans lack iGb3 due to the absence of functional iGb3-synthase: implications for NKT cell development and transplantation. PLoS Biol. 6, e172.

D'Alessandro, M., Mariani, P., Lomanto, D., Bachetoni, A., Speranza, V., 2002. Alterations in serum anti-α-galactosyl antibodies in patients with Crohn's disease and ulcerative colitis. Clin. Immunol. 103, 63–68.

Davin, J.C., Malaise, M., Foidart, J., Mahieu, P., 1987. Anti-α-galactosyl antibodies and immune complexes in children with Henoch–Schönlein purpura or IgA nephropathy. Kidney Int. 31, 1132–1139.

Der, C.J., Krontiris, T.G., Cooper, G.M., 1982. Transforming genes of human bladder and lung carcinoma cell lines are homologous to the *ras* genes of Harvey and Kirsten sarcoma viruses. Proc. Natl. Acad. Sci. U.S.A. 79, 3637–3640.

Deriy, L., Chen, Z.C., Gao, G.P., Galili, U., 2002. Expression of α-gal epitopes on HeLa cells transduced with adenovirus containing α1,3galactosyltransferase cDNA. Glycobiology 12, 135–144.

Endo, A., Rothfield, L., 1969. Studies of a phospholipid-requiring bacterial enzyme. I. Purification and properties of uridine diphosphate galactose:lipopolysaccharide α-3 galactosyltransferase. Biochemistry 8, 3500–3507.

Etienne-Decerf, J., Malaise, M., Mahieu, P., Winand, R., 1987. Elevated anti-α-galactosyl antibody titres. A marker of progression in autoimmune thyroid disorders and in endocrine ophthalmopathy? Acta Endocrinol. 115, 67–74.
Fullmer, J., Lindall, A., Bahn, R., Mariash, C.N., 2005. The possible contribution of anti-Gal to Graves' disease. Thyroid 15, 1239–1243.
Gabrielli, A., Candela, M., Pisani, E., Hermann, K., Wieslander, J., Danieli, G., et al., 1992. Antibodies against terminal galactosyl (α1-3) galactose epitopes in systemic sclerosis (scleroderma). Clin. Exp. Rheumatol. 10, 31–36.
Galili, U., 1989. Abnormal expression of α-galactosyl epitopes in man: a trigger for autoimmune processes? Lancet 2, 358–361.
Galili, U., 1999. Graves' disease as a model for anti-Gal involvement in autoimmune diseases pathogenesis. Subcell. Biochem. 32, 339–360.
Galili, U., Korkesh, A., Kahane, I., Rachmilewitz, A., 1983. Demonstration of a natural anti-galactosyl IgG antibody on thalassemic red blood cells. Blood 61, 1258–1264.
Galili, U., Rachmilewitz, E.A., Peleg, A., Flechner, I., 1984. A unique natural human IgG antibody with anti-α-galactosyl specificity. J. Exp. Med. 160, 1519–1531.
Galili, U., Flechner, I., Kniszinski, A., Danon, D., Rachmilewitz, E.A., 1986a. The natural anti-α-galactosyl IgG on human normal senescent red blood cells. Br. J. Haematol. 62, 317–324.
Galili, U., Clark, M.R., Shohet, S.B., 1986b. Excessive binding of the natural anti-α-galactosyl IgG to sickle red cells may contribute to extravascular cell destruction. J. Clin. Investig. 77, 27–33.
Galili, U., Clark, M.R., Shohet, S.B., Buehler, J., Macher, B.A., 1987. Evolutionary relationship between the anti-Gal antibody and the Galα1-3Gal epitope in primates. Proc. Natl. Acad. Sci. U.S.A. 84, 1369–1373.
Galili, U., Shohet, S.B., Kobrin, E., Stults, C.L.M., Macher, B.A., 1988a. Man, apes, and Old World monkeys differ from other mammals in the expression of α-galactosyl epitopes on nucleated cells. J. Biol. Chem. 263, 17755–17762.
Galili, U., Mandrell, R.E., Hamadeh, R.M., Shohet, S.B., Griffiss, J.M., 1988b. Interaction between human natural anti-α-galactosyl immunoglobulin G and bacteria of the human flora. Infect. Immun. 56, 1730–1737.
Galili, U., Anaraki, F., Thall, A., Hill-Black, C., Radic, M., 1993. One percent of circulating B lymphocytes are capable of producing the natural anti-Gal antibody. Blood 82, 2485–2493.
Galili, U., Wigglesworth, K., Abdel-Motal, U.M., 2007. Intratumoral injection of α-gal glycolipids induces xenograft-like destruction and conversion of lesions into endogenous vaccines. J. Immunol. 178, 4676–4687.
Hamadeh, R.M., Jarvis, G.A., Zhou, P., Cotleur, A.C., Griffiss, J.M., 1996. Bacterial enzymes can add galactose α1,3 to human erythrocytes and create a senescence-associated epitope. Infect. Immun. 64, 528–534.
Han, W., Cai, L., Wu, B., Li, L., Xiao, Z., Cheng, J., et al., 2012. The wciN gene encodes an α-1,3-galactosyltransferase involved in the biosynthesis of the capsule repeating unit of *Streptococcus pneumoniae* serotype 6B. Biochemistry 51, 5804–5810.
Higashi, H., Naiki, M., Matuo, S., Okouchi, K., 1977. Antigen of "serum sickness" type of heterophile antibodies in human sera: identification as gangliosides with *N*-glycolylneuraminic acid. Biochem. Biophys. Res. Commun. 79, 388–395.
Kagawa, Y., Takasaki, S., Utsumi, J., Hosoi, K., Shimizu, H., Kochibe, N., et al., 1988. Comparative study of the asparagine-linked sugar chains of natural human interferon-β1 and recombinant human interferon-β1 produced by three different mammalian cells. J. Biol. Chem. 267, 8012–8020.
Kannagi, R., Fukuda, M.N., Hakomori, S., 1982a. A new glycolipid antigen isolated from human erythrocyte membranes reacting with antibodies directed to globo-ZV-tetraosylceramide (globoside). J. Biol. Chem. 257, 4438–4442.
Kannagi, R., Levine, P., Watanabe, K., Hakomori, S.-I., 1982b. Recent studies of glycolipid and glycoprotein profiles and characterization of the major glycolipid antigen in gastric cancer of a patient of blood group genotype pp ("Tj") first studied in 1951. Cancer Res. 42, 5249–5254.
Kielczynski, W., Bartholomeusz, R.K., Harrison, L.C., 1994. Characterization of ganglioside associated with the thyrotropin receptor. Glycobiology 4, 791–796.
Klein, R., Timpl, R., Zanetti, F.R., Plester, D., Berg, P.A., 1989. High antibody levels against mouse laminin with specificity for galactosyl-(α1-3)galactose in patients with inner ear diseases. Ann. Otol. Rhinol. Laryngol. 98, 537–542.
Knobel, M., Umezawa, E.S., Cardia, M.S., Martins, M.J., Correa, M.L., Gianella-Neto, D., et al., 1999. Elevated anti-galactosyl antibody titers in endemic goiter. Thyroid 9, 493–498.
Kooyman, D.L., McClellan, S.B., Parker, W., Avissar, P.L., Velardo, M.A., Platt, J.L., et al., 1996. Identification and characterization of a galactosyl peptide mimetic. Implications for use in removing xenoreactive anti-α-Gal antibodies. Transplantation 61, 851–855.

Lang, J., Zhan, J., Xu, L., Yan, Z., 2006. Identification of peptide mimetics of xenoreactive α-Gal antigenic epitope by phage display. Biochem. Biophys. Res. Commun. 344, 214–220.

Leoni, P., Rupoli, S., Salvi, A., Sambo, P., Cinciripini, A., Gabrielli, A., 1993. Antibodies against terminal galactosyl α(1-3) galactose epitopes in patients with idiopathic myelofibrosis. Br. J. Haematol. 85, 313–319.

Lüderitz, O., Simmons, D.A., Westphal, G., 1965. The immunochemistry of *Salmonella* chemotype VI O-antigens. The structure of oligosaccharides from *Salmonella* group U (o 43) lipopolysaccharides. Biochem. J. 97, 820–826.

Malaise, M.G., Davin, J.C., Mahieu, P.R., Franchimont, P., 1986. Elevated anti-galactosyl antibody titers reflect renal injury after gold or D-penicillamine in rheumatoid arthritis. Clin. Immunol. Immunopathol. 40, 356–364.

Merrick, J.M., Zadarlik, K., Milgrom, F., 1978. Characterization of the Hanganutziu-Deicher (serum-sickness) antigen as gangliosides containing *N*-glycolylneuraminic acid. Int. Arch. Allergy Appl. Immunol. 57, 477–480.

Misrahi, M., Loosfelt, H., Atger, M., Sar, S., Guichon-Mantel, A., Milgrom, E., 1990. Cloning sequencing and expression of human TSH receptor. Biochem. Biophys. Res. Commun. 166, 394–403.

Nagayama, Y., Kaufmann, K.D., Seto, P., Rapoport, B., 1989. Molecular cloning, sequence and functional expression of the cDNA for the human thyrotropin receptor. Biochem. Biophys. Res. Commun. 165, 1184–1190.

Parada, L.F., Tabin, C.J., Shih, C., Weinberg, R.A., 1982. Human EJ bladder carcinoma oncogene is homologue of Harvey sarcoma virus *ras* gene. Nature 297, 474–478.

Parmentier, M., Libert, E., Maehaut, C., Lefort, A., Gerard, C., Perret, J., et al., 1989. Molecular cloning of the thyrotropin receptor. Science 245, 1620–1622.

Petryniak, J., Varani, J., Ervin, P.R., Goldstein, I.J., 1991. Differential expression of glycoproteins containing α-D-galactosyl groups on normal human breast epithelial cells and MCF-7 human breast carcinoma cells. Cancer Lett. 60, 59–65.

Posekany, K.J., Pittman, H.K., Bradfield, J.F., Haisch, C.E., Verbanac, K.M., 2002. Induction of cytolytic anti-Gal antibodies in α-1,3-galactosyltransferase gene knockout mice by oral inoculation with *Escherichia coli O86*:B7 bacteria. Infect. Immun. 70, 6215–6222.

Samraj, A.N., Pearce, O.M., Läubli, H., Crittenden, A.N., Bergfeld, A.K., Banda, K., et al., 2015. A red meat-derived glycan promotes inflammation and cancer progression. Proc. Natl. Acad. Sci. U.S.A. 112, 542–547.

Sandrin, M.S., Vaughan, H.A., Xing, P.X., McKenzie, I.F., 1997. Natural human anti-Gal α(1,3)Gal antibodies react with human mucin peptides. Glycoconj. J. 14, 97–105.

Santer, U.V., DeSantis, R., Hård, K.J., van Kuik, J.A., Vligenthart, J.F., Won, B., et al., 1989. *N*-linked oligosaccharide changes with oncogenic transformation required sialylation of multiantennae. Eur. J. Biochem. 181, 249–260.

Santos, E., Tronick, S.R., Aaronson, S.A., Pulciani, S., Barbacid, M., 1982. T24 human bladder carcinoma oncogene is an activated form of the normal human homologue of BALB- and Harvey-MSV transforming genes. Nature 298, 343–347.

Smith, D.F., Larsen, R.D., Mattox, S., Lowe, J.B., Cummings, R.D., 1990. Transfer and expression of a murine UDP-Gal:β-D-Gal-α1,3-galactosyltransferase gene in transfected Chinese hamster ovary cells. Competition reactions between the α1,3-galactosyltransferase and the endogenous α2,3-sialyltransferase. J. Biol. Chem. 265, 6225–6234.

Springer, G.F., Horton, R.E., 1969. Blood group isoantibody stimulation in man by feeding blood group-active bacteria. J. Clin. Investig. 48, 1280–1291.

Stehelin, D., Varmus, H.E., Bishop, J.M., Vogt, P.K., 1976. DNA related to the transforming gene(s) of avian sarcoma viruses is present in normal avian DNA. Nature 260, 170–173.

Sung, S.J., Sweely, C.C., 1979. The structure of canine intestinal trihexosylceramide. Biochim. Biophys. Acta 525, 295–298.

Takeuchi, M., Takasaki, S., Miyazaki, H., Kato, T., Hoshi, S., Kochibe, N., et al., 1988. Comparative study of the asparagine-linked sugar chains of human erythropoietins purified from urine and the culture medium of recombinant Chinese hamster ovary cells. J. Biol. Chem. 263, 3657–3663.

Tangvoranuntakul, P., Gagneux, P., Diaz, S., Bardor, M., Varki, N., Varki, A., et al., 2003. Human uptake and incorporation of an immunogenic nonhuman dietary sialic acid. Proc. Natl. Acad. Sci. U.S.A. 100, 12045–12050.

Taylor, S.G., McKenzie, I.F., Sandrin, M.S., 2003. Characterization of the rat α(1,3)galactosyltransferase: evidence for two independent genes encoding glycosyltransferases that synthesize Galα(1,3)Gal by two separate glycosylation pathways. Glycobiology 13, 327–337.

Taylor, R.E., Gregg, C.J., Padler-Karavani, V., Ghaderi, D., Yu, H., Huang, S., et al., 2010. Novel mechanism for the generation of human xeno-autoantibodies against the non-human sialic acid *N*-glycolylneuraminic acid. J. Exp. Med. 207, 1637–1646.

Teneberg, S., Lönnroth, I., Torres Lopez, J.F., Galili, U., Olwegard Halvarsson, M., Angstrom, J., et al., 1996. Molecular mimicry in the recognition of glycosphingolipids by Galα3Galß4GlcNAcß-binding *Clostridium difficile* toxin A, human natural anti-α-galactosyl IgG and the monoclonal antibody Gal-13: characterization of a binding-active human glycosphingolipid, non-identical with the animal receptor. Glycobiology 6, 599–609.
Thall, A., Etienne-Decerf, J., Winand, R.J., Galili, U., 1991. The α-galactosyl epitope on mammalian thyroid cells. Acta Endocrinol. 124, 692–699.
Varki, A., 2010. Colloquium paper: uniquely human evolution of sialic acid genetics and biology. Proc. Natl. Acad. Sci. U.S.A. 107, 8939–8946.
Whitfield, C., Richards, J.C., Perry, M.B., Clarke, B.R., MacLean, L.L., 1991. Expression of two structurally distinct D-galactan O antigens in the lipopolysaccharide of *Klebsiella pneumoniae* serotype O1. J. Bacteriol. 173, 1420–1431.
Winand, R.J., Kohn, L.D., 1975. Thyrotropin effects on thyroid cells in culture: effect of trypsin on the thyrotropin receptor and on thyrotropin-mediated cyclic 3′-5′ AMP changes. J. Biol. Chem. 250, 6534–6540.
Winand, R.J., Anaraki, F., Etienne-Decerf, J., Galili, U., 1993. Xenogeneic thyroid-stimulating hormone-like activity of the human natural anti-Gal antibody. Interaction of anti-Gal with porcine thyrocytes and with recombinant human thyroid-stimulating hormone receptors expressed on mouse cells. J. Immunol. 151, 3923–3934.
Winand, R.J., Winand-Devigne, J., Meurisse, M., Galili, U., 1994. Specific stimulation of Graves' disease thyrocytes by the natural anti-Gal antibody from normal and autologous serum. J. Immunol. 153, 1386–1395.
Zhan, J., Xia, Z., Xu, L., Yan, Z., Wang, K., 2003. A peptide mimetic of Gal-α1,3-Gal is able to block human natural antibodies. Biochem. Biophys. Res. Commun. 308, 19–22.
Zhu, A., Hurst, R., 2002. Anti-*N*-glycolylneuraminic acid antibodies identified in healthy human serum. Xenotransplantation 9, 376–381.

SECTION 3

ANTI-GAL AS FRIEND

CHAPTER

9

Anti-Gal-Mediated Amplification of Viral Vaccine Efficacy

OUTLINE

INTRODUCTION

When a viral vaccine, such as influenza (flu) vaccine, is injected intramuscular into the arm of the vaccinee, nothing happens at the injection site other than uptake of the vaccinating proteins via random pinocytosis by antigen-presenting cells (APC), such as dendritic cells and macrophages. The APC internalizing the vaccine transport the vaccinating protein molecules to the regional lymph nodes, process, and present the antigenic peptides on class I and class II MHC molecules. These processed peptides interact with the corresponding T cell receptors (TCR) and activate the peptide specific CD8+ and CD4+ T cells, respectively (Banchereau and Steinman, 1998; Zinkernagel et al., 1997). The clinical experience with several microbial vaccines such as hepatitis C virus vaccine, (Ahlén and Frelin, 2016), HIV vaccine (Goulder and Watkins, 2004; Lewis et al., 2014), and influenza virus vaccine (Chang et al., 2012) has indicated that their immunogenicity is suboptimal. One of the major prerequisites for a successful virus vaccine is effective uptake of the vaccinating material by APC. This is especially

The Natural Anti-Gal Antibody as Foe Turned Friend in Medicine
http://dx.doi.org/10.1016/B978-0-12-813362-0.00009-9

of significance with vaccine injections that contain small amounts of the vaccinating material such as flu vaccines, which contain only approximately 15–30 μg of each of the viruses' hemagglutinin (HA). Because the vaccinating proteins carry no markers ("labels") that target them to APC, the random pinocytosis of the vaccine results in uptake of only a small proportion of the vaccinating material by the APC and further processing and presentation of immunogenic peptides for T cell activation.

It is well established that immunogenicity of vaccines can be markedly increased by formation of immune complexes with the corresponding IgG antibody. Such antibody/vaccine immune complexes are targeted to APC for effective uptake because of the high-affinity interaction between the Fc portion of the immunocomplexed (opsonizing) IgG molecule and Fcγ receptors (FcγR) on the cell membrane of dendritic cells and macrophages (Ravetch and Clynes, 1998; Schuurhuis et al., 2002; Regnault et al., 1999). This Fc/FcγR interaction generates a signal that stimulates APC to internalize the antibody opsonized vaccine, resulting in marked increase in uptake, processing, and presentation of vaccines by APC. Immune response amplification following immunization with immunocomplexed vaccines has been known for more than 50 years. This amplification has been reported in vaccination studies with tetanus toxoid (Stoner and Terres, 1963; Manca et al., 1991; Fanger et al., 1996), hepatitis B envelope antigen (Celis and Chang, 1984), Eastern equine encephalitis virus (Houston et al., 1977), simian immunodeficiency virus (Villinger et al., 2003), and Trypanosoma (Stager et al., 2003). Immunization with each of these vaccines immunocomplexed with the corresponding antibody increased the immune response by 10- to 1000-fold in comparison with the same vaccine that was not immunocomplexed.

No large vaccination studies have been reported with vaccines immunocomplexed with their specific antibodies, because the need for purifying the corresponding antibody, complexing it *in vitro* with the vaccine and keeping the antibody active, all raise quality control regulatory issues that make the vaccine production cumbersome and difficult to standardize. In addition, antibodies from nonhuman species (e.g., monoclonal antibodies) would have to be humanized to prevent antibody production to the immunocomplexed antibody. Moreover, immunocomplexed antibodies may mask immunodominant peptide epitopes on the vaccinating antigen, which are important for specific T cell and B cell activation. These difficulties in using immunocomplexed vaccines may be overcome by *in vivo* formation of immune complexes between the natural anti-Gal antibody and vaccinating glycoproteins that carry α-gal epitopes (e.g., viral envelope glycoproteins) (Galili and LaTemple, 1997). This chapter describes the harnessing of anti-Gal for increasing immunogenicity of flu vaccine and of HIV-gp120 vaccine in anti-Gal producing α1,3galactosyltransferase (α1,3GT) knockout mice (GT-KO mice). In addition, the chapter describes a method for increasing immunogenicity of any vaccinating protein by formation of a fusion protein vaccine with gp120 of HIV engineered to carry multiple α-gal epitopes. This chapter further demonstrates the mechanism involved in anti-Gal-mediated increased immunogenicity of ovalbumin (OVA) as an experimental vaccine model in GT-KO mice. The method for amplification vaccine immunogenicity is likely to be feasible in humans because anti-Gal is produced in large amounts (~1% of immunoglobulins) in all humans that are not severely immunocompromised (Galili et al., 1984, 1985).

ANTI-GAL-MEDIATED AMPLIFICATION OF INFLUENZA VIRUS VACCINE IMMUNOGENICITY

Hypothesis on anti-Gal-mediated targeting of virus vaccines to antigen-presenting cells

Improving vaccine efficacy is of significance in flu vaccine development because of the suboptimal efficacy of present flu vaccines, in particular, in elderly populations in which 25%–50% of individuals may contract the disease during the flu season (Couch et al., 1997; Webster, 2000; Katz et al., 2004; Chang et al., 2012). In addition, an effective flu vaccine is required because of the constant danger of a flu pandemic. In the absence of an effective vaccine, such a pandemic is anticipated to cause damage and mortality even higher than that of the 1918 flu pandemic (Khanna et al., 2009). As discussed above, increased efficacy of flu vaccine may be achieved by active targeting of vaccinating inactivated influenza virus, or the influenza virus main envelope glycoprotein-HA subunit vaccines to dendritic cells and macrophages.

The HA molecule of the influenza virus envelope is a glycoprotein with 6–8 carbohydrate chains capped with *N*-acetyllactosamines (LacNAc epitopes) (Galβ1-4GlcNAc-R), as in the center carbohydrate chain in Fig. 1 (Matsumoto et al., 1983; Keil et al., 1985). α-Gal epitopes can be synthesized on the carbohydrate chains of HA (right carbohydrate chain in Fig. 1) of the vaccinating inactivated virus or on HA in the "subunit" vaccine by recombinant (r)α1,3GT, which links α1-3 the terminal galactose provided by the sugar donor uridine diphosphate galactose (UDP-Gal) (Fig. 1) (Henion et al., 1997). It could be hypothesized that immunization with whole virus flu vaccine or with subunit HA flu vaccine, which carry α-gal epitopes will result in *in vivo* formation of immune complexes between anti-Gal and such vaccines, at the vaccination site (Fig. 2). Binding of anti-Gal IgM to these vaccines will further activate the complement system and generate chemotactic complement cleavage peptides such as C5a and C3a, which recruit macrophages and dendritic cells to the vaccination site. This recruitment is further demonstrated in Chapters 10 and 12. The anti-Gal IgG molecules opsonizing the vaccine will target the vaccine for uptake by the recruited APC via the interaction between the Fc portion of anti-Gal bound to the vaccine and FcγR on these APC. This interaction generates a signal that induces the effective uptake of the vaccinating virus or HA bound to the FcγR. Following the uptake of the vaccine, the APC migrate to the regional lymph nodes, process, and present the vaccinating peptides in association with class I and II MHC molecules for effective activation of CD8+ and CD4+ T cells. Currently used flu vaccines lack α-gal epitopes that can actively target them to APC because the virus grown for vaccine preparation is propagated in embryonated chicken eggs or in Vero cells (from African Green monkey), both lacking α1,3GT that synthesizes α-gal epitopes in mammals (Chapters 1 and 2). Thus, the uptake of these vaccines is likely to be suboptimal and to be mediated only by random pinocytosis. This scenario of anti-Gal mediated targeting vaccines to APC is also applicable to other enveloped viruses or their envelope glycoproteins that are engineered to carry α-gal epitopes according to the enzymatic reactions illustrated in Fig. 1 (see studies on HIV-gp120 below).

The study of this hypothesis could be performed in GT-KO mice that lack α-gal epitopes because of the disruption (i.e., knockout) of the α1,3GT gene (Thall et al., 1995). As described in Chapter 5, these mice need to be immunized with xenoglycoproteins (e.g., pig

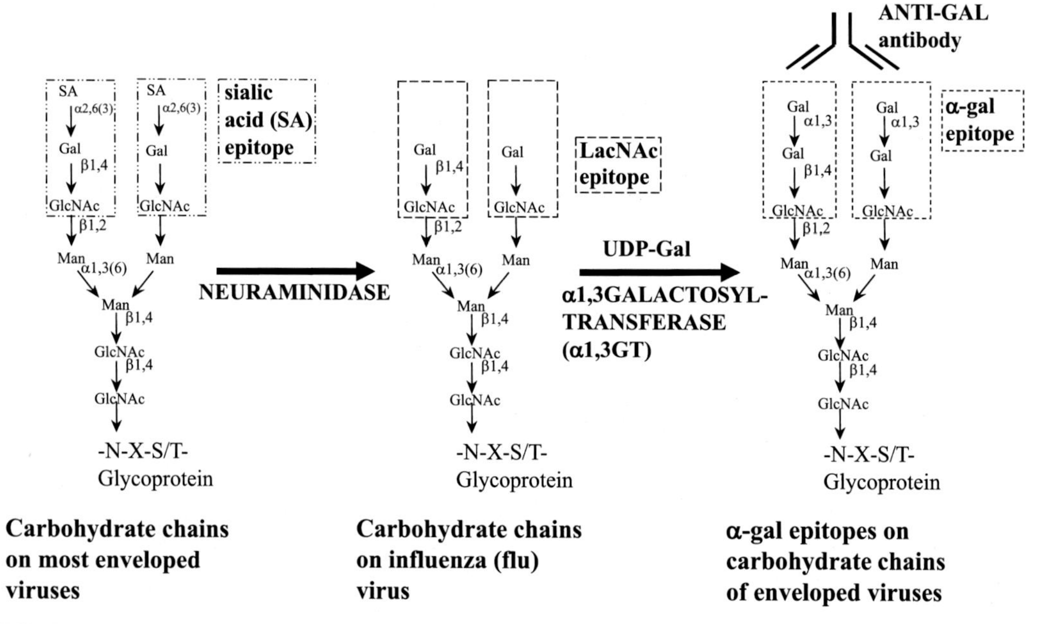

FIGURE 1 Synthesis of α-gal epitopes on viral envelope glycoproteins. Left chain—Carbohydrate chains of the complex type on viral envelope glycoproteins are synthesized on asparagine (N) in amino acid sequences of N-X-S/T-. Many of these carbohydrate chains are capped by sialic acid. Center chain—Sialic acid is removed from the carbohydrate chain by neuraminidase to expose the penultimate Galβ1-4GlcNAc-R called *N*-acetyllactosamine (LacNAc). In influenza virus, the sialic acid is removed by the virus neuraminidase upon arrival of hemagglutinin to the cell surface. Right chain—Incubation of inactivated virus or soluble envelope glycoproteins carrying the desialylated carbohydrate chain with rα1,3GT and with the sugar donor uridine diphosphate galactose (UDP-Gal) results in synthesis of α-gal epitopes (Galα1-3Galβ1-4GlcNAc-R) on the carbohydrate chain. These epitopes readily bind the anti-Gal antibody. *Adapted from Abdel-Motal, U., Wang, S., Lu, S., Wigglesworth, K., Galili, U., 2006. Increased immunogenicity of human immunodeficiency virus gp120 engineered to express Galα1-3Galβ1-4GlcNAc-R epitopes. J. Virol. 80, 6943–6951, with permission.*

kidney membranes homogenate), which carry multiple α-gal epitopes, to simulate the human immune system in production of the anti-Gal antibody.

Demonstration of anti-Gal-mediated increased uptake of influenza virus by antigen-presenting cells

Initial studies on anti-Gal-mediated targeting of influenza virus expressing α-gal epitopes to APC were performed with the influenza virus A/Puerto Rico/8/34-H1N1 (PR8 virus). The virus was propagated to express α-gal epitopes by growth in cells containing active α1,3GT (Galili et al., 1996). As discussed in Chapter 2, the carbohydrate chains on envelope glycoproteins of viruses are synthesized by the host cell glycosylation machinery. Thus, influenza virus produced in Madin–Darby bovine kidney (MDBK) cells and Madin–Darby canine kidney (MDCK) cells have α-gal epitopes on part of their carbohydrate chains because these

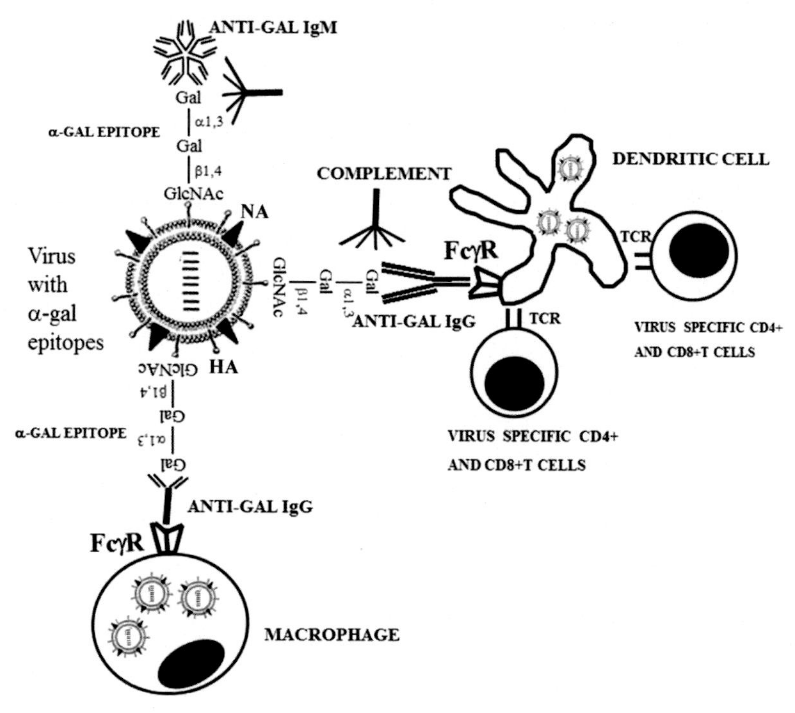

FIGURE 2 Suggested mechanism for amplification of viral vaccine immunogenicity by anti-Gal-mediated targeting of vaccinating virus to APC. Inactivated influenza virus presenting α-gal epitopes is used as vaccine example. Anti-Gal IgM and IgG binding to α-gal epitopes on the vaccinating virus activate the complement system, which generates complement cleavage chemotactic peptides that recruit APC such as dendritic cells and macrophages to the vaccination site. Anti-Gal IgG coating the virus targets it for active uptake of multiple virions by the recruited dendritic cells and macrophages via Fc/Fcγ receptors (FcγR) interaction. These APC transport the internalized virus vaccine to the regional lymph nodes and process the virus antigens. Within the lymph nodes, the APC present the immunogenic virus peptides on class I and class II MHC molecules for the activation of virus-specific CD8+ and CD4+ T cells, respectively. *HA*, hemagglutinin; *NA*, neuraminidase; *TCR*, T cell receptor.

nonprimate mammalian cells have endogenous α1,3GT synthesizing these epitopes. In contrast, PR8 virus propagated in embryonated chicken eggs lacks α-gal epitopes because chicken, like other birds, lacks α1,3GT, an enzyme found only in mammals (see Chapter 2). Uptake of inactivated influenza virus by APC could be assessed by activation of an HA-specific T cell clone by processed peptides presented on the APC. *In vitro* binding of anti-Gal to inactivated PR8 virus presenting α-gal epitopes resulted in increased uptake by APC. This was indicated by ~10-fold increase in subsequent activation of the HA-specific T cell clone by processed peptides presented on the APC in comparison with activation of these HA-specific T cells in the absence of anti-Gal (Galili et al., 1996). In contrast, incubation with anti-Gal of inactivated influenza virus produced in embryonated chicken eggs did not alter the extent of uptake, processing, and presentation of viral peptides by APC in comparison with that in the absence of anti-Gal.

Synthesis of α-gal epitopes on influenza virus by recombinant α1,3galactosyltransferase

Propagation of influenza virus in cells such as MDCK or MDBK results in synthesis of α-gal epitopes only on a proportion of the N-linked carbohydrate chains of HA. This proportion varies from one type of cell to the other. The extent of "capping" the N-linked carbohydrate chains by α-gal epitopes (Fig. 1) depends on the activity of α1,3GT as well as that of sialyltransferase (SiaT), which caps the carbohydrate chains with sialic acid (SA epitope on the left carbohydrate chain in Fig. 1). Both α1,3GT and SiaT reside in the same trans-Golgi compartment and when a glycoprotein such as HA carries the nascent N-linked carbohydrate chain into that compartment, the two enzymes compete for capping with the α-gal or SA (Smith et al., 1990). In most enveloped viruses, SA epitopes are the major epitopes on the carbohydrate chains of envelope glycoproteins. α-Gal epitope also is found on some of the carbohydrate chains, if the virus is propagated in cells with active α1,3GT (Repik et al., 1994; Galili et al., 1996). When HA of influenza virus reaches the surface of the host cell, the enzyme neuraminidase (sialidase), which is also a component of the virus envelope, cleaves the SA (center carbohydrate chain in Fig. 1). The elimination of the virus SA by neuraminidase prevents HA of one virion binding to SA on another virion. This enables the binding of HA only to SA receptors on cells that are to be infected. It was reasonable to assume that there may be a correlation between the number of α-gal epitopes on influenza virus and the ability of anti-Gal to effectively target the vaccinating inactivated influenza virus to APC via Fc/FcγR interaction (Fig. 2). Thus, it was of interest to determine whether it is possible to use rα1,3GT for synthesizing the maximum number of α-gal epitopes on inactivated PR8 virus. This was achieved by synthesizing α-gal epitopes on the N-linked carbohydrate chains of HA on PR8 virus produced in embryonated eggs. This synthesis was catalyzed by rα1,3GT with UDP-Gal as sugar donor, as illustrated in the second enzymatic reaction in Fig. 1 (Henion et al., 1997; Abdel-Motal et al., 2007). Quantification of the *de novo* synthesized α-gal epitopes indicated that ~3000 such epitopes were synthesized on each virion. The engineered PR8 virus carrying α-gal epitopes is referred to as $PR8_{\alpha gal}$.

Increased immunogenicity of $PR8_{\alpha gal}$ in comparison with PR8

The immunogenicity of the vaccinating $PR8_{\alpha gal}$ virus versus that of the parental PR8 lacking α-gal epitopes was evaluated in anti-Gal producing GT-KO mice. The mice received two immunizations in two-week interval with either 1 μg of inactivated $PR8_{\alpha gal}$ (corresponding to 400 hemagglutination units prior to inactivation) or a similar amount of inactivated PR8 virus. The immunization was performed with Ribi adjuvant. Production of anti-PR8 antibodies and activation of T cells against virus antigens were assayed 2 weeks after the second vaccination.

In four of the six mice immunized with $PR8_{\alpha gal}$ inactivated virus, the number of interferon-γ (IFNγ) secreting T cells per 10^6 spleen cells activated by PR8 peptides and assayed in enzyme-linked immunospot (ELISPOT) assay, was ~2200, whereas in mice immunized with PR8 inactivated virus, there were only ~600 IFNγ secreting T cells per 10^6 cells (Abdel-Motal et al., 2007). Intracellular staining for IFNγ analyzed by flow cytometry demonstrated ~21% of positive staining among CD8+ T cells of the four out of six $PR8_{\alpha gal}$ immunized mice versus

~3.5% of CD8+ cells in PR8 immunized mice. Similarly, ~13% of the CD4+ T cells from $PR8_{\alpha gal}$ immunized mice stained positively for intracellular IFNγ versus ~0.1% of CD4+ T cells in PR8 immunized mice (Abdel-Motal et al., 2007). In the remaining two mice immunized with $PR8_{\alpha gal}$, the cellular immune response was the same as that in PR8 immunized mice.

A marked increase in immunogenicity of $PR8_{\alpha gal}$ in comparison with PR8 was observed also in humoral immunity measured by production of antibodies to PR8 which served as solid-phase antigen in ELISA. The titer of anti-PR8 IgG antibodies (defined as serum dilution yielding 1.0 O.D. in ELISA) was ~100-fold higher in $PR8_{\alpha gal}$ immunized GT-KO mice than in the mice immunized with inactivated PR8 virus (Fig. 3A). Similar immunizations in wild-type (WT) mice, which lack the anti-Gal antibody, yielded low titers of anti-PR8 antibodies in both $PR8_{\alpha gal}$ immunized and PR8 immunized mice (Fig. 3B). Thus, in the absence of anti-Gal-mediated targeting of $PR8_{\alpha gal}$ vaccine to APC, the presence of α-gal epitopes on the virus vaccine did not increase immunogenicity of the virus. It is of interest to note that anti-PR8 IgA antibodies were found at high titers (1:800–1:6400) in serum of the four mice that also displayed T cell response to PR8 antigens, whereas mice immunized with PR8 displayed anti-PR8 IgA titers of 1:10–1:50 (Fig. 3C). Accordingly, GT-KO mice immunized with $PR8_{\alpha gal}$ were found to have anti-PR8 IgA antibodies in solutions prepared from lung homogenates, whereas GT-KO mice immunized with PR8 virus had no such antibodies in their lungs (Abdel-Motal et al., 2007).

Protection against challenge with live PR8 virus by $PR8_{\alpha gal}$ vaccine

The efficacy of $PR8_{\alpha gal}$ vaccine was further tested by its ability to protect against intranasal infection with a lethal dose of live PR8 virus. Anti-Gal producing GT-KO mice immunized with inactivated PR8 or $PR8_{\alpha gal}$ virus were studied for resistance to intranasal challenge with the lethal dose of 2000 plaque forming units (PFU) of live PR8 virus. As many as 89% of the mice immunized with inactivated PR8 virus succumbed to the infection and died by Day 10 post challenge. In contrast, only 11% of the mice immunized with inactivated $PR8_{\alpha gal}$ virus died following the challenge with the live virus (Fig. 4A) (Abdel-Motal et al., 2007). WT mice immunized with $PR8_{\alpha gal}$ virus displayed 70% death, indicating that the vaccine elicited no significant elevated immune protection in the absence of the anti-Gal antibody. Survival data on Day 30 were the same as on Day 15 post challenge. In parallel studies, mice immunized with $PR8_{\alpha gal}$ or with PR8 were euthanized 3 days post challenge with live PR8 virus, their lungs homogenized in PBS and supernatants of the homogenates were studied for virus titer by hemagglutination of chicken red blood cells. As shown in Fig. 4B, the titer of the virus in the lungs of PR8 immunized GT-KO mice was 10- to 100-fold higher than that measured in lungs of GT-KO mice immunized with $PR8_{\alpha gal}$ virus.

Overall, these observations suggest that immunization with influenza virus presenting multiple α-gal epitopes ($PR8_{\alpha gal}$) induces a protective immune response that is much higher than that of vaccinating virus lacking α-gal epitopes (PR8). This increase in the immune response to the vaccine seems to be the result of several sequential steps described in Fig. 2: Binding of anti-Gal to α-gal epitopes on the vaccinating virus opsonizes the vaccine and provides a specific "label" that induces effective uptake of the vaccinating virus by APC. Activation of the complement system by this antigen/antibody interaction further generates

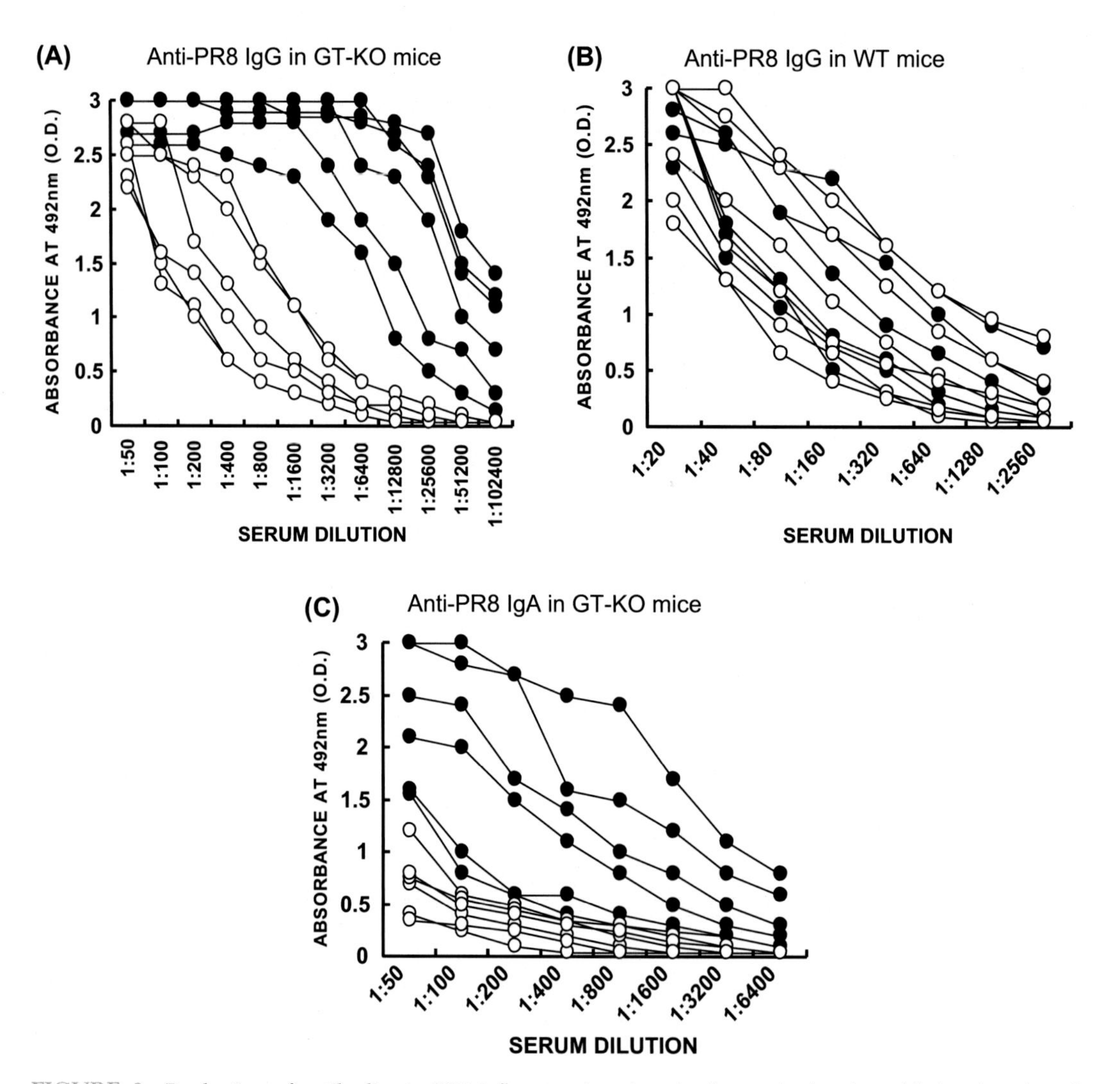

FIGURE 3 Production of antibodies to PR8 influenza virus in mice immunized twice with 1 μg inactivated $PR8_{\alpha gal}$ virus (●), or with inactivated PR8 virus (○). Antibody production was evaluated by ELISA with PR8 virus attached to ELISA wells as solid-phase antigen. (A) Anti-PR8 IgG production in GT-KO mice. (B) Anti-PR8 IgG production in WT mice. (C) Anti-PR8 IgA production in GT-KO mice (n=6 per group). *Reprinted from Abdel-Motal, U.M., Guay, H.M., Wigglesworth, K., Welsh, R.M., Galili, U., 2007. Immunogenicity of influenza virus vaccine is increased by anti-Gal-mediated targeting to antigen-presenting cells. J. Virol. 81, 9131–9141, with permission.*

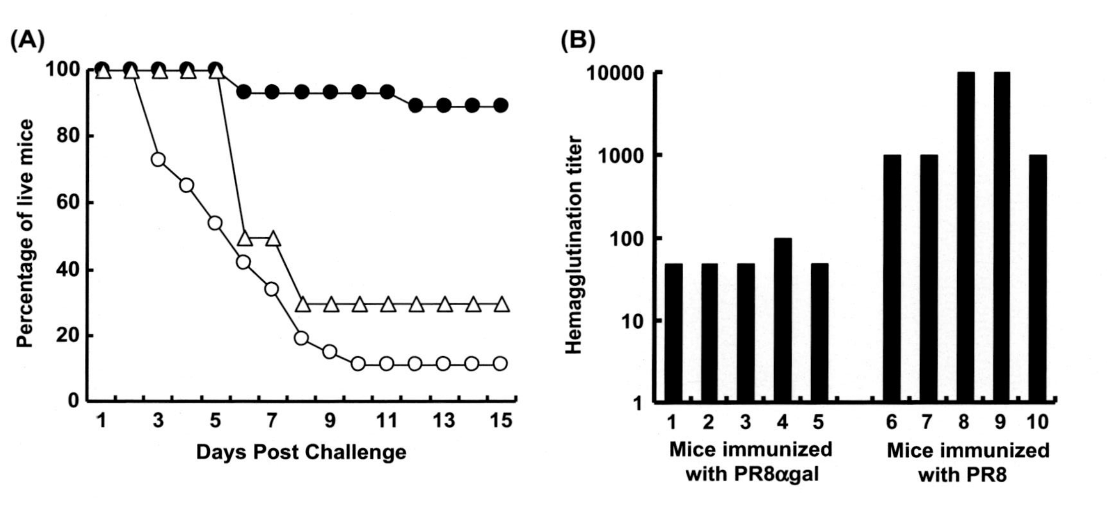

FIGURE 4 Results of intranasal challenge of GT-KO mice and WT mice with a lethal dose of PR8 virus in mice immunized with inactivated PR8 or $PR8_{\alpha gal}$ virus. (A) Survival of mice immunized twice with 1 μg inactivated PR8 vaccine (○) in GT-KO mice, $PR8_{\alpha gal}$ (●) vaccine in GT-KO mice, or with $PR8_{\alpha gal}$ (Δ) vaccine in WT mice and challenged with 2000 PFU of live PR8 in 50 μL (n = 25 per group). Survival data are presented as percentage of live mice at various days following the challenge. Survival data on Day 30 were as those on Day 15 after challenge. (B) PR8 virus titers in lungs of GT-KO mice, 3 days after challenge with live virus. The virus titers were assayed in supernatants of lung homogenates from the immunized mice by hemagglutination of chicken red blood cells (n = 5 per group). *Adapted from Abdel-Motal, U.M., Guay, H.M., Wigglesworth, K., Welsh, R.M., Galili, U., 2007. Immunogenicity of influenza virus vaccine is increased by anti-Gal-mediated targeting to antigen-presenting cells. J. Virol. 81, 9131–9141, with permission.*

chemotactic complement cleavage peptides that recruit APC to the vaccination site. Anti-Gal targeting of the vaccinating virus to the recruited APC is mediated by interaction between the Fc portion of the virus opsonizing antibody and FcγR on APC. The active uptake of the vaccine results in internalization of much higher number of vaccinating virions than the uptake by random pinocytosis of virions lacking α-gal epitopes. Moreover, endocytosis of such immune complexes induces maturation of the APC and improves their ability to present the processed immunogenic peptides (Ravetch and Clynes, 1998; Regnault et al., 1999; Schuurhuis et al., 2002). Within the endosome, the vaccinating virus undergoes proteolytic digestion, and its proteins are degraded by proteasomes into immunogenic peptides that are presented on class I MHC molecules for engaging TCR on CD8+ T cells. Such interaction activates these T cells into becoming cytotoxic T cells (CTL), which can kill virus infected target cells. Parallel TCR engaging immunogenic peptides on class II MHC molecules activate CD4+ T helper (Th) cells that provide help to B cells, which produce antibodies that neutralize the virus. This immune response is much more intense and faster than that of the immune system immunized with viruses lacking α-gal epitopes and, thus, can stop progression of viral infections in early stages of the infection.

An inherent difficulty in providing a long-term vaccine protection against influenza virus infection is the antigenic variation of the virus because of changes in both HA and neuraminidase sequences, which may occur from one infection season to the other. In contrast, the extracellular domain of the minor, virus-coded M2 protein is nearly invariant in all influenza A strains (Fiers et al., 2004). Because the M2 extracellular domain is comprised only of 24 amino acids, its immunogenicity as vaccine is very limited. Studies using vaccine of a fusion protein between M2 extracellular domain and hepatitis B virus core protein reported an increase in immunogenicity of this M2 peptide, as indicated by the production of neutralizing antibodies against influenza virus (Neirynck et al., 1999). In view of the high efficacy of anti-Gal-mediated targeting of PR8 carrying α-gal epitopes to APC, as described above, it would be of interest to determine whether fusion proteins between HA carrying α-gal epitopes and M2 extracellular domain may result in elevated immune response against M2, thereby providing protection against influenza virus with HA other than the immunizing one. Studies described below demonstrate a marked increase in immunogenicity of HIV recombinant P24 core (capsid) protein fused to gp120 carrying α-gal epitopes. These observations suggest that fusion vaccines between glycoproteins such as HA or gp120 carrying α-gal epitopes and nonglycosylated viral proteins, or other proteins, may elicit an extensive increase of immunogenicity of nonglycosylated proteins because of the effective anti-Gal-mediated targeting of the vaccines to APC.

INCREASED IMMUNOGENICITY OF HIV gp120 CARRYING α-GAL EPITOPES

Several viral vaccines are soluble protein or glycoprotein molecules that are prepared as subunit vaccines from envelope glycoproteins (e.g., flu vaccines prepared from HA produced by detergent disruption of influenza virus envelope proteins). Some vaccines are prepared as recombinant proteins produced in various expression systems. Thus, it was of interest to determine whether anti-Gal-mediated targeting to APC of soluble glycoproteins can increase the immunogenicity of such vaccines, in addition to the amplification of immunogenicity of inactivated whole virus vaccines, described above. The glycoprotein gp120 of HIV was chosen as a model for soluble vaccinating glycoproteins carrying α-gal epitopes. The study of this glycoprotein is of interest also because many of the studies on recombinant protein and DNA HIV vaccines in primate models or in clinical trials were reported to have insufficient or suboptimal efficacy in eliciting a sterilizing protective immune response against infection with HIV (Munier et al., 2011). The HIV envelope glycoprotein gp120 is the major component of the virus envelope, and it has 24 N-linked carbohydrate chains, of which 13–16 chains are of the complex type that are capped with SA (left carbohydrate chain in Fig. 1). The rest carbohydrate chains are of the high mannose type (Mizuochi et al., 1988, 1990; Leonard et al., 1990). The multiple carbohydrate chains on gp120 seem to contribute to the protection of the virus. Because they are hydrophilic, the carbohydrate chains protrude from gp120 molecule like the quills of a porcupine. The carbohydrate chains contribute to the protection of HIV against binding of neutralizing antibodies by forming a "glycan fence" or "glycan shield" on the virus (Wei et al., 2003). The increase in envelope carbohydrate chains on mutating gp120 in isolated virus clones of HIV patients further supports the notion that these carbohydrate chains protect the virus against the host immunologic assaults (Wei et al., 2003; Crooks et al., 2015). Much of this protection may be attributed to the electrostatic repulsion (ζ (zeta)-potential) between the multiple negative charges of SA units on gp120 and similar negative charges of SA on cells or on antibody molecules (Fig. 5). The conversion of SAs on gp120 into α-gal epitopes can alter the protective effects of the multiple N-linked carbohydrate chains on this glycoprotein into higher immunogenicity by anti-Gal-mediated targeting of the glycoprotein to APC, thereby increasing the immune response to $gp120_{\alpha gal}$ vaccine and the protection against HIV infections.

Synthesis of α-gal epitopes on gp120

α-Gal epitopes synthesis on gp120 can be achieved by enzymatic reaction with rα1,3GT as described above for influenza virus vaccine (Fig. 1). However, in contrast to envelope glycoproteins on influenza virus, which lack SA because of virus neuraminidase activity, envelope glycoproteins of most viruses carry SA that caps their carbohydrate chains within the trans-Golgi compartment (left carbohydrate chain in Fig. 1). The enzyme neuraminidase purified from *Vibrio cholera*, or from other sources, effectively removes SA from the viral glycoprotein (first enzymatic reaction in Fig. 1) and enables synthesis of α-gal epitopes by rα1,3GT on the carbohydrate chains devoid of sialic acid to generate $gp120_{\alpha gal}$ (second enzymatic reaction in Fig. 1). This two-step enzymatic reaction of sialic acid elimination and synthesis of α-gal epitopes on gp120 could be performed in one solution containing neuraminidase, rα1,3GT and UDP-Gal, as well as Mn^{++} ions, which are required for α1,3GT catalytic activity (Abdel-Motal et al., 2006). The recombinant gp120 was of the HIV_{BAL} strain and was produced in CHO cells that were transformed with the corresponding codon-optimized *env* gene. The synthesis of α-gal epitopes on gp120 enabled its subsequent isolation from the rest of reaction mixture components on an affinity agarose column of *Bandeiraea* (*Griffonia*) *simplicifolia* IB4 (BS lectin), which interacts specifically with α-gal epitopes (Wood et al., 1979). This purification process indicated that >97% of the gp120 was converted into $gp120_{\alpha gal}$, which readily bound anti-Gal (Abdel-Motal et al., 2006).

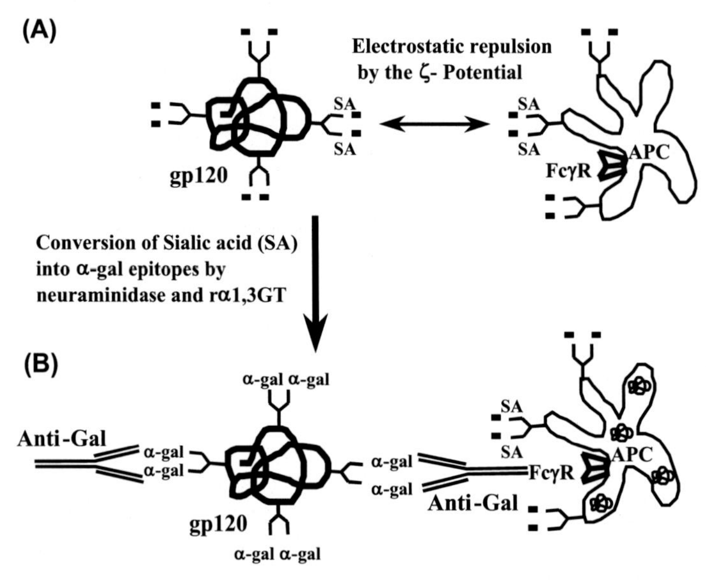

FIGURE 5 Anti-Gal-mediated targeting of gp120 to APC. (A) Uptake of gp120 by APC is greatly diminished because of the electrostatic repulsion (ζ [zeta]-potential) between the negative charges of sialic acids on gp120 and on APC. (B) Synthesis of α-gal epitopes on the carbohydrate chains of gp120 ($gp120_{\alpha gal}$) eliminates electrostatic repulsion and enables formation of immune complexes between $gp120_{\alpha gal}$ and the natural anti-Gal IgG antibody. Many of these immune complexes are effectively targeted for uptake by APC because of the binding of the immunocomplexed anti-Gal Fc portion to FcγR on APC.

Increased anti-gp120 antibody production following immunization with $gp120_{\alpha gal}$

The analysis of $gp120_{\alpha gal}$ immunogenicity versus that of gp120 was performed in anti-Gal producing GT-KO mice. These mice were immunized subcutaneously with 5 μg gp120 or $gp120_{\alpha gal}$ and Ribi adjuvant. These injections were repeated after 2 weeks, and anti-gp120 immune response was evaluated 17 days after the second injection. It was assumed that the immune complexes of anti-Gal/gp120 $_{\alpha gal}$ will bind via the Fc portion of anti-Gal to FcγR on dendritic cells and macrophages, thereby actively targeting the vaccine to APC. As argued above, it was further assumed that this binding is likely to induce differentiation and maturation of these APC into professional APC that present gp120 peptides, both on class I and class II MHC molecules for effective activation of gp120-specific CD8+ and CD4+ T cells, respectively (Clynes et al., 1998; Regnault et al., 1999; Schuurhuis et al., 2002; Rafiq et al., 2002).

Anti-gp120 antibody production was evaluated by ELISA with gp120 as solid-phase antigen and found to be ~200-fold higher in mice immunized with $gp120_{\alpha gal}$ in comparison with that measured in mice immunized with gp120 (Abdel-Motal et al., 2006). In contrast, immunization of WT mice (incapable of producing anti-Gal) resulted in production of anti-gp120 antibodies at low titers, both with $gp120_{\alpha gal}$ and gp120 as vaccinating antigens, further demonstrating the significance of anti-Gal in targeting $gp120_{\alpha gal}$ to APC. The produced anti-gp120 antibodies were further studied for their ability to neutralize HIV-1 strain MN (a strain that is convenient for manipulation in the lab) and thus, prevent infection of host cells. The neutralizing activity of these antibodies was evaluated by determining the extent of killing inhibition of the human T cell lymphoma MT-2 cells by the virus (Montefiori et al., 1996; Wang et al., 2005).

Whereas sera from gp120 immunized GT-KO mice displayed no significant neutralizing activity (Abdel-Motal et al., 2006), sera from the mice immunized with gp120$_{\alpha gal}$ displayed a very effective neutralization activity, as that of serum containing anti-HIV neutralizing antibodies, from a rabbit receiving multiple immunizations with gp120 (Wang et al., 2005). These data suggest that immunization of humans with gp120$_{\alpha gal}$ vaccine may induce the production of protective antibodies at high titers. These antibodies may contribute to preventing the progression of the infective process by neutralizing HIV released from infected cells.

Increased T cell response following immunization with gp120$_{\alpha gal}$

The gp120-specific T cell response following immunization with gp120$_{\alpha gal}$ was determined by quantification of IFNγ secreting cells among splenocytes in an ELISPOT assay with dendritic cells pulsed with gp120. The number of IFNγ secreting T cells in gp120$_{\alpha gal}$ immunized mice was ~15-fold higher than that in gp120 immunized mice (332 spots/10^6 cells and 23 spots/10^6 cells, respectively) (Abdel-Motal et al., 2006). These findings suggest that T cell activation against the gp120 peptides was much higher in gp120$_{\alpha gal}$ immunized mice than in gp120 immunized mice. Both anti-gp120 antibody and ELISPOT data demonstrate extensive amplification of the humoral and cellular immune responses against vaccinating viral antigens engineered to carry α-gal epitopes and which are effectively targeted to APC by the anti-Gal antibody.

TARGETING OF LOW IMMUNOGENICITY HIV PROTEINS BY FUSION WITH GP120$_{\alpha GAL}$

The envelope glycoprotein gp120 of HIV is capable of mutating during infection in humans. These mutations enable the virus to evade the detrimental effect of neutralizing antibodies. For this reason, vaccination only with gp120 does not induce a sufficient immune resistance to HIV infections in large populations (Goulder and Watkins, 2004; Lewis et al., 2014). In contrast, HIV internal proteins, such as tat, rev, p17, or p24, do not mutate in course of HIV infection and thus may serve as vaccines that elicit a cellular immune response, which can destroy HIV infected cells. However, because of poor targeting to APC, immunogenicity of these proteins may be low. In addition, these proteins are not glycosylated and thus, cannot be engineered to carry α-gal epitopes. Because gp120 is highly glycosylated and can be engineered by rα1,3GT to carry multiple α-gal epitopes, it was of interest to determine whether a vaccine made of a fusion protein between gp120$_{\alpha gal}$ and one of the nonglycosylated, nonmutating internal HIV proteins can increase the immunogenicity of such internal proteins. The core (capsid) protein p24 was chosen as a model protein for this analysis.

Production and immunogenicity of gp120$_{\alpha gal}$/p24 vaccine

A fusion protein gp120/p24 was produced by ligation of the gene regions of *env* coding for gp120 and of *gag* coding for p24 (Fig. 6A) (Abdel-Motal et al., 2010). The fused gene was inserted into a plasmid containing a strong promoter (cytomegalovirus), and the fusion product gp120/p24 was produced by transient transfection of 293 cells. The fusion protein secreted from the transfected cells was precipitated from the culture medium with ammonium sulfate. The fusion protein gp120/p24 was converted into gp120$_{\alpha gal}$/p24 vaccine carrying multiple

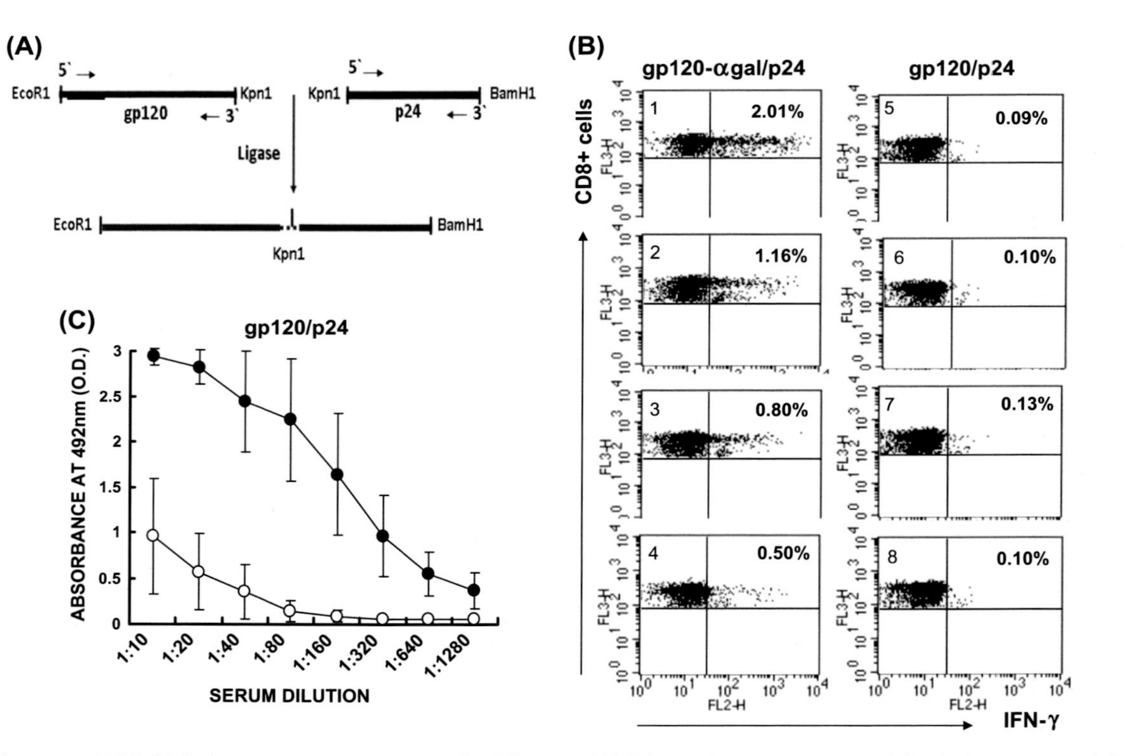

FIGURE 6 Studies on gp120/p24 fusion protein as vaccine in GT-KO mice. (A) Schematic representation of the fusion gene used for production of gp120/p24. The p24 gene was fused to the C-terminus of gp120 to keep the codon-optimized t-PA leader signal proximally up stream of the gp120 gene. (B) Measurement by flow cytometry of intracellular staining of IFNγ production in CD8+ T cells among splencocytes from mice immunized with $gp120_{\alpha gal}$/p24 (left panels, mice #1—#4) and from mice immunized with gp120/p24 (right panels, mice #5—#8) coincubated with APC pulsed with p24 peptides. (C) Production of anti-gp120 IgG antibodies in GT-KO mice immunized with $gp120_{\alpha gal}$/p24 (●), or with gp120/p24 (○), as measured by ELISA with gp120 solid-phase antigen. Mean ± Standard Deviation of results from five mice per group. *Reprinted from Abdel-Motal, U.M., Wang, S., Awad, A., Lu, S., Wigglesworth, K., Galili, U., 2010. Increased immunogenicity of HIV-1 p24 and gp120 following immunization with gp120/p24 fusion protein vaccine expressing α-gal epitopes. Vaccine 28, 1758–1765, with permission.*

α-gal epitopes by incubation with neuraminidase, rα1,3GT, and UDP-Gal, as performed above for production of $gp120_{\alpha gal}$ (Fig. 1).

Anti-Gal producing GT-KO mice received in two-week interval two subcutaneous immunizations of gp120/p24 or $gp120_{\alpha gal}$/p24 at 5 μg per injection with Ribi adjuvant. The splenocytes of the mice were obtained 2 weeks after the second injection and subjected to analysis of T cells specific to p24 by incubation with the p24 immunodominant peptide $p24_{189–207}$ (Qiu et al., 1999), which pulsed APC in the splenocytes mixture. Activation of the p24-specific T cells was determined by ELISPOT, which measured IFNγ secretion (Abdel-Motal et al., 2010). The number of T cells secreting IFNγ among lymphocytes from $gp120_{\alpha gal}$/p24 immunized mice was ~12-fold higher than the number of these cells in gp120/p24 immunized mice. Flow cytometry analysis of p24-specific CD8+ T cells by intracellular staining of IFNγ also demonstrated 5- to 20-fold higher number of such T cells in mice immunized with $gp120_{\alpha gal}$/p24 in comparison with mice immunized with gp120/p24 (Fig. 6B). A fifth mouse displaying an increase of twofold in CD8+ T cells producing IFNγ was observed but not included in Fig. 6B because of limited space (Abdel-Motal et al., 2010). On average, an increase of 10-fold in gp120-specific T cells was observed following immunization with the fusion vaccine carrying α-gal epitope in comparison with mice receiving the fusion protein vaccine lacking α-gal epitopes. The increased immunogenicity of gp120 in $gp120_{\alpha gal}$/p24 vaccine was further demonstrated in ELISA with gp120 as solid-phase antigen. A ~30-fold increase in anti-gp120 antibody titer was measured in serum of mice immunized with $gp120_{\alpha gal}$/p24 in comparison with mice receiving gp120/p24 vaccine (Fig. 6C). These observations suggest that the use of a fusion protein vaccine in which only gp120 portion carried α-gal epitopes resulted in anti-Gal-mediated increase in immunogenicity of both gp120 and p24, although the latter protein within the fusion protein vaccine lacks carbohydrate chains (Abdel-Motal et al., 2010).

Overall, the data suggest that $gp120_{\alpha gal}$ can serve as an effective platform for targeting other viral proteins fused with it because the enzymatic engineering of this glycoprotein results in the synthesis of 30–40 α-gal epitopes per molecule (each carbohydrate chain has 2–4 branches). Fusion of internal virus matrix or core nonglycosylated proteins such as nef, tat, and rev may result in similar enhancement of both humoral and cellular immune responses to candidate HIV vaccines. This proposed strategy of fusion proteins composed of envelope glycoproteins carrying α-gal epitopes and nonglycosylated internal proteins of a given virus may also be useful in preparing recombinant vaccinating fusion glycoproteins that elicit an effective protective immune response by B cells and T cells against other viral infections, as well as other microbial infections.

DEMONSTRATION OF ANTI-GAL-MEDIATED INCREASED UPTAKE, TRANSPORT, AND PROCESSING OF OVALBUMIN VACCINE BY ANTIGEN-PRESENTING CELLS

Ovalbumin liposomes as an experimental model vaccine

The studies with influenza virus and gp120 vaccines supported the hypothesis presented in Fig. 2 by demonstrating increased T and B cell response by linking α-gal epitopes to the vaccine. To determine whether the amplification of the immune response is the result of

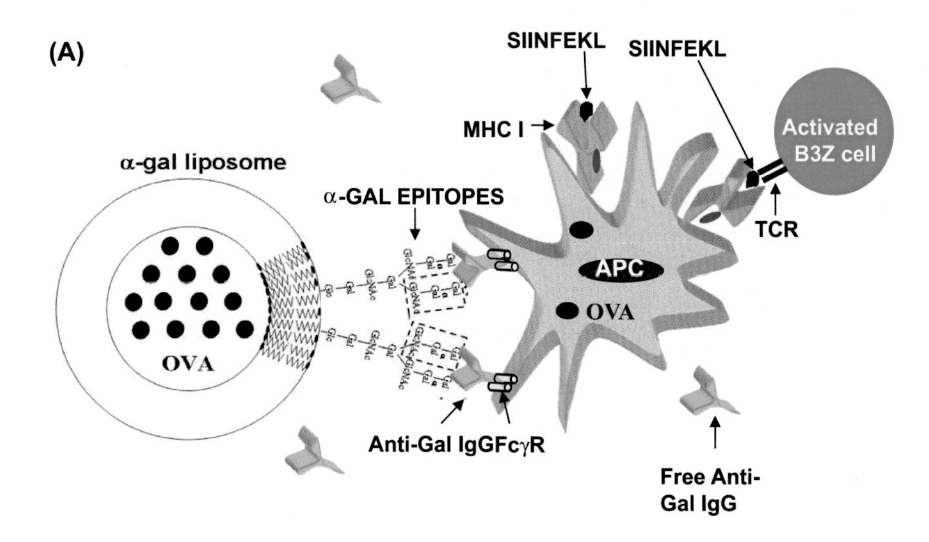
(A)
SIINFEKL
SIINFEKL
MHC I
Activated B3Z cell
α-gal liposome
α-GAL EPITOPES
TCR
APC
OVA
OVA
Anti-Gal IgGFcγR
Free Anti-Gal IgG

(B) GT-KO MICE WT MICE

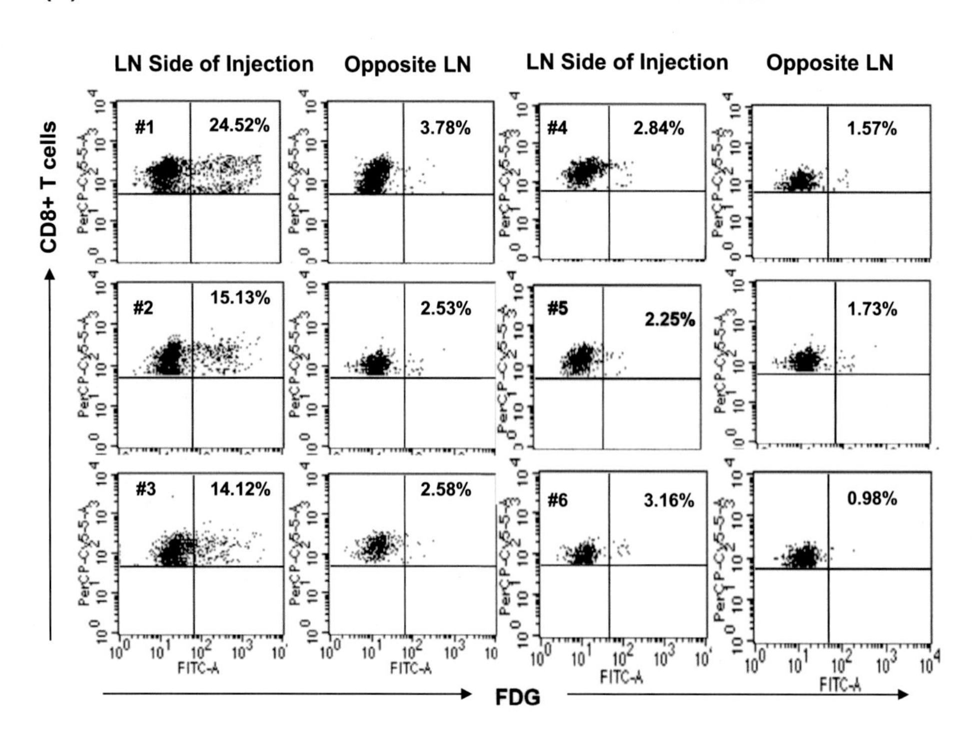
LN Side of Injection
Opposite LN
LN Side of Injection
Opposite LN
CD8+ T cells
#1 24.52%
3.78%
#4 2.84%
1.57%
#2 15.13%
2.53%
#5 2.25%
1.73%
#3 14.12%
2.58%
#6 3.16%
0.98%
FITC-A
FDG

increased anti-Gal-mediated uptake, processing, and presentation of immunogenic peptides by APC, chicken OVA was chosen as the vaccine model. OVA is a useful antigen for this purpose because the most immunogenic peptide of OVA for CD8+ T cells was identified as the 8-amino acid peptide SIINFEKL (Rötzschke et al., 1991). Moreover, a CD8+ T hybridoma cell line B3Z with a TCR specific for SIINFEKL was generated (Karttunen et al., 1992; Shastri and Gonzalez, 1993; Sanderson and Shastri, 1994). These B3Z cells are activated when they engage APC presenting SIINFEKL. Because B3Z cells have the β-galactosidase transgene *LacZ* under IL2 promoter, they produce this enzyme following interaction between SIINFEKL presented on MHC class I of H-2bK^b (called here H-2b) of APC and the TCR on B3Z cells. FITC-di-β-D-galactopyranoside (FDG) introduced into B3Z cells is hydrolyzed by β-galactosidase produced by the activated *LacZ* gene, causing the cells to be labeled by fluorescein, thus enabling detection of activated B3Z cells by flow cytometry.

OVA protein lacks N-linked carbohydrate chains that can be engineered to carry α-gal epitopes (unpublished data). Therefore, anti-Gal-mediated targeting of OVA to APC was studied with OVA encapsulated within liposomes that present multiple α-gal epitopes (Fig. 7A). As detailed in Chapter 12, these liposomes are prepared from phospholipids, cholesterol, and glycolipids extracted from rabbit red blood cell membranes. These liposomes present multiple α-gal epitopes (~10^{15} α-gal epitopes/mg) and thus, readily bind anti-Gal (Abdel-Motal et al., 2009; Wigglesworth et al., 2011). Studies with human tumor cells engineered to express α-gal epitopes demonstrated that human anti-Gal binding to α-gal epitopes on cells can further bind via its Fc portion to FcγR I (CD64) on human dendritic cells and macrophages and induce by this interaction effective uptake of the anti-Gal opsonized cells by these APC (Fig. 1B in Chapter 10) (Manches et al., 2005). A similar effective uptake could be demonstrated with anti-Gal opsonized α-gal liposomes (Abdel-Motal et al., 2009). Moreover, 24h following such *in vitro* internalization of the α-gal liposomes containing OVA (called OVA liposomes) by dendritic cells, these APC readily activated B3Z hybridoma T cells. This indicated that within 24h, the internalized OVA in OVA liposomes was processed and SIINFEKL was presented in association with MHC class I of H-2b on the APC, as illustrated in Fig. 7A (Abdel-Motal et al., 2009).

FIGURE 7 Anti-Gal-mediated targeting to APC of ovalbumin (OVA)-liposomes expressing α-gal epitopes, results in increased transport of OVA immunogenic peptides to lymph nodes draining the vaccination site. (A) Schematic illustration of an OVA-liposome encapsulating OVA and targeted by anti-Gal to APC. Anti-Gal binds to α-gal epitopes (*marked by dashed line rectangles*) on the vaccinating OVA-liposome. Opsonizing anti-Gal (colored blue) binds via its Fc portion to Fcγ receptors (FcγR) on APC (colored yellow). This interaction induces effective uptake of the liposome, followed by processing and presentation of OVA peptides on MHC molecules (e.g., the immunodominant peptide SIINFEKL on class I MHC molecules with their β-microglobulin colored red). B3Z T hybridoma cells engaging presented SIINFEKL with their TCRs are activated and detected by flow cytometry (colored green). (B) Detection of APC presenting SIINFEKL in the inguinal lymph nodes draining the OVA-liposomes vaccination site. APC presenting SIINFEKL were detected by activation of B3Z hybridoma (CD8+) T cell. Activated B3Z cells were quantified following incubation with cells from inguinal lymph nodes (LN) draining the immunization site in the right thigh (side of injection) and in left thigh (opposite LN), in GT-KO mice and WT mice. B3Z cells activated by SIINFEKL presented on APC were detected by flow cytometry of B3Z cells with FITC-di-β-D-galactopyranoside (FDG) hydrolyzed by endogenous β-galactosidase (staining activated B3Z cells, green). B3Z cells were also stained for CD8 detection by anti-CD8 antibody coupled to peridinin chlorophyll protein complex (PerCP). Lymph nodes were harvested 7days after the injection of 10mg OVA-liposomes into the right thigh muscle. Data from three mice per group. Adapted *from Abdel-Motal, U.M., Wigglesworth, K., Galili, U., 2009. Mechanism for increased immunogenicity of vaccines that form in vivo immune complexes with the natural anti-Gal antibody. Vaccine 27, 3072–3082, with permission.*

Anti-Gal-mediated increase in antigen-presenting cells transport of vaccinating ovalbumin to regional lymph nodes

The *in vivo* uptake and transport of OVA liposomes to regional lymph nodes by APC could be studied by subcutaneous injection of OVA liposomes into GT-KO mice near the right thigh. A week later the inguinal lymph nodes, which are the regional lymph nodes of this injection site, were harvested, minced, and the cells within them coincubated for 24h with B3Z cells. APC processing and presenting SIINFEKL within these lymph nodes were expected to activate the B3Z cells. Such activated B3Z cells could be detected by flow cytometry as cells double stained, green for hydrolyzed FDG and red for anti-CD8 antibody coupled to PerCP. Lymph nodes draining the vaccination site in three GT-KO mice displayed high numbers of SIINFEKL presenting APC, as indicated by the activation of 14%–24% B3Z cells, whereas the opposite (left) inguinal lymph node cells displayed activation of only 2.5%–3.7% B3Z cells (Fig. 7B) (Abdel-Motal et al., 2009). The presence of multiple APC presenting SIINFEKL in lymph nodes draining the vaccination site area depended on the presence of anti-Gal as further shown in Fig. 7B with WT mice. These mice are incapable of producing the anti-Gal antibody because they synthesize α-gal epitopes. Cells from lymph nodes draining the vaccination site in WT mice injected with OVA-liposomes activated only 2.2%–3.1% B3Z cells. This activation level was only slightly higher than that measured in the opposite leg inguinal lymph nodes. The data on B3Z activation by APC in lymph nodes draining the vaccination site suggest that anti-Gal-mediated targeting of the vaccine to APC greatly increases uptake and transport of the vaccine to regional lymph nodes, as well as the processing and presentation of vaccinating antigen by APC. This, in turn, amplifies the activation of T cells specific for the immunogenic vaccine's peptides. As shown in Abdel-Motal et al. (2009), increased activation of T cells could be demonstrated by ELISPOT with spleen lymphocytes coincubated with dendritic cells pulsed with SIINFEKL. There were on average 20-fold more SIINFEKL-specific T cells among splenocytes of GT-KO mice immunized with OVA-liposomes than among splenocytes obtained from WT mice receiving similar immunization. Accordingly, the proportion of CD8+ T cells binding pentamers of H-2b carrying SIINFEKL and the activity of CTL against target cells pulsed with this peptide was higher in the immunized GT-KO mice than in immunized WT mice. Similarly, the titer of anti-OVA antibodies (measured by ELISA with OVA as solid-phase antigen) was found to be ~30-fold higher in GT-KO mice immunized with OVA liposomes than in WT mice after similar immunization (Abdel-Motal et al., 2009). All these observations indicate that the elevated uptake of OVA-liposomes opsonized by anti-Gal and the increased transport of the internalized OVA by APC to regional lymph nodes, ultimately result in amplification of the cellular and humoral immune responses against the vaccinating protein.

CONCLUSIONS

The natural anti-Gal antibody may be harnessed for increasing the efficacy of viral vaccines as well as other vaccines, if the vaccine is engineered to carry the α-gal epitope. The binding of anti-Gal, present in large amounts in humans, to the α-gal epitopes on the vaccinating virus or on the glycoprotein vaccines results in formation of immune complexes. These immune complexes activate the complement system and generate complement cleavage chemotactic factors that recruit macrophages and dendritic cells to the immunization site. These APC bind via their Fc receptors, the Fc portion of anti-Gal immunocomplexed with the vaccine and are induced to effectively internalize the vaccine, transport it to the regional lymph nodes, process, and present the immunogenic peptides for effective activation of CTL and helper T cells. Studies on immunization of anti-Gal producing GT-KO mice with inactivated influenza virus presenting α-gal epitopes and with gp120 of HIV carrying α-gal epitopes indicated that anti-Gal-mediated targeting of vaccines to APC results in amplification of the antibody response by 30- to 200-fold and the cellular immune response by 10- to 30-fold. Synthesis of α-gal epitopes on carbohydrate chains of enveloped virus glycoproteins or on recombinant virus glycoproteins is feasible by using rα1,3GT and UDP-Gal. If the carbohydrate chains are capped with sialic acid, addition of neuraminidase removes the sialic acid and exposes the penultimate LacNAc (Galβ1-4GlcNAc-R), which is the acceptor for synthesis of α-gal epitopes by α1,3GT. Increased immunogenicity of vaccinating proteins lacking carbohydrate chains (e.g., core and matrix viral proteins) can be achieved by producing fusion proteins with glycoproteins carrying multiple α-gal epitopes, such as gp120 of HIV or HA of influenza virus.

References

Abdel-Motal, U., Wang, S., Lu, S., Wigglesworth, K., Galili, U., 2006. Increased immunogenicity of human immunodeficiency virus gp120 engineered to express Galα1-3Galβ1-4GlcNAc-R epitopes. J. Virol. 80, 6943–6951.

Abdel-Motal, U.M., Guay, H.M., Wigglesworth, K., Welsh, R.M., Galili, U., 2007. Immunogenicity of influenza virus vaccine is increased by anti-Gal-mediated targeting to antigen-presenting cells. J. Virol. 81, 9131–9141.

Abdel-Motal, U.M., Wigglesworth, K., Galili, U., 2009. Mechanism for increased immunogenicity of vaccines that form in vivo immune complexes with the natural anti-Gal antibody. Vaccine 27, 3072–3082.

Abdel-Motal, U.M., Wang, S., Awad, A., Lu, S., Wigglesworth, K., Galili, U., 2010. Increased immunogenicity of HIV-1 p24 and gp120 following immunization with gp120/p24 fusion protein vaccine expressing α-gal epitopes. Vaccine 28, 1758–1765.

Ahlén, G., Frelin, L., 2016. Methods to evaluate novel hepatitis C virus vaccines. Methods Mol. Biol. 1403, 221–244.

Banchereau, J., Steinman, R.M., 1998. Dendritic cells and the control of immunity. Nature 392, 245–252.

Celis, E., Chang, T.W., 1984. Antibodies to hepatitis B surface antigen potentiate the response of human T lymphocyte clones to the same antigen. Science 224, 297–299.

Chang, Y.T., Guo, C.Y., Tsai, M.S., Cheng, Y.Y., Lin, M.T., Chen, C.H., et al., 2012. Poor immune response to a standard single dose non-adjuvanted vaccination against 2009 pandemic H1N1 influenza virus A in the adult and elder hemodialysis patients. Vaccine 30, 5009–5018.

Clynes, R., Takechi, Y., Moroi, Y., Houghton, A., Ravetch, J.V., 1998. Fc receptors are required in passive and active immunity to melanoma. Proc. Natl. Acad. Sci. U.S.A. 95, 652–656.

Couch, R.B., Keitel, W.A., Cate, T.R., 1997. Improvement of inactivated influenza virus vaccines. J. Infect. Dis. 176 (Suppl. 1), S38–S44.

Crooks, E.T., Tong, T., Chakrabarti, B., Narayan, K., Georgiev, I.S., Menis, S., 2015. Vaccine elicited tier 2 HIV-1 neutralizing antibodies bind to quaternary epitopes involving glycan-deficient patches proximal to the CD4 binding site. PLoS Pathog. 11, e1004932.

Fanger, N.A., Wardwell, K., Shen, L., Tedder, T.F., Guyre, P.M., 1996. Type I (CD64) and type II (CD32) Fc gamma receptor-mediated phagocytosis by human blood dendritic cells. J. Immunol. 157, 541–548.

Fiers, W., De Filette, M., Birkett, A., Neirynck, S., Min Jou, W., 2004. A "universal" human influenza A vaccine. Virus Res. 103, 173–176.

Galili, U., Rachmilewitz, E.A., Peleg, A., Flechner, I., 1984. A unique natural human IgG antibody with anti-α-galactosyl specificity. J. Exp. Med. 160, 1519–1531.

Galili, U., Macher, B.A., Buehler, J., Shohet, S.B., 1985. Human natural anti-α-galactosyl IgG. II. The specific recognition of α(1-3)-linked galactose residues. J. Exp. Med. 162, 573–582.

Galili, U., Repik, P.M., Anaraki, F., Mozdzanowska, K., Washko, G., Gerhard, W., 1996. Enhancement of antigen presentation of influenza virus hemagglutinin by the natural human anti-Gal antibody. Vaccine 14, 321–328.

Galili, U., LaTemple, D.C., 1997. Natural anti-Gal antibody as a universal augmenter of autologous tumor vaccine immunogenicity. Immunol. Today 18, 281–285.

Goulder, P.J., Watkins, D.I., 2004. HIV and SIV CTL escape: implications for vaccine design. Nat. Rev. Immunol. 4, 630–640.

Henion, T.R., Gerhard, W., Anaraki, F., Galili, U., 1997. Synthesis of α-gal epitopes on influenza virus vaccines, by recombinant α1,3galactosyltransferase, enables the formation of immune complexes with the natural anti-Gal antibody. Vaccine 15, 1174–1182.

Houston, W.E., Kremer, R.J., Crabbs, C.L., Spertzel, R.O., 1977. Inactivated Venezuelan equine encephalomyelitis virus vaccine complexed with specific antibody: enhanced primary immune response and altered pattern of antibody class elicited. J. Infect. Dis. 135, 600–610.

Karttunen, J., Sanderson, S., Shastri, N., 1992. Detection of rare antigen-presenting cells by the lacZ T-cell activation assay suggests an expression cloning strategy for T-cell antigens. Proc. Natl. Acad. Sci. U.S.A. 89, 6020–6024.

Katz, J.M., Plowden, J., Renshaw-Hoelscher, M., Lu, X., Tumpey, T.M., Sambhara, S., 2004. Immunity to influenza: the challenges of protecting an aging population. Immunol. Res. 29, 113–124.

Keil, W., Geyer, R., Dabrowski, J., Dabrowski, U., Niemann, H., Stirm, S., et al., 1985. Carbohydrates of influenza virus. Structural elucidation of the individual glycans of the FPV hemagglutinin by two-dimensional 1H n.m.r. and methylation analysis. EMBO J. 4, 2711–2720.

Khanna, M., Gupta, N., Gupta, A., Vijayan, V.K., 2009. Influenza A (H1N1) 2009: a pandemic alarm. J. Biosci. 34, 481–489.

Lewis, G.K., DeVico, A.L., Gallo, R.C., 2014. Antibody persistence and T-cell balance: two key factors confronting HIV vaccine development. Proc. Natl. Acad. Sci. U.S.A. 111, 15614–15621.

Leonard, C.K., Spellman, M.W., Riddle, L., Harris, R.J., Thomas, J.N., Gregory, T.J., 1990. Assignment of intrachain disulfide bonds and characterization of potential glycosylation sites of the type 1 recombinant human immunodeficiency virus envelope glycoprotein (gp120) expressed in Chinese hamster ovary cells. J. Biol. Chem. 265, 10373–10382.

Manca, F., Fenoglio, D., Li Pira, G., Kunkl, A., Celada, F., 1991. Effect of antigen/antibody ratio on macrophage uptake, processing, and presentation to T cells of antigen complexed with polyclonal antibodies. J. Exp. Med. 173, 37–48.

Manches, O., Plumas, J., Lui, G., Chaperot, L., Molens, J.P., Sotto, J.J., et al., 2005. Anti-Gal-mediated targeting of human B lymphoma cells to antigen-presenting cells: a potential method for immunotherapy using autologous tumor cells. Haematologica 90, 625–634.

Matsumoto, A., Yoshima, H., Kobata, A., 1983. Carbohydrates of influenza virus hemagglutinin: structures of the whole neutral sugar chains. Biochemistry 22, 188–196.

Mizuochi, T., Spellman, M.W., Larkin, M., Solomon, J., Basa, L.J., Feizi, T., 1988. Carbohydrate structures of the human-immunodeficiency-virus (HIV) recombinant envelope glycoprotein gp120 produced in Chinese-hamster ovary cells. Biochem. J. 254, 599–603.

Mizuochi, T., Matthews, T.J., Kato, M., Hamako, J., Titani, K., Solomon, J., Feizi, T., 1990. Diversity of oligosaccharide structures on the envelope glycoprotein gp120 of human immunodeficiency virus 1 from the lymphoblastoid cell line H9. Presence of complex-type oligosaccharides with bisecting *N*-acetylglucosamine residues. J. Biol. Chem. 265, 8519–8524.

Montefiori, D.C., Pantaleo, G., Fink, L.M., Zhou, J.T., Zhou, J.Y., Bilska, M., et al., 1996. Neutralizing and infection-enhancing antibody responses to human immunodeficiency virus type 1 in long-term nonprogressors. J. Infect. Dis. 173, 60–77.

Munier, C.M., Andersen, C.R., Kelleher, A.D., 2011. HIV vaccines: progress to date. Drugs 71, 387–414.

Neirynck, S., Deroo, T., Saelens, X., Vanlandschoot, P., Jou, W.M., Fiers, W., 1999. A universal influenza A vaccine based on the extracellular domain of the M2 protein. Nat. Med. 5, 1157–1163.

Qiu, J.T., Song, R., Dettenhofer, M., Tian, C., August, T., Felber, B.K., et al., 1999. Evaluation of novel human immunodeficiency virus type 1 Gag DNA vaccines for protein expression in mammalian cells and induction of immune responses. J. Virol. 73, 9145–9152.

Rafiq, K., Bergtold, A., Clynes, R., 2002. Immune complex-mediated antigen presentation induces tumor immunity. J. Clin. Investig. 110, 71–79.

Ravetch, J.V., Clynes, R.A., 1998. Divergent roles for Fc receptors and complement in vivo. Annu. Rev. Immunol. 16, 421–432.

Regnault, A., Lankar, D., Lacabanne, V., Rodriguez, A., Théry, C., Rescigno, M., 1999. Fcγ receptor-mediated induction of dendritic cell maturation and major histocompatibility complex class I-restricted antigen presentation after immune complex internalization. J. Exp. Med. 189, 371–380.

Repik, P.M., Strizki, J.M., Galili, U., 1994. Differential host-dependent expression of α-galactosyl epitopes on viral glycoproteins: a study of eastern equine encephalitis virus as a model. J. Gen. Virol. 75, 1177–1181.

Rötzschke, O., Falk, K., Stevanović, S., Jung, G., Walden, P., Rammensee, H.G., 1991. Exact prediction of a natural T cell epitope. Eur. J. Immunol. 21, 2891–2894.

Sanderson, S., Shastri, N., 1994. LacZ inducible, antigen/MHC-specific T cell hybrids. Int. Immunol. 6, 369–376.

Schuurhuis, D.H., Ioan-Facsinay, A., Nagelkerken, B., van Schip, J.J., Sedlik, C., Melief, C.J., 2002. Antigen-antibody immune complexes empower dendritic cells to efficiently prime specific CD8+ CTL responses in vivo. J. Immunol. 168, 2240–2246.

Shastri, N., Gonzalez, F., 1993. Endogenous generation and presentation of the ovalbumin peptide/Kb complex to T cells. J. Immunol. 150, 2724–2736.

Smith, D.F., Larsen, R.D., Mattox, S., Lowe, J.B., Cummings, R.D., 1990. Transfer and expression of a murine UDP-Gal:β-D-Gal- α1,3-galactosyltransferase gene in transfected Chinese hamster ovary cells. Competition reactions between the α1,3-galactosyltransferase and the endogenous α2,3-sialyltransferase. J. Biol. Chem. 265, 6225–6234.

Stäger, S., Alexander, J., Kirby, A.C., Botto, M., Rooijen, N.V., Smith, D.F., et al., 2003. Natural antibodies and complement are endogenous adjuvants for vaccine-induced CD8+ T-cell responses. Nat. Med. 9, 1287–1292.

Stoner, R.D., Terres, G., 1963. Enhanced antitoxin responses in irradiated mice elicited by complexes of tetanus toxoid and specific antibody. J. Immunol. 91, 761–770.

Thall, A.D., Maly, P., Lowe, J.B., 1995. Oocyte Galα1,3Gal epitopes implicated in sperm adhesion to the zona pellucida glycoprotein ZP3 are not required for fertilization in the mouse. J. Biol. Chem. 270, 21437–21440.

Villinger, F., Mayne, A.E., Bostik, P., Mori, K., Jensen, P.E., Ahmed, R., 2003. Evidence for antibody-mediated enhancement of simian immunodeficiency virus (SIV) Gag antigen processing and cross presentation in SIV-infected rhesus macaques. J. Virol. 77, 10–24.

Wang, S., Arthos, J., Lawrence, J.M., Van Ryk, D., Mboudjeka, I., Shen, S., et al., 2005. Enhanced immunogenicity of gp120 protein when combined with recombinant DNA priming to generate antibodies that neutralize the JR-FL primary isolate of human immunodeficiency virus type 1. J. Virol. 79, 7933–7937.

Webster, R.G., 2000. Immunity to influenza in the elderly. Vaccine 18, 1686–1689.

Wei, X., Decker, J.M., Wang, S., Hui, H., Kappes, J.C., Wu, X., et al., 2003. Antibody neutralization and escape by HIV-1. Nature 422, 307–312.

Wigglesworth, K., Racki, W.J., Mishra, R., Szomolany-Tsuda, E., Greiner, D.L., Galili, U., 2011. Rapid recruitment and activation of macrophages by anti-Gal/α-gal liposome interaction accelerates wound healing. J. Immunol. 186, 4422–4432.

Wood, C., Kabat, E.A., Murphy, L.A., Goldstein, I.J., 1979. Immunochemical studies of the combining sites of the two isolectins, A4 and B4, isolated from *Bandeiraea simplicifolia*. Arch. Biochem. Biophys. 198, 1–11.

Zinkernagel, R.M., Ehl, S., Aichele, P., Oehen, S., Kündig, T., Hengartner, H., 1997. Antigen localisation regulates immune responses in a dose- and time-dependent fashion: a geographical view of immune reactivity. Immunol. Rev. 156, 199–209.

CHAPTER

10

Cancer Immunotherapy by Anti-Gal-Mediated *In Situ* Conversion of Tumors Into Autologous Vaccines

OUTLINE

INTRODUCTION

A major challenge in treatment of patients with solid cancer lesions is the destruction of micrometastases. Many of the patients who fully recover are tumor free when a cancer lesion that has not released metastatic cells into the circulation is detected and removed by resection or ablation. However, in a large proportion of cancer patients, the solid tumor releases individual cancer cells, or small groups of metastatic cells, called micrometastases, which develop into lethal tumor lesions after the primary tumor has been eliminated. Micrometastases cannot be detected because of their microscopic size and thus cannot be removed by surgery or destroyed

The Natural Anti-Gal Antibody as Foe Turned Friend in Medicine
http://dx.doi.org/10.1016/B978-0-12-813362-0.00010-5

by radiotherapy. Destruction of metastatic tumor cells by chemotherapy also may be insufficient in a proportion of the patients. The reason is that some of the metastatic tumor cells may reside "far enough" from the nearest capillary, so that the concentration of the chemotherapy drug is too low for affecting the tumor cells. A promising therapeutic approach that has been extensively studied and used in the clinic in the recent decade is immunotherapy against tumor-associated antigens (TAA) that are generated by the multiple mutations within tumor cells.

IMMUNOTHERAPY AGAINST TUMOR-ASSOCIATED ANTIGENS

Whole genome sequencing of tumors from various cancer patients has demonstrated the presence of multiple-coding mutations introduced into the cells because of genomic instability (Mumberg et al., 1996; Wood et al., 2007; Campbell et al., 2010; Vogelstein et al., 2013). Many of these mutations are nonsynonymous (missense) mutations that change individual amino acids within various proteins. Most of those mutations are specific to the patient and are not found in other individuals with the same kind of tumor. A proportion of these mutations may convert proteins in a way that provides growth advantage to the tumor cells, whereas other mutations are neutral because they do not alter the mutated protein function within tumor cells. Nevertheless, both types of mutated proteins might be the immunological "Achilles' heel" of cancer cells if the immune system can be "educated" to recognize and destroy cells that carry these tumor specific mutated proteins. These proteins, which may become recognizable by the immune system as "foreign" antigens, are referred to as TAA or tumor antigens.

The ability of the immune system to recognize TAA and destroy the tumor cells presenting them is reflected in the correlation between the extent of CD8+ and Th1 memory T cell infiltration into tumors and the prognosis of patients with solid tumors such as ovarian carcinoma, and colorectal carcinoma (Zhang et al., 2003; Galon et al., 2006, 2007; Mlecnik et al., 2011). These retrospective studies demonstrated correlation between cancer patients' survival and the extent of T cell infiltration observed in resected primary tumors. Patients with tumors displaying high infiltration of these T cells had much longer survival and low or no metastases formation. The good prognosis was observed even in patients with tumor cells in the regional lymph nodes of the primary tumor. In contrast, patients with tumors lacking T cell infiltration had poor prognosis, even if the regional lymph nodes displayed no infiltration of metastatic tumor cells. In patients with good prognosis, the infiltration of T cells was exclusively observed within the malignant tissue and not in normal tissues. This implies that the antigens recognized by the tumor infiltrating T cells are specific to tumor cells and are not present on normal cells, i.e., these antigens are the TAA.

The findings of resected tumors with no or very few infiltrating T cells suggest that the immune system in these patients is not "nimble" enough to recognize the TAA as foreign antigens. Alternatively, the tumors "conceal" TAA from the immune system by secreting cytokines that are immunosuppressive and induce tolerance, anergy, or lymphocyte death (Pawelec et al., 2000; Ribas, 2015). Thus, it is reasonable to assume that immunotherapy which activates the immune system to react against the TAA of the individual patient may be beneficial in enabling the immune system to recognize metastatic cells by their TAA and destroy them. One type of immunotherapy aimed to achieve such activation has been infusion of checkpoint inhibitors. Those are antibodies that prevent suppression of T cells following their activation by tumor-specific peptides presented on antigen presenting cells (APC). Among checkpoint inhibitors are antibodies against CTLA4 (ipilimumab) and against PD1 (nivolumab and pembrolizumab), which prevent the corresponding interactions between activated T cells and their ligands (Korman et al., 2006; Intlekofer and Thompson, 2013; Callahan and Wolchok, 2013). In the absence of such antibodies, these T cells/APC interactions suppress further activation and clonal expansion of activated T cells, thereby preventing over reactivity that may destroy not only tumor cells but also may lead to autoimmune events. Although checkpoint inhibitor antibodies are effective in a substantial proportion of the treated patients in improving prognosis of patients with metastases, such treatments have high incidence of negative side effects, primarily autoimmune reactions because of the nonspecific activation of TAA-specific T cells as well as T cells that can mediate autoimmune reactivity against normal tissue antigens.

As with viral or other microbial vaccine, an effective immune response against autologous TAA requires effective uptake of the tumor cells and cell membranes by APC, such as dendritic cells and macrophages. However, because tumor cells are larger than microbial vaccines, they cannot be internalized randomly by pinocytosis into APC. Uptake of tumor vaccines can be induced if the vaccinating tumor cells present markers recognizable by APC such as Fc portion of opsonizing IgG or C3b deposits on the tumor cell membranes. Similar to the processing of viral vaccines by APC, described in Chapter 9, tumor cells internalized by APC are transported to the regional lymph nodes, processed, and their immunogenic TAA peptides presented on class I and class II MHC molecules. These peptides activate tumor-specific CD8+ cytotoxic T cells and CD4+ Th1 cells, respectively (Maass et al., 1995; Zinkernagel et al., 1997; Dunn et al., 2004). In the absence of markers on tumor cells that induce uptake by APC, these cells are "ignored" by the immune system, as indicated by the ability of tumor cells to reside in lymph nodes without being affected by the immune system. Moreover, TAA are usually "concealed" from the immune system because the tumor cytokine milieu is often suppressive toward immune function and induces tolerance, anergy, or lymphocyte death (Staveley-O'Carroll et al., 1998; Malmberg, 2004; Furumoto et al., 2004). This prompted the development of methods for recruiting APC into tumor lesions by intratumoral administration of immunomodulators such as GM-CSF or CpG oligonucleotides (Mastrangelo et al., 1999; Furumoto et al., 2004). APC recruited into tumors cannot identify tumor cells within the lesion as cells that "ought" to be internalized because the tumor cells lack identifying markers that label them for uptake by APC. Thus, uptake of tumor cells by recruited APC is suboptimal (Galili, 2004). In view of these considerations, it is reasonable to assume that *in situ* conversion of tumor cells within autologous lesions into vaccines requires methods that recruit APC into the treated tumor, as well as induce uptake of tumor cells by APC.

As described in Chapter 9, one of the most effective mechanisms, by which macrophages and dendritic cells can internalize vaccinating antigens and process them for T cell activation, is the formation of immune complexes with the vaccinating antigens. The interaction of the Fc portion of the immunocomplexed antibody molecules with Fcγ receptors (FcγR) on APC generates signals for the uptake of the immune complex (Manca et al., 1991; Gosselin et al., 1992; Fanger et al., 1996; Ravetch and Clynes, 1998). In addition, interaction of immune complexes with FcγR of dendritic cells induces their maturation into "professional" APC that elicit effective immune response against the immunocomplexed vaccine (Schuurhuis et al., 2002; Regnault et al., 1999). This principle of amplifying the immunogenicity of a vaccine by formation of immune complexes targeted to APC via Fc/FcγR interaction was described in

Chapter 9 for viral vaccines. This principle was further demonstrated with tumor antigens and tumor cells forming immune complexes with their corresponding antibodies (Berlyn et al., 2001; Rafiq et al., 2002; Dhodapkar et al., 2002; Groh et al., 2005). Targeting tumor cells to APC by immunocomplexing them with their corresponding antibody is not practical at present because in most human tumors, the identity of such antibodies and their customized production for individual patients are technically challenging. However, human tumors can be targeted to APC by engineering them to present α-gal epitopes, thereby enabling them to form immune complexes with the natural anti-Gal antibody.

Hypothesis on cancer immunotherapy by anti-Gal-mediated targeting of tumor cells to antigen presenting cells

The production of anti-Gal in response of immunizing gastrointestinal bacteria is a very potent process occurring in all humans who are not severely immunocompromised (Galili et al., 1984, 1988), even under immunosuppressive treatments that prevent rejection of allografts (Galili et al., 1995). Several studies further reported that anti-Gal titers in patients with various malignancies do not differ significantly from those in healthy individuals (Galili et al., 1984; Tremont-Lukats et al., 1997; Qiu et al., 2011, 2013; Hamanová et al., 2014). Thus, similar to the rationale described in Chapter 9 for anti-Gal-mediated targeting of viral vaccines to APC, anti-Gal may serve as an antibody capable of targeting tumor cells presenting α-gal epitopes to APC to increase the immunogenicity of TAA of the immunizing tumor cells (Galili and LaTemple, 1997). It was hypothesized that tumor cells or cell membranes engineered or processed to present multiple α-gal epitopes readily bind the natural anti-Gal antibody. If this interaction occurs *in vivo*, then it results in activation of the complement system and generation of complement chemotactic factors such as C5a and C3a, which recruit APC as dendritic cells and macrophages (Fig. 1A). Anti-Gal bound to the tumor cells and cell membranes

FIGURE 1 Targeting of tumor cells presenting α-gal epitopes to antigen presenting cells (APC) by the natural anti-Gal antibody. (A) α-Gal epitopes are expressed on tumor cells following intratumoral injection of α-gal glycolipids. Representative α-gal glycolipid with two branches (antennae) capped by α-gal epitopes (*dashed line rectangles*) inserted into the tumor cell membranes. Natural anti-Gal IgM and IgG antibodies bind to α-gal epitopes and activate the complement system, which causes lysis of tumor cells, as well as generation of the complement cleavage chemotactic peptides C5a and C3a that induce rapid recruitment of APC (dendritic cells and macrophages) into the treated tumor. Anti-Gal IgG further opsonizes the tumor cells by binding of its Fc portion to Fcγ receptors (FcγR) on APC. This interaction induces uptake of opsonized tumor cells by APC. Phagocytosis of intact or lysed tumor cells results in internalization of the tumor-associated antigens (TAA) of these cells. Internalized TAA are transported by the APC to regional lymph nodes, processed and the immunogenic TAA peptides (●, ■, ▲) are presented on class I and class II MHC molecules. The presented TAA peptides engage the corresponding TCR and activate TAA specific cytotoxic and helper T cell clones. The activated T cells proliferate, leave the lymph nodes and mediate protective anti-tumor immune responses. (B) *In vitro* demonstration of anti-Gal mediated uptake of human B lymphoma cells by autologous APC. α-Gal epitopes were synthesized on these cells by incubation with neuraminidase, rα1,3GT and UDP-Gal. Lymphoma cells with or without α-gal epitopes were incubated with autologous anti-Gal for 30 min, then for 2 h at 37°C with autologous macrophages or dendritic cells. Triangles mark the nuclei of the APC. Note the uptake of five lymphoma cells presenting α-gal epitopes by the macrophage and one lymphoma cell by the dendritic cell. No uptake of lymphoma cells lacking α-gal epitopes was observed (×1000). *(A) Reprinted with permission from Galili, U., 2013. Anti-Gal: an abundant human natural antibody of multiple pathogeneses and clinical benefits. Immunology 140, 1–11. (B) Adapted with permission from Manches, O., Plumas, J., Lui, G., Chaperot, L., Molens, J.P., Sotto, J.J., et al., 2005. Anti-Gal-mediated targeting of human B lymphoma cells to antigen-presenting cells: a potential method for immunotherapy using autologous tumor cells. Haematologica 90, 625–634.*

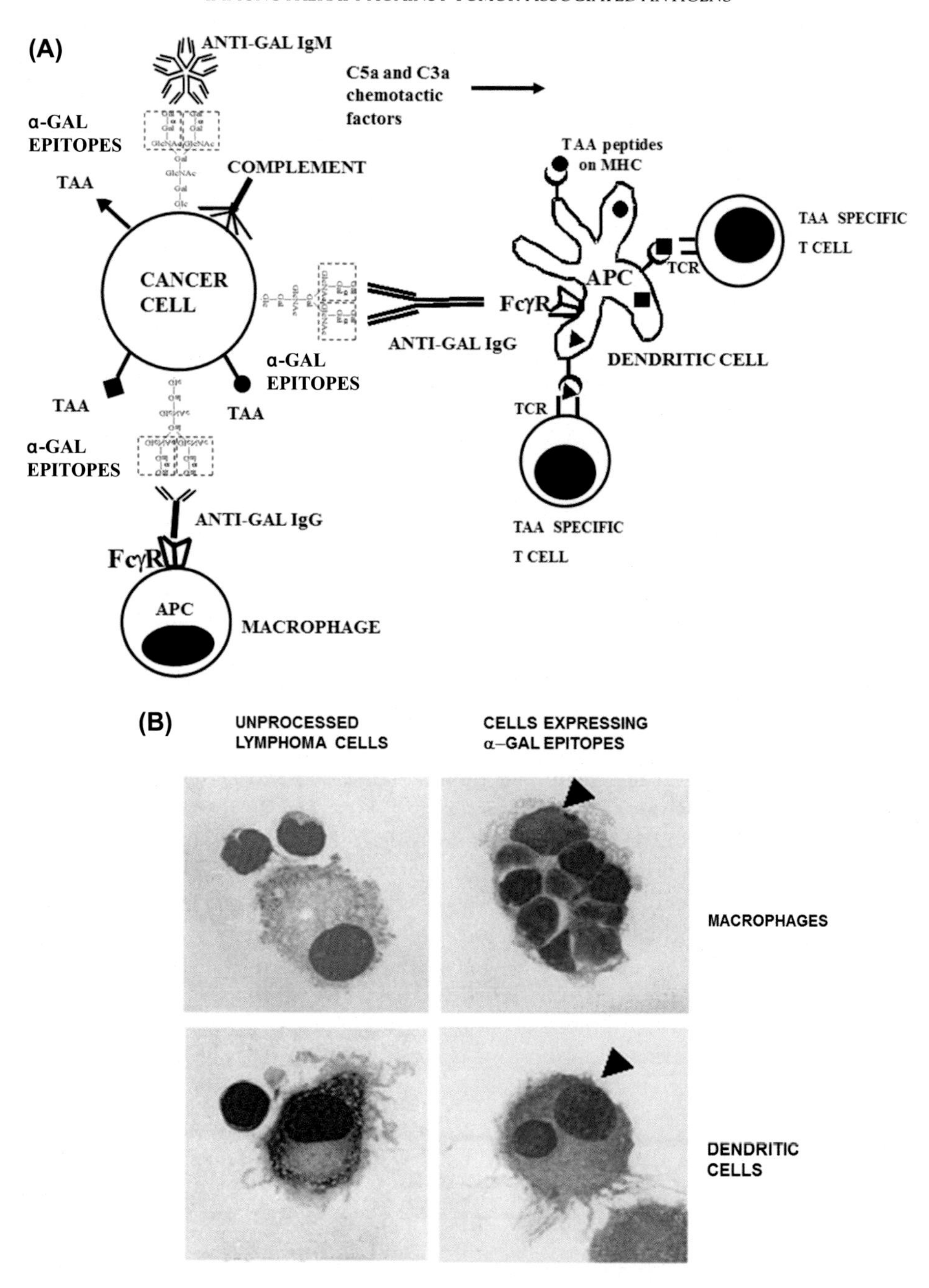
(A)
ANTI-GAL IgM
C5a and C3a
chemotactic
factors
α-GAL
EPITOPES
COMPLEMENT
TAA
CANCER
CELL
TAA peptides
on MHC
TAA SPECIFIC
T CELL
TCR
APC
FcγR
ANTI-GAL IgG
DENDRITIC CELL
α-GAL
EPITOPES
TAA
TAA
TCR
α-GAL
EPITOPES
ANTI-GAL IgG
TAA SPECIFIC
T CELL
FcγR
APC
MACROPHAGE
(B)
UNPROCESSED
LYMPHOMA CELLS
CELLS EXPRESSING
α–GAL EPITOPES
MACROPHAGES
DENDRITIC
CELLS

further serves as a "label" that targets the tumor cells and their TAA for effective uptake by the recruited APC. This uptake is mediated by the Fc portion anti-Gal opsonizing tumor cells and cell membranes, which interacts with the FcγR on APC, and thus stimulates the APC to phagocytose the tumor cells. C3b/C3b receptor interaction may have similar effects.

One example for this basic principle of effective uptake of human tumor cells by autologous APC is described in Fig. 1B (Manches et al., 2005). Human lymphoma cells obtained from a B cell lymphoma patient were engineered to present α-gal epitope by incubation with neuraminidase, recombinant α1,3galactosyltransferase (rα1,3GT), and UDP-Gal, like the method used for synthesis of α-gal epitopes on gp120 of HIV described in Fig. 1 of Chapter 9. Subsequently, lymphoma cells presenting α-gal epitopes were incubated with macrophages and dendritic cells prepared from cultured monocytes of the patient and in the presence of autologous anti-Gal. Coincubation of anti-Gal opsonized lymphoma cells with autologous macrophages or dendritic cells resulted in uptake of the tumor cells by the APC, whereas in absence of α-gal epitopes no such phagocytosis could be observed (Fig. 1B). Similarly, no phagocytosis of tumor cells presenting α-gal epitopes was observed without anti-Gal (Manches et al., 2005).

APC that internalize the tumor cells or cell membranes, process the TAA, and transport them to the regional lymph nodes where TAA peptides presented on class I and class II MHC molecules activate TAA-specific CD8+ and CD4+ T cells. These activated CD8+ T cells proliferate, leave the lymph nodes, circulate in the body, and seek and destroy metastatic tumor cells that present the autologous TAA peptides. Activated CD4+ T helper (Th) cells provide help for CD8+ T cells and for tumor specific B cells to mount a protective cellular and humoral immune response against the tumor cells. This immune response may enable the destruction of micrometastases that are too small to be identified and destroyed by resection or ablation. It was further suggested that development of methods that result in *in situ* presentation of α-gal epitopes on tumor lesions may convert the treated lesions into autologous vaccines that bind anti-Gal, recruit APC, and target the tumor cells for uptake by these APC. This may result in eliciting a protective anti-tumor immune response against the multiple autologous TAA without the need to identify these TAA in the individual cancer patient (Galili et al., 2007). This chapter describes studies on the increased anti-tumor immune protection following the *in vitro* or *in vivo* manipulation of vaccinating autologous tumor cells to present α-gal epitopes. It should be noted that conversion of autologous tumor cells to vaccine was also proposed by using tumor cells engineered to secrete immune modulating cytokines such as GM-CSF, thereby recruiting APC with this cytokine (Dranoff et al., 1993; Simons et al., 1997). It is not clear, however, whether GM-CSF or other APC recruiting cytokines can increase uptake of the tumor cells, because such tumor cells lack any marker that actively targets them to be internalized by the recruited APC.

INCREASED IMMUNOGENICITY OF TUMOR CELLS WITH α1,3GALACTOSYLTRANSFERASE TRANSGENE

Analysis of anti-Gal-mediated targeting of vaccinating tumor cells to APC could be performed in anti-Gal producing α1,3GT knockout mice (GT-KO) mice using syngeneic tumor cells, which lack α-gal epitopes. Wild-type (WT) mice could not be used for this purpose because they do not produce anti-Gal. Most mouse tumor cell lines also are not suitable because they present α-gal epitopes and thus, they are destroyed by mechanisms like those causing anti-Gal-mediated rejection of xenografts (see Chapter 6). The mouse melanoma cell lines B16, JBRH, and JBDS are the only known mouse cell lines that lack α-gal epitopes (Gorelik et al., 1995), because of suppression of the α1,3GT gene (*GGTA1*) expression. The highly tumorigenic B16 cell line was chosen as a model for determining anti-Gal-mediated increase in anti-tumor immune response. B16 tumor lesions usually double their size every 4–8 days. Because GT-KO mice are kept in sterile environment and fed sterilized food, they do not develop the gastrointestinal flora, which stimulates the immune system to produce the natural anti-Gal antibody. These mice produce anti-Gal following 3–4 weekly immunizations with rabbit red blood cells (RBC) (LaTemple et al., 1999) or with porcine kidney membranes homogenate, which present multiple α-gal epitopes (Galili et al., 2007).

For the preparation of B16 melanoma cells presenting α-gal epitopes, the cells underwent transfection with the mouse α1,3GT gene (*GGTA1*) and were selected for stable expression of the gene based on the constant synthesis of α-gal epitopes (LaTemple et al., 1999). These cells, designated $B16_{\alpha GT}$, were found to present ~10^6 α-gal epitopes per cell. Two million irradiated $B16_{\alpha GT}$ cells were injected as vaccine into anti-Gal producing GT-KO mice, whereas control GT-KO mice were immunized with irradiated parental B16 cells lacking α-gal epitopes. Two weeks following the immunization, the mice were challenged subcutaneously with 0.2×10^6 live parental B16 cells lacking α-gal epitopes. The mice were monitored for 60 days for tumor development, which could be identified as a black lesion in the skin. As many as 60% of the mice immunized with $B16_{\alpha GT}$ cells were protected from the tumor challenge and did not develop a skin lesion. In contrast, only 20% of mice immunized with the parental B16 cells lacking α-gal epitopes displayed such protection (LaTemple et al., 1999).

Histologic analysis of the tumors developing in 80% of the mice immunized with B16 cells displayed no mononuclear cell infiltrates. The tumors in these mice were surrounded by a thin layer of connective tissue. The melanoma cells displayed extensive proliferation, indicated by relatively small size of the tumor cells, only few melanin granules and basophilic cytoplasm, characteristic of high protein synthesis in dividing cells (Fig. 2A). In contrast, the tumors developing in 40% of the mice immunized with $B16_{\alpha GT}$ cells exhibited distinct mononuclear cell infiltrates that surrounded the tumors (Fig. 2B). Many of the tumor cells adjacent to the mononuclear cell infiltrate were large with pale cytoplasm and multiple melanin granules, suggesting that these cells stopped dividing. Immunohistochemistry analysis of the mononuclear cell infiltrates indicated that as many as 70% of the cells were T cells stained by anti-CD3 antibodies and 30% were macrophages. No B cells were found in these infiltrates (LaTemple et al., 1999). The mononuclear infiltrates surrounding tumors developing in mice immunized with $B16_{\alpha gal}$ cells suggest that these mice developed an anti-tumor immune response. In 40% of these immunized mice, the anti-tumor immune response failed to destroy the fast-growing tumor lesion but slowed its growth. However, in 60% of the mice, the elicited anti-tumor immune response succeeded in destroying the parental B16 cells administered in the challenge, before they could develop into a tumor lesion.

When the number of challenging tumor cells was increased to 0.5×10^6 live B16 cells, all mice immunized with parental B16 cells lacking α-gal epitopes developed tumors within 21–25 days. However, as many as 35% of mice immunized with $B16_{\alpha GT}$ cells displayed protection against any tumor development (LaTemple et al., 1999). These findings suggested that immunization with mouse melanoma cells engineered to present α-gal epitopes induces a protective immune response against the tumor TAA that may prevent the development

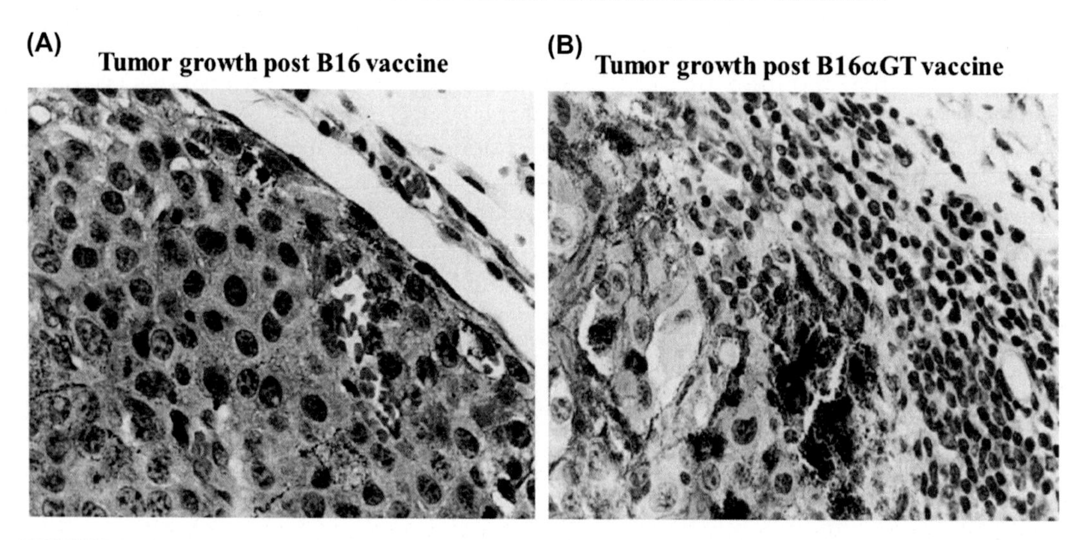

FIGURE 2 The immune response against B16 tumors in GT-KO mice immunized with B16 melanoma cells (A), or with $B16_{\alpha GT}$ melanoma cells that present multiple α-gal epitopes (B). Tumor lesions were removed 2 weeks post challenge, sectioned and subjected to H&E staining. (A) Tumor from mice immunized with irradiated B16 cells displays no cell infiltrates and is surrounded by a thin connective tissue. (B) In mice immunized with irradiated $B16_{\alpha GT}$ cells the tumor is surrounded by many mononuclear cells and the melanoma cells bordering these mononuclear cells are large vacuolated cells, filled with melanin granules (×400). *Reprinted with permission from LaTemple, D.C., Abrams, J.T., Zhang, S.Y., Galili, U., 1999. Increased immunogenicity of tumor vaccines complexed with anti-Gal: studies in knockout mice for α1,3galactosyltranferase. Cancer Res. 59, 3417–3423.*

of tumor lesions following challenge with tumor cells lacking these epitopes. None of the immunized mice developed any autoimmune response, suggesting that anti-Gal-mediated targeting of the tumor cells to APC did not cause a breakdown in immune tolerance to normal antigens on the vaccinating cells.

A similar protective anti-Gal-mediated immune response against melanoma cells in GT-KO mice was observed in studies using B16 cells engineered to synthesize α-gal epitopes by transduction of the α1,3GT gene within an adenovirus vector (Deriy et al., 2005) or a retrovirus vector (Rossi et al., 2005a, 2005b). In the latter studies, immune protection was also observed when the challenge with tumor cells lacking α-gal epitopes was performed prior to the immunization with tumor cells presenting these epitopes. A protective effect of tumor cells expressing α-gal epitopes has been demonstrated also in a pancreatic adenocarcinoma model of GT-KO mice immunized with MUC1 proteoglycan (overexpressed in most pancreatic carcinoma, thus may function as TAA) expressing α-gal epitopes (Deguchi et al., 2010; Tanemura et al., 2015). Similarly, increased immune response against TAA of a human pancreatic cancer cell line (PANC1) engineered to express α-gal epitopes was observed in anti-Gal producing GT-KO mice immunized with lysates of this tumor in comparison with vaccinating lysate of PANC1 tumor lacking α-gal epitopes (Tanida et al., 2015). The analysis of the anti-tumor immune response was performed in NOD/SCID mice carrying this tumor and receiving by adoptive transfer lymphocytes from GT-KO mice immunized with the tumor lysates presenting or lacking α-gal epitopes.

SYNTHESIS OF α-GAL EPITOPES ON HUMAN TUMOR LYSATES BY RECOMBINANT α1,3GALACTOSYLTRANSFERASE

Most TAA in human tumors are encoded by mutations in various genes. These mutations are specific to the individual patient and differ from one patient to the other (Wood et al., 2007; Campbell et al., 2010; Vogelstein et al., 2013). Moreover, the genomic instability of tumors results in appearance of mutations in various metastases that are not found in the primary tumor lesion (Yachida et al., 2010). Thus, immunotherapy for eliciting a protective immune response against the multiple TAA of the individual patient requires the use of the tumor itself in each patient as the source of immunizing TAA. It is difficult to transfect or transduce freshly obtained tumor cells with a vector containing the α1,3GT gene in order to induce expression of α-gal epitopes, because most tumor cells from resected tumors do not proliferate *in vitro*. Thus, the transgene cannot integrate into the genome. In addition, freshly obtained tumor cells are usually difficult to grow in large enough numbers required for vaccine preparation.

One possible approach for harnessing the anti-Gal antibody for targeting autologous tumor vaccines to APC is to use primary or metastatic tumor lesions, which are resected and processed into tumor lysates. These lysates can be subjected to synthesis of α-gal epitopes on glycolipids and glycoproteins of the tumor cell membranes. Such synthesis is achieved by incubation of the lysate with neuraminidase, rα1,3GT and UDP-Gal, like the combined enzymatic reactions illustrated in Fig. 1 of Chapter 9 (LaTemple et al., 1996; Galili et al., 2003; Galili, 2004). Neuraminidase removes sialic acid and exposes the *N*-acetyllactosamine residues on the carbohydrate chains of glycoproteins and on gangliosides (glycolipids with carbohydrate chains capped by sialic acid). Subsequently, rα1,3GT links galactose from the UDP-Gal sugar donor to the exposed *N*-acetyllactosamine residues to synthesize $1–27 \times 10^6$ α-gal epitopes/cell (Ogawa and Galili, 2006). Pancreatic carcinoma tumor lysate expressing α-gal epitopes was found to be much more active in eliciting a protective anti-tumor immune response in anti-Gal producing GT-KO mice than tumor lysates lacking this epitope (Tanida et al., 2015).

The method of conversion of a tumor lysate into a vaccine by synthesis of α-gal epitopes with rα1,3GT was also studied in clinical trials in patients with resected hepatocellular carcinoma (Qiu et al., 2011) and with pancreatic carcinoma (Qiu et al., 2013). The resected tumors in both groups were dispersed by incubation of the minced tissue in EDTA followed by passing through a narrow needle and subjected to enzymatic synthesis of α-gal epitopes on tumor cell membranes with neuraminidase and bovine rα1,3GT. These cells and cell membranes were washed and coincubated with the patient's dendritic cells in presence of anti-Gal, followed by coincubation with *in vitro* expanded patient's T cells. Subsequently, this mixture was infused into the treated patients. The researchers found the procedure to be safe with no serious side effects. This procedure induced a significant elevation in tumor-specific immune response as measured by skin test and by *in vitro* IFNγ secretion by activated T cells, as measured in enzyme-linked immunospot (ELISPOT) (Qiu et al., 2011, 2013). This immunotherapy was reported to significantly prolong the survival of treated patients with hepatocellular carcinoma as compared with the controls (~17 vs. 10 months) (Qiu et al., 2011).

CONVERSION OF TUMORS INTO VACCINES BY INTRATUMORAL INJECTION OF α-GAL GLYCOLIPIDS

The use of tumor lysates is work intensive and requires specialized facilities for preparing autologous tumor vaccines with α-gal epitopes. Thus, it was of interest to develop methods for *in situ* conversion of solid tumors into vaccines presenting α-gal epitopes. The efficacy of various agents injected intratumoral to induce α-gal epitope expression on tumor cells was assessed in the experimental model of B16 melanoma tumors grown subcutaneously in GT-KO mice.

The first two methods tested were as follows: (1) intratumoral injection of a mixture of neuraminidase, rα1,3GT, and UDP-Gal with the purpose of synthesizing α-gal epitopes on the cells within the tumor, and (2) injection of replication defective adenovirus vector containing the α1,3GT gene. This adenovirus vector was shown to effectively transduce *in vitro* mouse B16 and human HeLa cells and induces synthesis and expression of α-gal epitopes (Deriy et al., 2002, 2005). Both methods displayed low *in vivo* efficacy, possibly because the effects of these agents were limited to the "walls of the bubble" in which they were injected and because of poor diffusion within the tumor (unpublished observations). As described below, the third method that consisted of intratumoral injection of α-gal glycolipids proved to be successful (Galili et al., 2007).

Extraction of α-gal glycolipids from rabbit red blood cells membranes

α-Gal glycolipids are glycolipids consisting of a ceramide lipid tail and a carbohydrate chain with 1–8 branches (antennae) and capped by the α-gal epitope (Galili et al., 2007) (Fig. 3). The richest known natural source of α-gal glycolipids is rabbit RBC membranes. These RBC have ~2×10^6 α-gal epitopes/cell (Galili et al., 1998), many of which are on glycolipids and the rest on glycoproteins. The shortest α-gal glycolipid in rabbit RBC has five carbohydrates (Eto et al., 1968; Stellner et al., 1973) and is shown in Fig. 3A and B. α-Gal glycolipids with longer carbohydrate chains increase in size by increment of 5 carbohydrate (i.e., 10, 15, 20, 25, and up to 40 carbohydrates) except for a glycolipid with 7 carbohydrates and one α-gal epitope (Fig. 3) (Honma et al., 1981; Dabrowski et al., 1984; Egge et al., 1985; Hanfland et al., 1988). Each increment also forms a new branch capped with an α-gal epitope (Fig. 3B). Prior to extraction of glycolipids, rabbit RBC are lysed by hypotonic shock, and the cell membranes are washed repeatedly to remove the hemoglobin. The glycolipids, phospholipids, and cholesterol are extracted from these RBC membranes for 2h in chloroform:methanol 1:1 solution followed by overnight extraction in 1:2 solution of chloroform:methanol and filtration for removal of the precipitating proteins (Fig. 3A) (Galili et al., 2007). The hydrophobic cholesterol, phospholipids, and a glycolipid with a three-carbohydrate chain called ceramide trihexoside (CTH—Galα1-4Galβ1-4Glc-R, which does not have the α-gal epitope, but it is also present in human RBC) are removed. This removal from the hydrophilic glycolipids with ≥5-carbohydrate in each chain (chains carrying α-gal epitopes) (Fig. 3A), is performed by the process of Folch partition (Folch et al., 1957). In this process, gradual addition of water to the chloroform:methanol extraction solution results in partition between a lower organic phase containing cholesterol, phospholipids, and CTH and an upper aqueous phase containing α-gal glycolipids with >5 carbohydrates (Fig. 3A). The structure of the α-gal glycolipids with

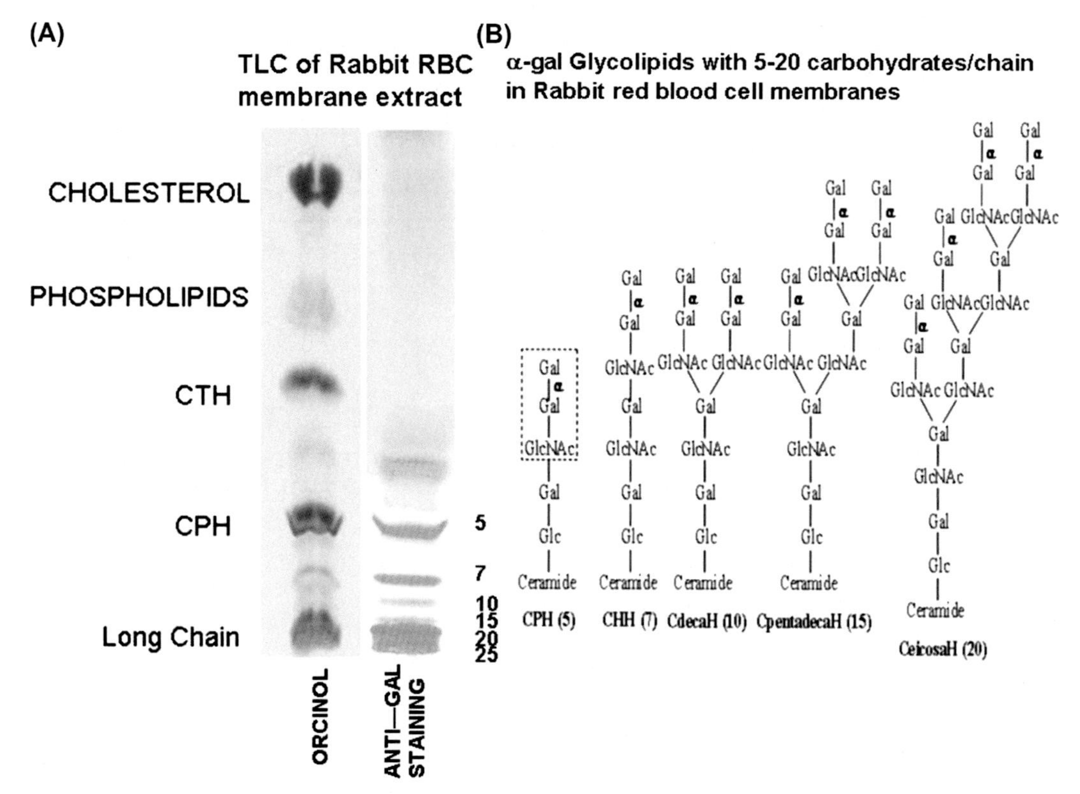

FIGURE 3 Rabbit RBC α-gal glycolipids. (A) Chloroform:methanol extract of rabbit RBC membranes separated on TLC plate and stained nonspecifically with orcinol (left lane) or immunostained with human anti-Gal (right lane). CTH—ceramide trihexoside with three carbohydrates lacks α-gal epitopes. CPH—ceramide pentahexoside with five carbohydrates. CPH and larger glycolipids (carbohydrate numbers per chain are indicated in the right side of figure), all have α-gal epitopes as indicated by binding of anti-Gal in the immunostained right lane. (B) Structures of the α-gal glycolipids with 5–20 carbohydrates (CHH—ceramide heptahexoside). The α-gal epitope on CPH is marked by the *dashed line rectangle* and is found on all other rabbit RBC α-gal glycolipids. *Adapted with permission from Galili, U., Wigglesworth, K., Abdel-Motal, U.M., 2007. Intratumoral injection of α-gal glycolipids induces xenograft-like destruction and conversion of lesions into endogenous vaccines. J. Immunol. 178, 4676–4687.*

5–20 carbohydrates is illustrated in Fig. 3B. The extracted α-gal glycolipids are dried and then dissolve in water or PBS as ball-like structures called micelles. In these micelles, the hydrophobic portion of the glycolipid (i.e., the lipid tail) is in the core of the micelle, whereas the hydrophilic carbohydrate chain protrudes into the aqueous surrounding. A two-dimensional illustration of a micelle made of representative α-gal glycolipid with 10 carbohydrate units (ceramide decahexoside—CDH), each with two branches capped with α-gal epitopes, is schematically described in Fig. 4A.

Insertion of α-gal glycolipids into tumor cell membranes

When α-gal glycolipid micelles are incubated with tumor cells at 37°C, there is a spontaneous insertion of these glycolipids into the tumor cell membranes (Fig. 4A). The reason

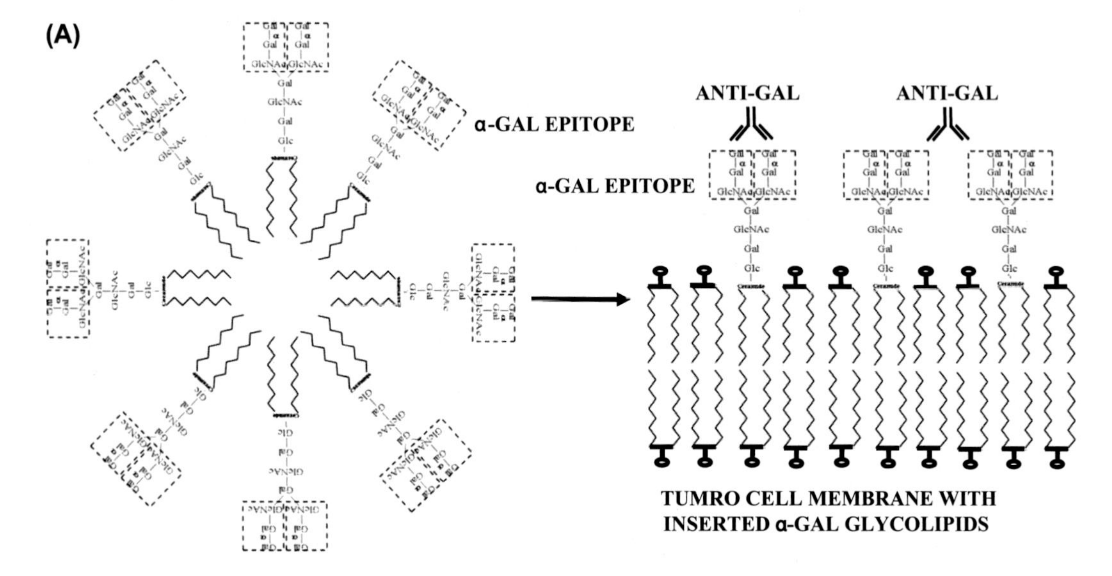

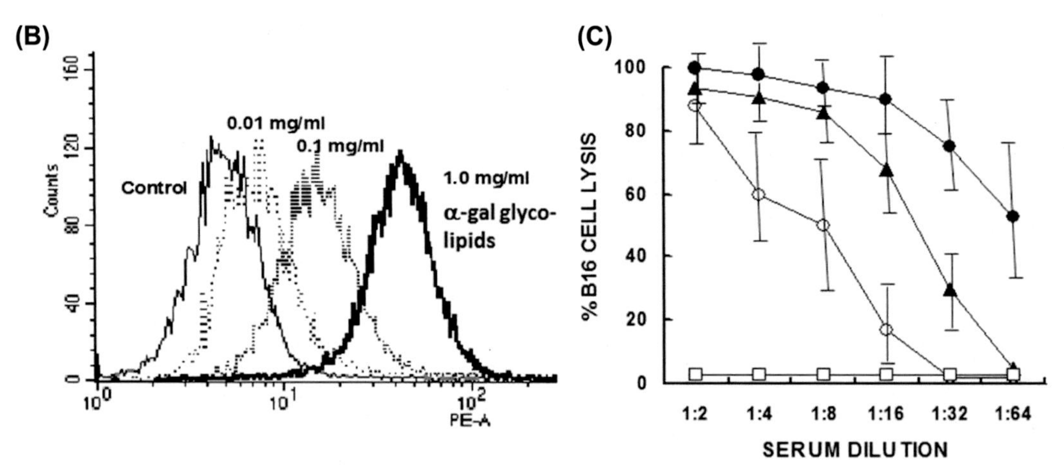

FIGURE 4 Insertion of α-gal glycolipids into tumor cell membranes. (A) Micelles composed of α-gal glycolipids have the hydrophobic ceramide tails in the core of the micelle and the hydrophilic carbohydrate chains protruding into the surrounding aqueous environment. When injected into tumors, the glycolipids spontaneously insert into the outer phospholipid leaflet of tumor cell membranes. The insertion occurs because the hydrophobic ceramide tails are in a more stable energetic state when surrounded by the fatty acid tails of the cell membrane phospholipids than in the core of micelles, surrounded by water molecules. This insertion of α-gal glycolipids is followed by binding of the natural anti-Gal antibody to α-gal epitopes presented *de novo* on the tumor cell membranes. The representative α-gal glycolipid in this illustration is ceramide decahexoside (CdecaH) with 10 carbohydrate units and two branches (antennae), each capped by an α-gal epitope marked by a dashed line rectangle. (B) Flow cytometry analysis of α-gal epitope expression on B16 cells incubated for 2h at 37°C with various concentrations of α-gal glycolipids, or with no glycolipids, and subsequently with monoclonal anti-Gal M86 antibody, then with secondary FITC—anti-mouse IgM antibody. (C) Complement dependent cytotolysis (CDC) of B16 cells that were incubated for 2h with 1mg/mL α-gal glycolipids. Mouse serum containing anti-Gal and rabbit complement (●); mouse serum without rabbit complement (○); human serum (▲); control B16 cells lacking α-gal epitopes and incubated with mouse serum and rabbit complement, or with human serum (□) (mean ± SD of four experiments). *(A) Reprinted with permission from Galili, U., 2013. Anti-Gal: an abundant human natural antibody of multiple pathogeneses and clinical benefits. Immunology 140, 1–11. (B and C) were reprinted with permission from Galili, U., Wigglesworth, K., Abdel-Motal, U.M., 2007. Intratumoral injection of α-gal glycolipids induces xenograft-like destruction and conversion of lesions into endogenous vaccines. J. Immunol. 178, 4676–4687.*

for this insertion is that the hydrophobic lipid tail of glycolipids is energetically much more stable when surrounded by phospholipids of the lipid bilayer in the cell membrane than when surrounded by water molecules in the micelle. Because the aggregation of glycolipids into micelles is a reversible process, lipid tails of glycolipids that "leave" the micelle tend to insert into the outer leaflet of the lipid bilayer of tumor cell membranes. α-Gal glycolipids inserted into tumor cell membranes present their multiple α-gal epitopes, which bind anti-Gal to the tumor cell membrane (Fig. 4A). Thus, incubation of B16 melanoma cells with α-gal glycolipids for 2h at 37°C results in expression of multiple α-gal epitopes on the tumor cells, as indicated by subsequent binding of anti-Gal to the cells (Fig. 4B). This further suggests that a similar insertion may occur within tumor lesions injected with α-gal glycolipids. Indeed, injection of 1mg α-gal glycolipids into B16 lesions resulted in *in situ* insertion of these glycolipids into tumor cell membranes throughout the injected lesion, as demonstrated by binding of the α-gal epitope specific lectin *Bandeiraea* (*Griffonia*) *simplicifolia* IB4 to histologic sections of the injected tumor (Galili et al., 2007). Insertion of α-gal glycolipids into tumor cell membranes, in a dose response manner, could be demonstrated both with mouse B16 melanoma cells (Fig. 4B) and with human melanoma and mammary carcinoma cells (Abdel-Motal et al., 2009; Whalen et al., 2012). Incubation of both mouse and human melanoma cells with inserted α-gal glycolipids in the presence of anti-Gal and complement resulted in cytolysis of the cells because of complement activation by anti-Gal binding to these tumor cells, as shown in Fig. 4C with B16 melanoma cells (Galili et al., 2007; Abdel-Motal et al., 2009).

Induction of natural antibodies binding to tumor cells also was reported to be feasible with natural anti-rhamnose antibodies (Chen et al., 2011; Sheridan et al., 2014; Long et al., 2014; Li et al., 2016). These studies indicated that natural anti-rhamnose antibodies are found in human sera, and that incubation of tumor cells with rhamnose capped glycolipids results in insertion of these glycolipids into tumor cell membranes, binding of the natural anti-rhamnose antibodies to the rhamnose carried by these glycolipids, and induction of complement-mediated cytolysis by this interaction.

Effects of α-gal glycolipids injected into tumors

Intratumoral injection of α-gal glycolipids is followed by rapid interaction between these glycolipids and anti-Gal released from capillaries that are ruptured by the injecting needle. This interaction results in activation of the complement system, which further initiates the following sequential processes (illustrated in Fig. 1A):

1. *Complement-mediated cytolysis and vasodilation within the tumor*—Anti-Gal binding to α-gal epitopes of the glycolipids activates the complement cascade resulting in cytolysis of cells with inserted α-gal glycolipids (Fig. 4C). This antigen/antibody interaction further results in generation of complement cleavage peptides such as C5a and C3a, which induces vasodilation of the blood vessels within the tumor. Serum fluid and proteins are released into the tumor, as indicated by swelling of the treated tumor within 24–72h post injection. The combination of complement-mediated cytolysis of tumor cells and release of serum fluid and proteins enable further dispersion of the α-gal glycolipid micelles into the tumor. Thus, many of the injected B16 tumor lesions stopped expanding and some even regressed or slowed in their growth after injection of 1mg α-gal glycolipids in 2–3

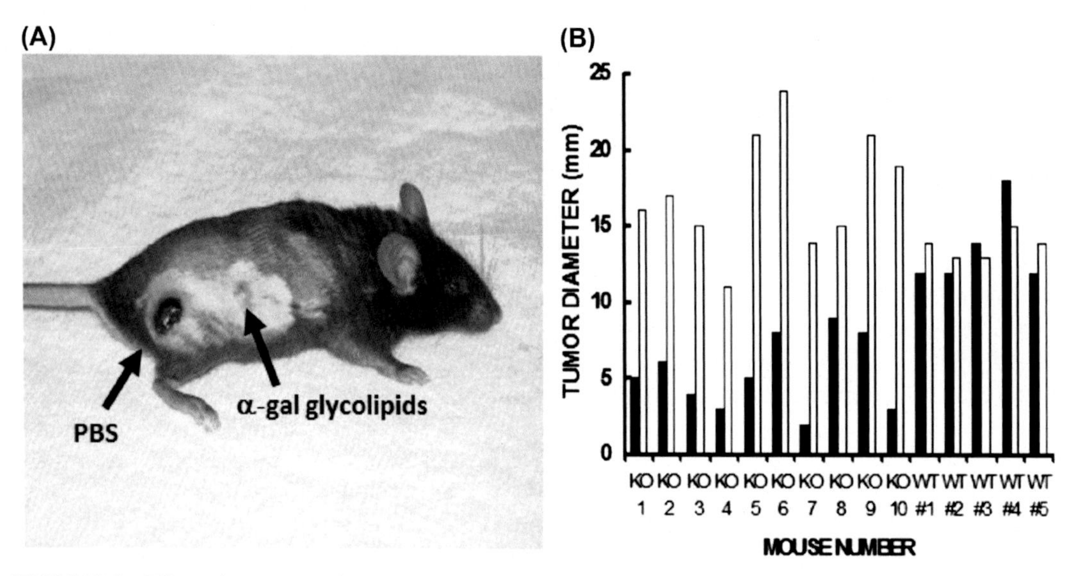

FIGURE 5 Effects of intratumoral injection of α-gal glycolipids on growth of cutaneous B16 melanoma in anti-Gal producing GT-KO mice. (A) Melanoma lesions 10 days after intratumoral injection of 1 mg α-gal glycolipids or PBS in a representative mouse. Treated tumors had an initial diameter of ~5 mm. (B) Comparison of tumor size, 10 days following injection of 1 mg α-gal glycolipids (*closed columns*), or PBS (*open columns*) in pairs of melanoma lesions growing in the same mouse (n = 10). Similar study, performed in WT mice (n = 5) demonstrated no significant differences between α-gal glycolipids and PBS intratumoral injections in the absence of the anti-Gal antibody. *Reprinted with permission from Galili, U., Wigglesworth, K., Abdel-Motal, U.M., 2007. Intratumoral injection of α-gal glycolipids induces xenograft-like destruction and conversion of lesions into endogenous vaccines. J. Immunol. 178, 4676–4687.*

sites within the tumor in comparison with PBS injected tumors on the same mouse (Fig. 5) (Galili et al., 2007). Similar analysis in WT mice resulted in continuous growth of both α-gal glycolipids and PBS-injected tumors. It should be stressed that complete regression of the injected tumors was observed in only a third of the treated mice and slow of tumor growth but no regression in most of the remaining mice. It is possible that the rapid proliferation of B16 cells *in vivo* (B16 tumors double their size every 4–8 days) enables even small numbers of tumor cells that remain alive following the intratumoral injection to cause tumor growth.

2. *Chemotactic Recruitment of APC into the tumor*—The complement cleavage peptides C5a and C3a are effective chemotactic factors that induce recruitment of APC such as macrophages and dendritic cells into the treated tumor (Galili et al., 2007). This chemotactic recruitment can be observed already within 2 days in perivascular areas as extravasation of the mononuclear cells from capillaries (Fig. 6A). Migration of the cells recruited into the tumor further increases after 7 days (Fig. 6B). In many of the injected tumors, the B16 melanoma cells are killed *in situ* by anti-Gal binding to the tumor cells, activating complement, and inducing cell lysis by complement-dependent cytolysis (CDC). The high efficacy of anti-Gal IgG in mediating antibody dependent cell cytolysis of xenograft cells (ADCC) (Galili, 1993) suggests that this mechanism also may contribute to the killing of tumor cells with inserted α-gal glycolipids. These CDC and ADCC mechanisms induce the regression of treated tumors as shown in Figs. 5 and 6C, in which

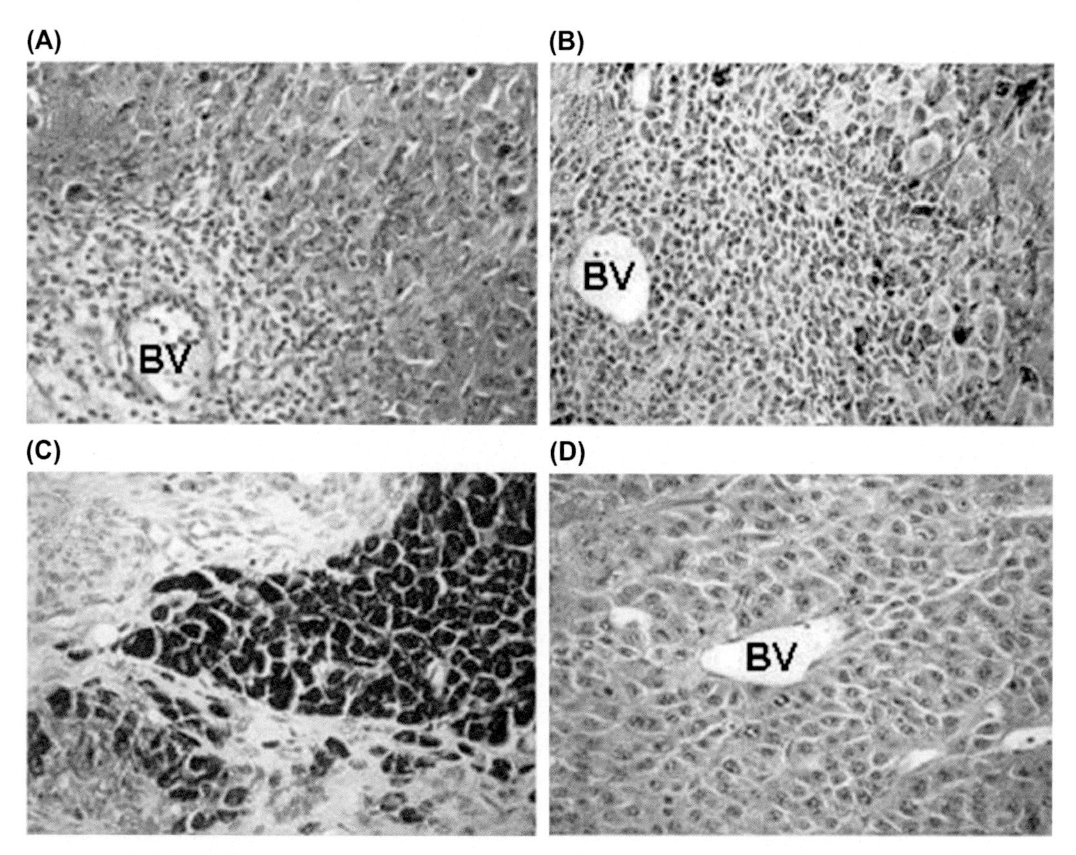

FIGURE 6 Induction of mononuclear cell extravasation within ~5 mm B16 melanoma lesions in anti-Gal producing GT-KO mice that were injected intratumoral with 1mg α-gal glycolipids. (A) Tumors resected on Day 2 post injection display extravasation of mononuclear cells from blood vessels (BV). (B) Tumors resected on Day 7 display a more extensive extravasation of mononuclear cells than on Day 2. (C) Tumor resected on Day 20 displays pyknotic B16 cells, suggesting destruction of the melanoma lesion. (D) Control melanoma lesions injected with PBS and studied on Day 7. No mononuclear cell extravasation is observed in the tumor (H&E, ×200). *Adapted with permission from Galili, U., Albertini, A., Sondel, P., Wigglesworth, K., Sullivan, M., Whalen, G., 2010. In situ conversion of melanoma lesions into autologous vaccine by intratumoral injections of α-gal glycolipids. Cancers 2, 773–793.*

the tumor cells die as indicated by their pyknotic appearance. Control tumors injected with PBS displayed no recruitment of cells into the tumor (Fig. 6D).

3. *Uptake of tumor cells by APC for increased immunogenicity of autologous TAA*—As discussed above, in many cancer patients the immune system is "oblivious" to the developing tumor, and thus there is no uptake of tumor cells or cell membranes by APC, and no infiltration of T cells is observed in tumors. In these patients, the absence of an anti-TAA immune response further enables metastatic tumor cells to spread from the primary tumor to various organs. The opsonization by anti-Gal bound to α-gal epitopes on tumor cells in treated lesions overcomes this deficiency. Opsonizing anti-Gal IgG molecules binds via their Fc portion to FcγR on dendritic cells and macrophages and stimulates these APC to internalize the tumor cells and cell membranes with the autologous TAA (Fig. 1).

4. *Transport of TAA to draining (regional) lymph nodes*—The TAA are transported by APC internalizing them to regional lymph nodes. These APC further process and present immunogenic TAA peptides for the activation of tumor-specific T cells. Activated T cells proliferate within the lymph nodes, then migrate into the circulation and seek and destroy tumor cells in micrometastases, which express the immunizing TAA. The general efficacy of anti-Gal in increasing the immunogenicity of antigens by targeting them to APC was demonstrated in Chapter 9, which describes manyfold increase in the immunogenicity of ovalbumin (OVA), flu virus and gp120 of HIV carrying α-gal epitopes. As described below, OVA was further studied as a surrogate TAA to evaluate the effect of intratumoral injection of α-gal glycolipids on the APC mediated transport of TAA to lymph nodes (Fig. 7) and thus, on the immunogenicity of autologous TAA.

Increased immunogenicity of ovalbumin as surrogate tumor-associated antigen in B16 tumors injected with α-gal glycolipids

B16 cells producing OVA as a surrogate TAA (designated B16/OVA cells) were generated by stable transfection of B16 cells with the OVA gene (Brown et al., 2001; Lugade et al., 2005). The processing and presentation of the most immunogenic OVA peptide SIINFEKL, presented to CD8+ T cells on class I MHC molecules, was evaluated by the use of the B3Z T cell hybridoma with the T cell receptor (TCR) for SIINFEKL (Karttunen et al., 1992; Shastri and Gonzalez, 1993; Sanderson and Shastri, 1994) as detailed in Chapter 9. When SIINFEKL is presented by APC that internalized B16/OVA tumor cells, it binds to the TCR of the cocultured B3Z cells, resulting in the activation of the β-galactosidase gene and fluorescent detection of the catalytic activity of the produced enzyme.

Intratumoral injection of α-gal glycolipids into B16/OVA lesions on the right thigh of anti-Gal producing GT-KO mice was found to be followed by effective transport and presentation of SIINFEKL by APC in the draining inguinal lymph nodes. This was indicated by an average of approximately five-fold higher activation of B3Z cells coincubated with cells from inguinal lymph nodes draining B16/OVA lesions injected with α-gal glycolipids than the activation of B3Z cells by inguinal lymph node cells from mice with PBS injected tumors (Fig. 7). This, in turn, resulted in a much higher number of SIINFEKL specific activated CD8+ T cells in mice with α-gal glycolipids injected tumors than in mice with PBS injected tumors. The elevated activity could be demonstrated both by ELISPOT and by cytolysis of target cells presenting SIINFEKL on class I MHC molecules (Galili et al., 2007).

INDUCTION OF ANTI-TUMOR PROTECTIVE IMMUNE RESPONSE AGAINST DISTANT LESIONS BY α-GAL GLYCOLIPIDS

The putative ability of activated tumor specific T cells to destroy B16 tumor cells was first studied by adoptive transfer with spleen cells from mice with B16 tumors that were injected with α-gal glycolipids. GT-KO mice bearing B16 tumors received two intratumoral injections of 1 mg α-gal glycolipids or PBS in 1 week interval. One week after the second injection, the spleens were harvested, and 40×10^6 splenocytes were infused into naïve GT-KO mice that

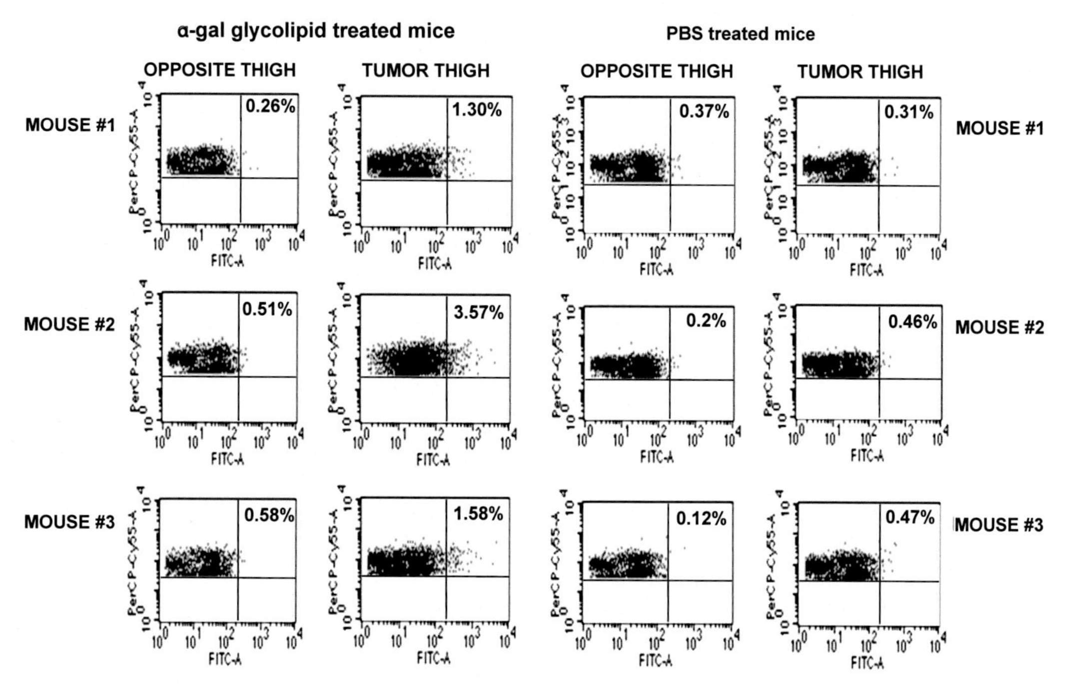

FIGURE 7 Transport of ovalbumin (OVA) as a surrogate TAA from B16/OVA melanoma lesions injected with α-gal glycolipids to inguinal draining lymph nodes and in anti-Gal producing GT-KO mice. Lymph nodes were harvested 2 weeks post intratumoral injection of α-gal glycolipids. Antigen presenting cells presenting the OVA immunodominant SIINFEKL peptide were detected by flow cytometry of activated CD8+ B3Z T hybridoma cells containing *lac-Z* transgene under IL2 promoter. These T hybridoma cells have TCR for SIINFEKL presented in association with class I MHC molecules. The activated B3Z cells were detected by double staining for CD8 expression with PerCP-anti-CD8 antibody (red in the flow cytometer) and FITC-di-β–galactoside hydrolyzed by β-galactosidase, produced by activated *lac-Z* (green in the flow cytometer). Data from three representative mice out of eight mice in each group. *Reprinted with permission from Galili, U., Wigglesworth, K., Abdel-Motal, U.M., 2007. Intratumoral injection of α-gal glycolipids induces xenograft-like destruction and conversion of lesions into endogenous vaccines. J. Immunol. 178, 4676–4687.*

were challenged subcutaneously with 3×10^5 live B16 cells. This challenge was performed 24 h prior to the administration of the splenocytes. The recipients were monitored for growth of the challenging tumor cells. Recipients of lymphocytes from α-gal glycolipids–treated B16 tumors display no growth or very slow growth of tumors in 80% of the mice. In contrast, recipients of lymphocytes from mice with PBS-injected tumors displayed no growth or slow growth of the challenging tumor cells in only 20% of the mice, whereas the rest of the mice displayed normal tumor growth (Galili et al., 2007). These observations suggested that injection of B16 tumors with α-gal glycolipids activates lymphocytes that can recognize and kill B16 melanoma cells. These activated lymphocytes could be detected by their adoptive transfer into naïve mice that were challenged with B16 tumor cells. The transferred lymphocytes killed the challenging tumor cells, thereby prevented the tumor cells from developing into tumor lesions in the naïve recipients.

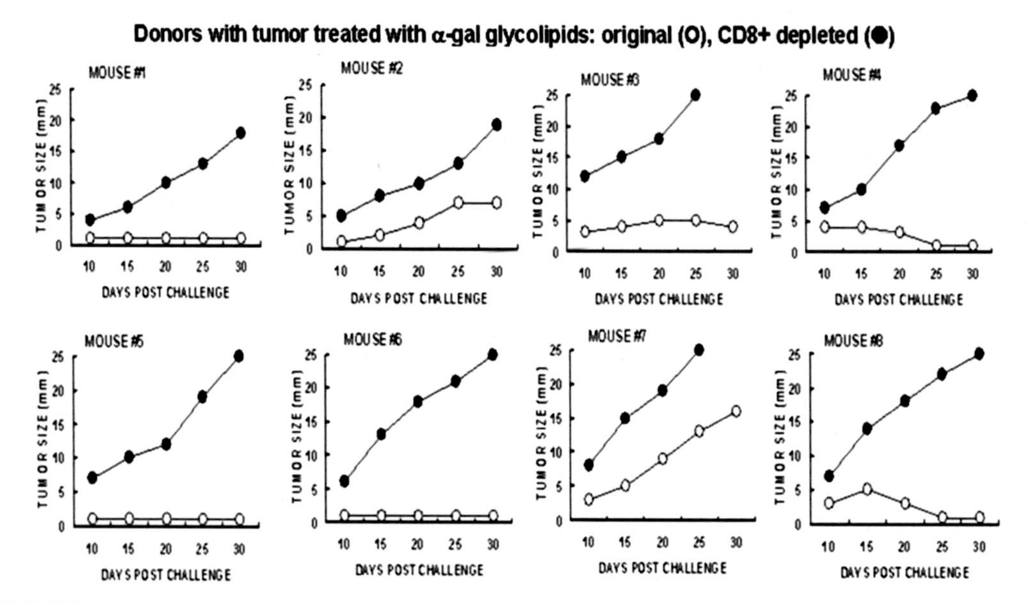

FIGURE 8 Tumor growth in GT-KO naïve recipient mice of spleen lymphocytes from donor mice with B16 melanoma lesions injected with α-gal glycolipids. Naïve recipients were challenged subcutaneously with 0.5×10^6 live B16 cells, 24h prior to transfer of lymphocytes from tumor bearing mice treated with α-gal glycolipids. Adoptive transfer of lymphocytes from each donor spleen was performed in pairs of recipients, each receiving 40×10^6 cells. (○)—total lymphocytes, or (●)—lymphocytes depleted of CD8+ T cells. The size of the growing tumors is described at various time points up to 30 days post adoptive transfer. *Reprinted with permission from Abdel-Motal, U.M., Wigglesworth, K., Galili, U., 2009. Intratumoral injection of α-gal glycolipids induces a protective anti-tumor T cell response which overcomes Treg activity. Cancer Immunol. Immunother. 58, 1545–1556.*

Identification of the lymphocyte population that protects against the challenging B16 tumor cells was performed by repeating the adoptive transfer studies with splenocytes from donors carrying B16 tumors that were injected with α-gal glycolipids. Half of the splenocytes from each donor were subjected to elimination of CD8+ T cells by magnetic microbeads coated with the corresponding antibody. Pairs of recipient naïve mice, challenged subcutaneously with B16 cells, received 40×10^6 splenocytes that were either unseparated or splenocytes devoid of CD8+ T cells. The adoptive transfer was performed 24h post challenge with B16 cells. Monitoring of the tumor growth in the recipient mice indicated that the unseparated splenocytes introduced by adoptive transfer completely prevented the development of tumor lesions in 75% of the recipients (Fig. 8). However, mice receiving the splenocytes from the same donor, which were depleted of CD8+ T cells, displayed no immune protection, as indicated by the tumor growth in them (Fig. 8) (Abdel-Motal et al., 2009). These observations indicate that B16 specific CD8+ T cells are likely to mediate much of the protection against tumor growth.

In accord with the protection observed in adoptive transfer studies, 65% of anti-Gal producing GT-KO mice bearing B16 tumors and receiving three weekly injections of 1 mg α-gal glycolipids were protected from challenge by 0.5×10^6 B16 cells administered subcutaneously in the contralateral flank (Fig. 9A). The rest of the mice displayed slowed growth of the challenging cells into tumor lesions. In contrast, mice with tumors that were ablated by intratumoral injection of ethanol displayed no protection against the challenging tumor cells in the contralateral flank (Abdel-Motal et al., 2009) (Fig. 9B). These observations suggested that

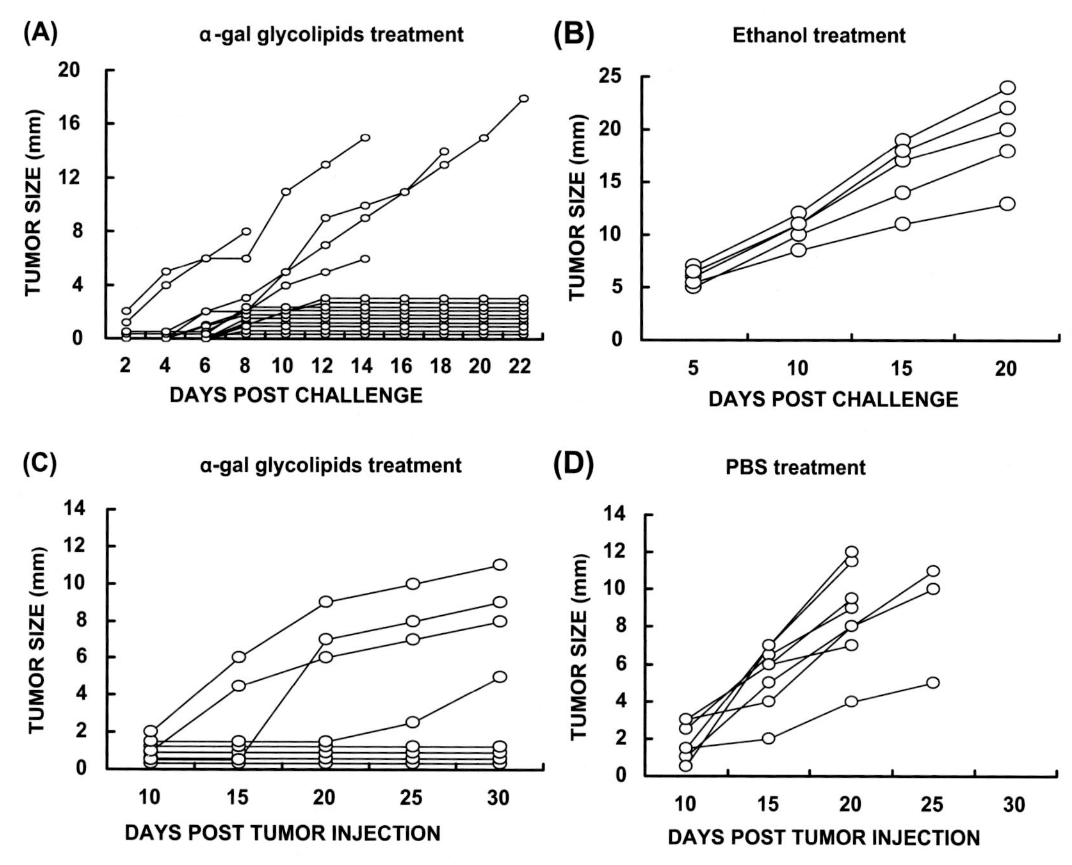

FIGURE 9 Protection against tumor challenge and from distant lesion growth by intratumoral injection of α-gal glycolipids. (A) GT-KO mice with B16 tumors received 3 weekly intratumoral injections of 1 mg α-gal glycolipids. The mice were subsequently challenged with 0.5×10^6 B16 cells in the contralateral flank and tumor growth of the challenge monitored. 10 of 15 mice injected with α-gal glycolipids displayed no development of the challenging B16 cells into melanoma lesions. (B) Tumor growth following challenge of mice in which the tumor was destroyed by ablation with ethanol injection. The challenge was performed 3 weeks after ethanol ablation. All five challenged mice developed melanoma lesions. (C) Prevention of distant tumor growth by intratumoral injection of α-gal glycolipids into the "primary" tumor. B16 melanoma cells were administered subcutaneously into anti-Gal producing GT-KO mice as10^4 cells in the left flank ("distant" lesion) and 10^6 cells in the right flank ("primary" tumor). After 5 days, the right flank tumor received an intratumoral injection of 1 mg α-gal glycolipids. The injection was repeated a week later and the growth of the left flank tumor was monitored. (D) Tumor cells were administered as in Fig. 9C, however, the right flank tumor received two PBS injections (n = 8 mice/group). Mice were euthanized when the right flank "primary" tumor reached the size of 20 mm. *Reprinted with permission from Abdel-Motal, U.M., Wigglesworth, K., Galili, U., 2009. Intratumoral injection of α-gal glycolipids induces a protective anti-tumor T cell response which overcomes Treg activity. Cancer Immunol. Immunother. 58, 1545–1556.*

dead tumor cells, as in ethanol ablation, do not induce a protective immune response and that active targeting to APC, such as that mediated by anti-Gal, is required for eliciting protective anti-tumor immune response.

Simulation of the protective immune response against distant "established metastases" by intratumoral injection of α-gal glycolipids was studied in mice receiving subcutaneously 10^6 tumor cells in the right flank simulating the "primary tumor" and

10^4 tumor cells in the left flank simulating a "distant metastasis." After 5–6 days, the primary tumors reaching the size of ~5 mm were injected with 1 mg α-gal glycolipids or with PBS. At that time, the left flank tumors were not visible. The injections were repeated after a week, and tumor growth in the left flank "distant metastasis" was monitored. In all PBS-treated mice, tumor lesions developed in the left flank were visible as 2–8 mm tumors by Day 15 post"primary" tumor injection (Fig. 9D). However, in 50% of mice receiving α-gal glycolipids treatment no tumor growth was observed in the left flank even after a month, whereas the remaining mice displayed slower tumor growth than most PBS-treated mice (Fig. 9C) (Abdel-Motal et al., 2009).

Analysis of T cell response to immunogenic melanoma peptides TRP2 and gp100 was measured by IFNγ secretion in ELISPOT. This analysis indicated that in mice with B16 tumors injected with α-gal glycolipids, there are approximately six-fold more T cells activated by $TRP_{180\text{-}188}$ peptide and $gp100_{25\text{-}33}$ peptide than in mice receiving intratumoral injection of PBS (Abdel-Motal et al., 2009). The increased numbers of tumor-specific T cells in spleens and the higher protective immune response against B16 tumor cells in mice with lesions injected with α-gal glycolipids suggest that this treatment amplifies the immune response to autologous TAA, which otherwise may be "ignored" by the immune system.

α-Gal glycolipids injections do not elicit autoimmune reactions

Immunotherapy by intratumoral injection of α-gal glycolipids raised the concern that anti-Gal-mediated uptake by APC also includes normal antigens present on tumor cells. In addition, α-gal glycolipids are likely to insert into cell membranes of normal cells within tumors (e.g., fibroblasts and endothelial cells). This may result in uptake of normal cells by APC via opsonizing anti-Gal interaction with FcγR on APC. None of the mice with B16 tumors injected with α-gal glycolipids displayed autoimmune reactivity. The possibility of autoimmunity induction was further studied as part of the studies required for IND 12946 submission to the FDA for a clinical trial (see below). Anti-Gal producing GT-KO mice received two intradermal injections of 1 mg α-gal glycolipids in 1 week interval. Fifteen different tissues from each of the injected mice were studied for any inflammatory infiltrates, 13 and 30 days post injection. All tissues studied displayed no such infiltrates. In addition, no autoimmunity induction was observed in anti-Gal producing GT-KO mice that were repeatedly injected with kidney or liver homogenates presenting α-gal epitopes (i.e., tissue homogenates from WT mice). All these observations suggest that the immune tolerance to normal antigens is maintained, despite anti-Gal-mediated targeting of normal cell membranes presenting α-gal epitopes, or of normal antigens on tumor cells presenting these epitopes.

PHASE I CLINICAL TRIALS WITH α-GAL GLYCOLIPIDS IMMUNOTHERAPY IN CANCER PATIENTS

In vitro studies demonstrated insertion of α-gal glycolipids into human melanoma and mammary carcinoma cells as that observed with mouse B16 melanoma cells (Abdel-Motal et al., 2009; Whalen et al., 2012). These observations and the observations on conversion of B16 melanoma in anti-Gal producing GT-KO mice into autologous vaccines led to phase I clinical trials on the safety of immunotherapy with these glycolipids. Two such studies were performed as follows: (1) dose escalation study in patients with internal solid tumors having advanced disease (Whalen et al., 2012) and (2) dose escalation in patients with advance cutaneous melanoma (Albertini et al., 2016).

Phase I clinical trial in patients with internal solid tumors

Eleven patients (age range 40–80 years) with a variety of solid tumors were treated in this phase I trial at the UMass Medical Center (Whalen et al., 2012). These studies were performed after approval by a local Human Investigations Committee and in accordance with an assurance filed with and approved by the Department of Health and Human Services, (FDA IND 12946). Informed consent was obtained from each of the participants. Eight patients had metastatic cancers including colon, pancreatic neuroendocrine, prostate, renal, ovarian, and mucinous appendiceal tumors and were heavily pretreated. The three patients, who did not have metastatic disease when they were injected with α-gal glycolipids, had locally advanced pancreatic adenocarcinoma. The patients were treated with standard therapy and experienced recurrence of the disease prior to the clinical trial. The patients received intratumorally 0.1, 1, or 10 mg α-gal glycolipids and kept under observation for 24 h. α-Gal glycolipids were injected into the target tumor lesions under ultrasound guidance including endoscopic ultrasound, CT guidance, or under manual and visual control in patients who had large palpable tumors. The endpoint of the study for monitoring toxicity was 4-weeks post injection. Subsequently, the patients were monitored at regular intervals for a period of 2 years. None of the patients developed clinical or laboratory signs of toxicity, symptoms of allergic responses, autoimmune conditions, or anti-nuclear antibodies. No autoantibodies to normal tissue antigens were found when assayed by ELISA with human colon or skeletal muscle homogenates as solid-phase antigens (Whalen et al., 2012). Thus, like the observations in GT-KO mice, immune tolerance to self-normal antigens was maintained following intratumoral injection of α-gal glycolipids and did not enable the emergence of autoimmune responses. Injected tumors did not regress and patients developed evidence of disease progression at various time points after the 4-week endpoint. However, several patients were alive with disease for 13+ to 48+ months, even though disease progression was evident by imaging. In addition, two patients with advanced pancreatic adenocarcinoma had an unexpectedly long survival of 18 and 23 months (Whalen et al., 2012).

Phase I clinical trial in patients with cutaneous melanoma

A second phase I clinical trial was performed with nine patients with advanced melanoma at the University of Wisconsin Carbone Cancer Center under FDA IND 12946 (Albertini et al., 2016). All patients had advanced unresectable melanoma (recurrent stage III or stage IV) that was refractory to therapy, with at least one cutaneous metastasis. Patients received two intratumoral injections of 0.1, 1.0, or 10 mg α-gal glycolipids administered 4 weeks apart. Because of the observations on the possible presence of an allergic anti-Gal IgE antibody in a small proportion of various populations (see Chapter 7), patients received prior to the second treatment, an intradermal injection in an extremity with 10 μg of α-gal glycolipids and then observed for 1 h to test for allergic reaction to α-gal glycolipids. The sizes of treated and untreated lesions were monitored during regular clinic visits. Two patients maintained stable disease for 8 and 7 months after treatment, and the remaining seven patients had disease progression. Three of the injected lesions were stable in size, 4 weeks after the second of α-gal glycolipids injection, and the remaining six lesions had at least a 20% increase in that period. Tumor biopsies demonstrated increased tumor necrosis in the injected lesion in five of nine patients, 2 weeks after the first injection, and nontreated tumor nodules in two of four evaluable patients showed necrosis, as well (Albertini et al., 2016). In attempt to determine whether the treatment affects tumor-specific T cell populations, blood lymphocytes from eight of nine treated patients, who were HLA-A2$^+$, were studied for binding of pooled melanoma TAA pentamers. TAA-specific T cells are expected to bind TAA peptides presented on the biotin-labeled HLA-A2 of the pentamers. An approximate two-fold to five-fold increase in the frequency of TAA pentamer+ cells posttreatment versus pretreatment was observed in three of eight patients. In addition, the fluorescence intensity of T cells-binding TAA pentamers increased posttreatment in three of the patients studied. These observations suggest that in a proportion of the patients, intratumoral injection of α-gal glycolipids induced the expansion of tumor-specific T cells.

Overall, the observations in cancer patients studied in the two phase I clinical trials indicated that the treatment is safe. However, no cure or distinct regression of tumors was observed under the α-gal glycolipids doses used for treatment of patients in the advanced stage of the disease. It is possible that increased doses may be more effective in eliciting a protective immune response in patients with large tumor burden. Evidently, the α-gal glycolipids should be administered into the tumor and not infused into the blood, to ensure that the injected α-gal glycolipids reach the malignant tissue in the treated patient. It should be further stressed that although the α-gal glycolipids are likely to induce CDC and ADCC mediated killing of tumor cells in the injected tumors, this treatment should not be considered as a therapy for direct destruction of treated tumor lesions because the injected glycolipids may not reach all tumor cells in the treated lesion. The primary objective of the proposed treatment with α-gal glycolipids is to convert the injected tumor into a vaccine, which provides immunogenic autologous TAA for eliciting an immune response against micrometastases. Resection or ablation is likely to be much more effective methods for direct destruction of tumor lesions. As discussed below, in addition to the use of α-gal glycolipids in patients who are refractory to other therapies, these glycolipids may be considered for use as neoadjuvant immunotherapy prior to the resection of primary tumors and they may synergize with other immunotherapies such as checkpoint inhibitors.

FUTURE DIRECTIONS IN α-GAL GLYCOLIPIDS IMMUNOTHERAPY RESEARCH

Possible use of α-gal glycolipids in neoadjuvant immunotherapy

In addition to immunotherapeutic effects of α-gal glycolipids in cancer patients with advanced disease, this immunotherapy may be of significance as neoadjuvant immunotherapy. In such immunotherapy, the patient undergoes treatment for eliciting protective anti-TAA immune response prior to the resection or ablation of the primary tumor. Monitoring the processing and transport of TAA internalized by APC in GT-KO mice with melanoma lesions injected with α-gal glycolipids indicated that TAA within APC reach the regional lymph within a period shorter than 2 weeks (Galili et al., 2007) (Fig. 7). These observations suggest that if a primary tumor (e.g., mammary carcinoma and colon carcinoma) is injected with α-gal glycolipids upon detection, by the time the tumor is resected 2–4 weeks later; APC internalizing the TAA might migrate out of the tumor toward the draining lymph nodes. Therefore, a window of time of at least 2 weeks between injection and resection may allow for the temporary conversion of the tumor into an autologous TAA vaccine. By the time the tumor is resected, as part of the standard treatment protocol of the cancer patient, the APC with internalized TAA have left the treated tumor and migrated toward the draining lymph nodes to activate the immune system against the immunogenic TAA peptides. TAA-specific T cells activated by these APC may mediate a protective immune surveillance against tumor cells presenting TAA in micrometastases. This neoadjuvant therapy may prevent recurrence of the disease because of development of invisible micrometastases into lethal tumor lesions. The TAA in micrometastases may further serve as a boost, restimulating the immune system to maintain a long lasting anti-TAA immunological memory, which can improve the prognosis of high risk patients.

Synergism of α-gal glycolipids with checkpoint inhibitors

A successful application of cancer immunotherapy in the recent decade has been the use of checkpoint inhibitors such as anti-CTLA1 monoclonal antibody (ipilimumab) and anti-PD1 monoclonal antibody (nivolumab and pembrolizumab). These antibodies prevent apoptosis of T cells following their activation and thus enable the expansion of activated T cell clones, many of which are directed against tumor cells presenting TAA peptides (Korman et al., 2006; Intlekofer and Thompson, 2013; Callahan and Wolchok, 2013). This immunotherapy was found to delay tumor growth and increase survival. However, many of the patients display a wide range of autoimmune adverse events, which seem to be associated with hyperactivated T-cell response with reactivity directed against normal tissue antigens (Weber et al., 2015; Day and Hansen, 2016; Kostine et al., 2017). In checkpoint inhibitors immunotherapy, the activation of T cells is not specific to the T cell clones with TAA specificity. Thus, the autoimmune adverse events may represent expansion of multiple T cell clones that cross-react with normal tissue antigens, resulting in organ damage. In view of these considerations, it would be of interest to determine whether combining the checkpoint inhibitors treatment with α-gal glycolipids immunotherapy can decrease toxicity of the former treatment and increase its efficacy. In could be assumed that such a combined treatment may consist of initial administration of α-gal glycolipids followed by administration of checkpoint inhibitors. The initial injection of α-gal glycolipids will induce activation of TAA-specific T cells, whereas the subsequent administration of checkpoint inhibitors, few weeks later, will enable the specific expansion of these activated T cell clones. Such a combined immunotherapy may lower the dose of the checkpoint inhibitors, so they enable the selective expansion of the TAA activated T cell clones without affecting other T cell clones. Thus, doses of α-gal glycolipids and of checkpoint inhibitors, found optimal for this combined immunotherapy, may decrease the likelihood of expansion of T cells specific to normal tissue antigens, while inducing selective expansion of protective TAA specific activated T cells.

CONCLUSIONS

Invisible micrometastases may develop into lethal metastases after the resection or ablation of detectable lesions. Destruction of micrometastases may be feasible by activating the immune system to react against tumor associated antgens (TAA) of the patient's tumor cells. Most TAA are generated because of genomic instability in tumor cells, which causes the formation of multiple mutations that are unique to the tumor of each patient. It is not feasible at present to identify the wide range of immunogenic TAA peptides in tumor lesions in order to prepare customized vaccine for each individual patient. Therefore, the autologous tumor cells may be considered as the source for vaccinating TAA. The immunogenicity of TAA in patients developing metastases is very low. Increasing the immunogenicity of TAA of autologous tumor cells requires effective targeting of these cells for uptake by APC. Such targeting may be achieved by *in situ* manipulation of the tumor lesions to present α-gal epitopes on their cells. Binding of the natural anti-Gal antibody to these epitopes activates the complement system that induces complement-mediated lysis of the tumor cells and generates chemotactic complement cleavage peptides that recruit APC such as dendritic cells and macrophages into the tumor. Binding of the Fc portion of anti-Gal opsonizing the tumor cells to FcγR on APC induces effective uptake of tumor cells and cell membranes by the APC and transport of the internalized TAA by the APC to the regional lymph nodes. These APC further process the immunogenic TAA peptides and present them on class I and II MHC molecules for the activation of TAA-specific CD8+ and CD4+ T cells, respectively. The activated T cells mediate a protective anti-tumor immune response against metastatic cells. Analysis of various methods for inducing α-gal epitope expression indicated that intratumoral injection of α-gal glycolipid micelles is the most effective method tested. The method was found to be safe in cancer patients with advanced disease. It is suggested that in addition to the use of this immunotherapy method in patients who are likely to have micrometastases, this method may elicit a protective immune response as a neoadjuvant immunotherapy prior to the resection of the primary tumor. This method may also synergize with checkpoint inhibitors therapy by amplifying the anti-tumor specific immune response and reducing the risk of autoimmune adverse events.

References

Abdel-Motal, U.M., Wigglesworth, K., Galili, U., 2009. Intratumoral injection of α-gal glycolipids induces a protective anti-tumor T cell response which overcomes Treg activity. Cancer Immunol. Immunother. 58, 1545–1556.

Albertini, M.R., Ranheim, E.A., Zuleger, C.L., Sondel, P.M., Hank, J.A., Bridges, A., et al., 2016. Phase I study to evaluate toxicity and feasibility of intratumoral injection of α-gal glycolipids in patients with advanced melanoma. Cancer Immunol. Immunother. 65, 897–907.

Berlyn, K.A., Schultes, B., Leveugle, B., et al., 2001. Generation of CD4(+) and CD8(+) T lymphocyte responses by dendritic cells armed with PSA/anti-PSA (antigen/antibody) complexes. Clin. Immunol. 101, 276–283.

Brown, D.M., Fisher, T.L., Wei, C., Frelinger, J.G., Lord, E.M., 2001. Tumours can act as adjuvants for humoral immunity. Immunology 102, 486–497.

Callahan, M.K., Wolchok, J.D., 2013. At the bedside: CTLA-4- and PD-1-blocking antibodies in cancer immunotherapy. J. Leukoc. Biol. 94, 41–53.

Campbell, P.J., Yachida, S., Mudie, L.J., Stephens, P.J., Pleasance, E.D., Stebbings, L.A., et al., 2010. The patterns and dynamics of genomic instability in metastatic pancreatic cancer. Nature 467, 1109–1113.

Chen, W., Gu, L., Zhang, W., Motari, E., Cai, L., Styslinger, T.J., et al., 2011. L-rhamnose antigen: a promising alternative to α-gal for cancer immunotherapies. ACS Chem. Biol. 6, 185–191.

Dabrowski, U., Hanfland, P., Egge, H., Kuhn, S., Dabrowski, J., 1984. Immunochemistry of I/i-active oligo- and polyglycosylceramides from rabbit erythrocyte membranes. Determination of branching patterns of a ceramide pentadecasaccharide by 1H nuclear magnetic resonance. J. Biol. Chem. 259, 7648–7651.

Day, D., Hansen, A.R., 2016. Immune-related adverse events associated with immune checkpoint inhibitors. BioDrugs 30, 571–584.

Deguchi, T., Tanemura, M., Miyoshi, E., Nagano, H., Machida, T., Ohmura, Y., et al., 2010. Increased immunogenicity of tumor-associated antigen, mucin 1, engineered to express α-gal epitopes: a novel approach to immunotherapy in pancreatic cancer. Cancer Res. 70, 5259–5269.

Deriy, L., Chen, Z., Gao, G., Galili, U., 2002. Expression of α-Gal epitopes (Galα1-3Galβ1-4GlcNAc-R) on human cells following transduction with adenovirus vector containing α1,3galactosyltransferase cDNA. Glycobiology 12, 135–144.

Deriy, L., Ogawa, H., Gao, G., Galili, U., 2005. In vivo targeting of vaccinating tumor cells to antigen presenting cells by a gene therapy method with adenovirus containing the α1,3galactosyltransferase gene. Cancer Gene Ther. 12, 528–539.

Dhodapkar, K.M., Krasovsky, J., Williamson, B., Dhodapkar, M.V., 2002. Antitumor monoclonal antibodies enhance cross-presentation of cellular antigens and the generation of myeloma-specific killer T cells by dendritic cells. J. Exp. Med. 195, 125–133.

Dranoff, G., Jaffe, E., Lazenby, A., Golumbek, P., Levitsky, H., Brose, K., et al., 1993. Vaccination with irradiated tumor cells engineered to secrete murine granulocyte-macrophage colony-stimulating factor stimulate potent specific and long-lasting anti-tumor immunity. Proc. Natl. Acad. Sci. U.S.A. 90, 3539–3543.

Dunn, G.P., Old, L.J., Schreiber, R.D., 2004. The immunobiology of cancer immunosurveilance and immunoediting. Immunity 21, 137–148.

Egge, H., Kordowicz, M., Peter-Katalinic, J., Hanfland, P., 1985. Immunochemistry of I/i-active oligo- and polyglycosylceramides from rabbit erythrocyte membranes. J. Biol. Chem. 260, 4927–4935.

Eto, T., Ichikawa, Y., Nishimura, K., Ando, S., Yamakawa, T., 1968. Chemistry of lipids of the posthemolytic residue or stroma of erythrocytes. XVI. Occurrence of ceramide pentasaccharide in the membrane of erythrocytes and reticulocytes in rabbit. J. Biochem. (Tokyo) 64, 205–213.

Fanger, N.A., Wardwell, K., Shen, L., Tedder, T.F., Guyre, P.M., 1996. Type I (CD64) and type II (CD32) Fcγ receptor-mediated phagocytosis by human blood dendritic cells. J. Immunol. 157, 541–548.

Folch, J., Lees, M., Sloane Stanley, G.H., 1957. A simple method for the isolation and purification of total lipids from animal tissues. J. Biol. Chem. 226, 497–509.

Furumoto, K., Soares, L., Engleman, E.G., Merad, M., 2004. Induction of potent antitumor immunity by in situ targeting of intratumoral DCs. J. Clin. Investig. 113, 774–783.

Galili, U., 1993. Interaction of the natural anti-Gal antibody with α-galactosyl epitopes: a major obstacle for xenotransplantation in humans. Immunol. Today 14, 480–482.

Galili, U., 2004. Autologous tumor vaccines processed to express α-gal epitopes: a practical approach to immunotherapy in cancer. Cancer Immunol. Immunother. 53, 935–945.

Galili, U., 2013. Anti-Gal: an abundant human natural antibody of multiple pathogeneses and clinical benefits. Immunology 140, 1–11.

Galili, U., Rachmilewitz, E.A., Peleg, A., Flechner, I., 1984. A unique natural human IgG antibody with anti-α-galactosyl specificity. J. Exp. Med. 160, 1519–1531.

Galili, U., Mandrell, R.E., Hamadeh, R.M., Shohet, S.B., Griffiss, J.M., 1988. Interaction between human natural anti-α-galactosyl immunoglobulin G and bacteria of the human flora. Infect. Immun. 56, 1730–1737.

Galili, U., Tibell, A., Samuelsson, B., Rydberg, L., Groth, C.G., 1995. Increased anti-Gal activity in diabetic patients transplanted with fetal porcine islet cell clusters. Transplantation 59, 1549–1556.

Galili, U., LaTemple, D.C., 1997. The natural anti-Gal antibody as a universal augmenter of autologous tumor vaccine immunogenicity. Immunol. Today 18, 281–285.

Galili, U., LaTemple, D.C., Radic, M.Z., 1998. A sensitive assay for measuring α-gal epitope expression on cells by a monoclonal anti-Gal antibody. Transplantation 65, 1129–1132.

Galili, U., Chen, Z., DeGeest, K., 2003. Expression of α-gal epitopes on ovarian carcinoma membranes to be used as a novel autologous tumor vaccine. Gynecol. Oncol. 90, 100–108.

Galili, U., Wigglesworth, K., Abdel-Motal, U.M., 2007. Intratumoral injection of α-gal glycolipids induces xenograft-like destruction and conversion of lesions into endogenous vaccines. J. Immunol. 178, 4676–4687.

Galili, U., Albertini, A., Sondel, P., Wigglesworth, K., Sullivan, M., Whalen, G., 2010. In situ conversion of melanoma lesions into autologous vaccine by intratumoral injections of α-gal glycolipids. Cancers 2, 773–793.

Galon, J., Costes, A., Sanchez-Cabo, F., Kirilovsky, A., Mlecnik, B., Lagorce-Pagès, C., et al., 2006. Type, density, and location of immune cells within human colorectal tumors predict clinical outcome. Science 313, 1960–1964.

Galon, J., Fridman, W.H., Pagès, F., 2007. The adaptive immunologic microenvironment in colorectal cancer: a novel perspective. Cancer Res. 67, 1883–1886.

Gorelik, E., Duty, L., Anaraki, F., Galili, U., 1995. Alterations of cell surface carbohydrates and inhibition of metastatic property of murine melanomas by α1,3galactosyltransferase gene transfection. Cancer Res. 55, 4168–4173.

Gosselin, E.J., Wardwell, K., Gosselin, D.R., Alter, N., Fisher, J.L., Guyre, P.M., 1992. Enhanced antigen presentation using human Fcγ receptor (monocyte/macrophage)-specific immunogens. J. Immunol. 149, 3477–3481.

Groh, V., Li, Y.Q., Cioca, D., Hunder, N.N., Wang, W., Riddell, S.R., et al., 2005. Efficient cross-priming of tumor antigen-specific T cells by dendritic cells sensitized with diverse anti-MICA opsonized tumor cells. Proc. Natl. Acad. Sci. U.S.A. 102, 6461–6466.

Hamanová, M., Zdražilová Dubská, L., Valík, D., Lokaj, J., 2014. Natural antibodies against α(1,3) galactosyl epitope in the serum of cancer patients. Epidemiol. Mikrobiol. Imunol. 63, 130–133.

Hanfland, P., Kordowicz, M., Peter-Katalinic, J., Egge, H., Dabrowski, J., Dabrowski, U., 1988. Structure elucidation of blood group B-like and I-active ceramide eicosa- and pentacosasaccharides from rabbit erythrocyte membranes by combined gas chromatography-mass spectrometry; electron-impact and fast atom-bombardment mass spectrometry; and two-dimensional correlated, relayed-coherence transfer, and nuclear overhauser effect 500-MHz 1H-n.m.r. spectroscopy. Carbohydr. Res. 178, 1–21.

Honma, K., Manabe, H., Tomita, M., Hamada, A., 1981. Isolation and partial structural characterization of macroglycolipid from rabbit erythrocyte membranes. J. Biochem. 90, 1187–1196.

Intlekofer, A.M., Thompson, C.B., 2013. At the bench: preclinical rationale for CTLA-4 and PD-1 blockade as cancer immunotherapy. J. Leukoc. Biol. 94, 25–39.

Karttunen, J., Sanderson, S., Shastri, N., 1992. Detection of rare antigen-presenting cells by the lacZ T-cell activation assay suggests an expression cloning strategy for T-cell antigens. Proc. Natl. Acad. Sci. U.S.A. 89, 6020–6024.

Korman, A.J., Peggs, K.S., Allison, J.P., 2006. Checkpoint blockade in cancer immunotherapy. Adv. Immunol. 90, 297–339.

Kostine, M., Chiche, L., Lazaro, E., Halfon, P., Charpin, C., Arniaud, D., et al., 2017. Opportunistic autoimmunity secondary to cancer immunotherapy (OASI): an emerging challenge. Rev. Med. Interne 38, 513–525 .

LaTemple, D.C., Henion, T.R., Anaraki, F., Galili, U., 1996. Synthesis of α-galactosyl epitopes by recombinant α1,3galactosyltransferase for opsonization of human tumor cell vaccines by anti-Gal. Cancer Res. 56, 3069–3074.

LaTemple, D.C., Abrams, J.T., Zhang, S.Y., Galili, U., 1999. Increased immunogenicity of tumor vaccines complexed with anti-Gal: studies in knockout mice for α1,3galactosyltranferase. Cancer Res. 59, 3417–3423.

Li, X., Rao, X., Cai, L., Liu, X., Wang, H., Wu, W., et al., 2016. Targeting tumor cells by natural anti-carbohydrate antibodies using rhamnose-functionalized liposomes. ACS Chem. Biol. 11, 1205–1209.

Long, D.E., Karmakar, P., Wall, K.A., Sucheck, S.J., 2014. Synthesis of α-l-rhamnosyl ceramide and evaluation of its binding with anti-rhamnose antibodies. Bioorg. Med. Chem. 22, 5279–5289.

Lugade, A.A., Moran, J.P., Gerber, S.A., Rose, R.C., Frelinger, J.G., Lord, E.M., 2005. Local radiation therapy of B16 melanoma tumors increases the generation of tumor antigen-specific effector cells that traffic to the tumor. J. Immunol. 174, 7516–7523.

Maass, G., Schmidt, W., Berger, M., Schmidt, W., Herbst, E., Zatloukal, K., et al., 1995. Priming of tumor-specific T cells in the draining lymph nodes after immunization with interleukin-2 secreting tumor cells: three consecutive stages may be required for successful tumor vaccination. Proc. Natl. Acad. Sci. U.S.A. 92, 5540–5544.

Malmberg, K.J., 2004. Effective immunotherapy against cancer: a question of overcoming immune suppression and immune escape? Cancer Immunol. Immunother. 53, 879–891.

Manca, F., Fenoglio, D., Li Pira, G., Kunk, A., Celada, F., 1991. Effect of antigen/antibody ratio on macrophage uptake, processing, and presentation to T cells of antigen complexed with polyclonal antibodies. J. Exp. Med. 173, 37–48.

Manches, O., Plumas, J., Lui, G., Chaperot, L., Molens, J.P., Sotto, J.J., et al., 2005. Anti-Gal-mediated targeting of human B lymphoma cells to antigen-presenting cells: a potential method for immunotherapy using autologous tumor cells. Haematologica 90, 625–634.

Mastrangelo, M.J., Maguire Jr., H.C., Eisenlohr, L.C., Laughlin, C.E., Monken, C.E., McCue, P.A., et al., 1999. Intratumoral recombinant GM-CSF-encoding virus as gene therapy in patients with cutaneous melanoma. Cancer Gene Ther. 6, 409–422.

Mlecnik, B., Tosolini, M., Kirilovsky, A., Berger, A., Bindea, G., Meatchi, T., et al., 2011. Histopathologic-based prognostic factors of colorectal cancers are associated with the state of the local immune reaction. J. Clin. Oncol. 29, 610–618.

Mumberg, D., Wick, M., Schreiber, H., 1996. Unique tumor antigens redefined as mutant tumor-specific antigens. Semin. Immunol. 8, 289–293.

Ogawa, H., Galili, U., 2006. Profiling terminal *N*-acetyllactoamines of glycans on mammalian cells by an immunoenzymatic assay. Glycoconj. J. 23, 663–674.

Pawelec, G., Heinzel, S., Kiessling, R., Müller, L., Ouyang, Q., Zeuthen, J., 2000. Escape mechanisms in tumor immunity: a year 2000 update. Crit. Rev. Oncog. 11, 97–133.

Qiu, Y., Xu, M.B., Yun, M.M., Wang, Y.Z., Zhang, R.M., Meng, X.K., et al., 2011. Hepatocellular carcinoma-specific immunotherapy with synthesized α1,3-galactosyl epitope-pulsed dendritic cells and cytokine-induced killer cells. World J. Gastroenterol. 17, 5260–5266.

Qiu, Y., Yun, M.M., Xu, M.B., Wang, Y.Z., Yun, S., 2013. Pancreatic carcinoma-specific immunotherapy using synthesized α-galactosyl epitope-activated immune responders: findings from a pilot study. Int. J. Clin. Oncol. 18, 657–665.

Rafiq, K., Bergtold, A., Clynes, R., 2002. Immune complex-mediated antigen presentation induces tumor immunity. J. Clin. Investig. 110, 71–79.

Ravetch, J.V., Clynes, R.A., 1998. Divergent roles for Fc receptors and complement in vivo. Annu. Rev. Immunol. 16, 421–432.

Regnault, A., Lankar, D., Lacabanne, V., Rodriguez, A., Théry, C., Rescigno, M., et al., 1999. Fcγ receptor-mediated induction of dendritic cell maturation and major histocompatibility complex class I-restricted antigen presentation after immune complex internalization. J. Exp. Med. 189, 371–380.

Ribas, A., 2015. Adaptive immune resistance: how cancer protects from immune attack. Cancer Discov. 5, 915–919.

Rossi, G.R., Mautino, M.R., Awwad, D.Z., Husske, K., Lejukole, H., Koenigsfeld, M., et al., 2005a. Effective treatment of preexisting melanoma with whole cell vaccines expressing α(1,3)-galactosyl epitopes. Cancer Res. 65, 10555–10561.

Rossi, G.R., Unfer, R.C., Seregina, T., Link, C.J., 2005b. Complete protection against melanoma in absence of autoimmune depigmentation after rejection of melanoma cells expressing α(1,3)galactosyl epitopes. Cancer Immunol. Immunother. 54, 999–1009.

Sanderson, S., Shastri, N., 1994. LacZ inducible, antigen/MHC-specific T cell hybrids. Int. Immunol. 6, 369–376.

Schuurhuis, D.H., Ioan-Facsinay, A., Nagelkerken, B., van Schip, J.J., Sedlik, C., Melief, C.J., 2002. Antigen-antibody immune complexes empower dendritic cells to efficiently prime specific CD8+ CTL responses in vivo. J. Immunol. 168, 2240–2246.

Shastri, N., Gonzalez, F., 1993. Endogenous generation and presentation of the ovalbumin peptide/K^b complex to T cells. J. Immunol. 150, 2724–2736.

Sheridan, R.T., Hudon, J., Hank, J.A., Sondel, P.M., Kiessling, L.L., 2014. Rhamnose glycoconjugates for the recruitment of endogenous anti-carbohydrate antibodies to tumor cells. Chembiochem 15, 1393–1398.

Simons, J.W., Jaffee, E.M., Weber, C.E., Levitsky, H.I., Nelson, W.G., Carducci, M.A., et al., 1997. Bioactivity of autologous irradiated renal cell carcinoma vaccines generated by ex vivo granulocyte-macrophage colony-stimulating factor gene transfer. Cancer Res. 57, 1537–1546.

Staveley-O'Carroll, K., Sotomayor, E., Montgomery, J., Borrello, I., Hwang, L., Fein, S., et al., 1998. Induction of antigen-specific T cell anergy: an early event in the course of tumor progression. Proc. Natl. Acad. Sci. U.S.A. 95, 1178–1183.

Stellner, K., Saito, H., Hakomori, S., 1973. Determination of aminosugar linkage in glycolipids by methylation. Aminosugar linkage of ceramide pentasaccharides of rabbit erythrocytes and of Forssman antigen. Arch. Biochem. Biophys. 133, 464–472.

Tanemura, M., Miyoshi, E., Nagano, H., Eguchi, H., Matsunami, K., Taniyama, K., et al., 2015. Cancer immunotherapy for pancreatic cancer utilizing α-gal epitope/natural anti-Gal antibody reaction. World J. Gastroenterol. 21, 11396–11410.

Tanida, T., Tanemura, M., Miyoshi, E., Nagano, H., Furukawa, K., Nonaka, Y., et al., 2015. Pancreatic cancer immunotherapy using a tumor lysate vaccine, engineered to express α-gal epitopes, targets pancreatic cancer stem cells. Int. J. Oncol. 46, 78–90.

Tremont-Lukats, I.W., Avila, J.L., Hernández, D., Vásquez, J., Teixeira, G.M., Rojas, M., 1997. Antibody levels against α-galactosyl epitopes in sera of patients with squamous intraepithelial lesions and early invasive cervical carcinoma. Gynecol. Oncol. 64, 207–212.

Vogelstein, B., Papadopoulos, N., Velculescu, V.E., Zhou, S., Diaz Jr., L.A., Kinzler, K.W., 2013. Cancer genome landscapes. Science 339, 1546–1558.

Weber, J.S., Yang, J.C., Atkins, M.B., Disis, M.L., 2015. Toxicities of immunotherapy for the practitioner. J. Clin. Oncol. 33, 2092–2099.

Whalen, G.F., Sullivan, M., Piperdi, B., Wasseff, W., Galili, U., 2012. Cancer immunotherapy by intratumoral injection of α-gal glycolipids. Anticancer Res. 32, 3861–3868.

Wood, L.D., Parsons, D.W., Jones, S., Lin, J., Sjöblom, T., Leary, R.J., et al., 2007. The genomic landscapes of human breast and colorectal cancers. Science 318, 1108–1113.

Yachida, S., Jones, S., Bozic, I., Antal, T., Leary, R., Fu, B., et al., 2010. Distant metastasis occurs late during the genetic evolution of pancreatic cancer. Nature 467, 1114–1117.

Zhang, L., Conejo-Garcia, J.R., Katsaros, D., Gimotty, P.A., Massobrio, M., Regnani, G., et al., 2003. Intratumoral T cells, recurrence, and survival in epithelial ovarian cancer. New Engl. J. Med. 348, 203–213.

Zinkernagel, R.M., Ehl, S., Aichele, P., Oehen, S., Kündig, T., Hengartner, H., 1997. Antigen localization regulates immune responses in a dose- and time-dependent fashion: a geographical view of immune reactivity. Immunol. Rev. 156, 199–209.

CHAPTER

11

Anti-Gal as Cancer Cell Destroying Antibody and as Antibiotics Targeted by α-Gal Bifunctional Molecules

OUTLINE

INTRODUCTION

The harnessing of anti-Gal described in Chapters 9 and 10 depends on expression of α-gal epitopes on the administered vaccine or on tumor cells converted into an autologous vaccine. Anti-Gal may be further harnessed for destruction of pathogens such as viruses, bacteria, or of tumor cells, by α-gal bifunctional molecules. The α-gal epitope in one arm of the bifunctional molecule is linked to a ligand in the second arm that can bind to a target receptor on cells (e.g., tumor cells), bacteria, or a target viral protein. The ligand may be a molecule or part of a molecule that binds with high affinity to the corresponding receptor, or it can be an aptamer (i.e., a short sequence of single strand (ss) DNA or RNA) that is selected out of a large aptamer library, based on its ability to bind to the target molecule. Development of α-gal bifunctional molecules was based on research on different antibodies directed to targets by bifunctional molecules, also called "antibody recruiting molecules". These molecules include two arms, one is the antibody-binding arm and the other is the target-binding arm (cf. McEnaney et al., 2012). The principle of this method is described in the next section,

The Natural Anti-Gal Antibody as Foe Turned Friend in Medicine
http://dx.doi.org/10.1016/B978-0-12-813362-0.00011-7

using tumor cells as an example. This example is also applicable to bacteria and viruses, as described in another section of the chapter. An additional section describes the possible use of the highly versatile aptamers tagged with to α-gal epitopes (alphamers), as a method that converts anti-Gal into antibiotics that may be effective in destruction of a wide range of pathogens. Although the extent of research on α-gal bifunctional molecules has been limited, the potential of this method may be substantial both in cancer therapy and in development of alternatives to classical antibiotics.

KILLING OF TUMOR CELLS BY α-GAL BIFUNCTIONAL MOLECULES

The principle of α-gal bifunctional molecules activity is illustrated in Fig. 1 as an example of Arg–Gly–Asp (RGD) interaction with αVβ3 integrin, based on studies of Owen et al. (2007) and Carlson et al. (2007). The bifunctional molecule has an α-gal epitope as its antibody-binding arm and RGD tripeptide as the target-binding arm, which binds with high affinity to its receptor, the αVβ3 integrin. Incubation of these α-gal bifunctional molecules with tumor cells that have multiple RGD receptors on their cell membranes results in binding of many of these molecules to the receptors. This interaction, in turn, leads to the binding of many anti-Gal

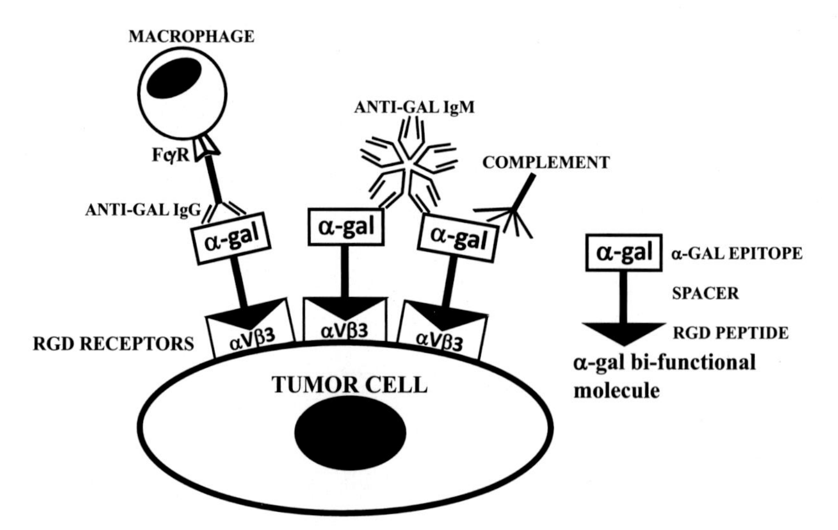

FIGURE 1 Schematic illustration of an α-gal bifunctional molecule in cancer immunotherapy, using the example of an RGD peptide as the ligand binding to its receptor-integrin αVβ3 on tumor cells. The α-gal epitope on the antibody binding arm of α-gal bifunctional molecules is tagged via a linker to the RGD amino acid sequence. α-Gal bifunctional molecules binding to tumor cells via RGD/αVβ3 interaction, further bind natural anti-Gal IgM and IgG antibodies via the α-gal epitope on the antibody binding arm. Anti-Gal IgM and to a lesser extent anti-Gal IgG activate the complement system resulting in complement mediated cytolysis of the tumor cells. The chemotactic cleavage complement peptides C5a and C3a recruit macrophages that adhere via their Fcγ receptors (FcγR) to the Fc portion of anti-Gal IgG bound to α-gal epitopes. This interaction results in destruction of tumor cells by ADCC. *Based on Carlson, C.B., Mowery, P., Owen, R.M., Dykhuizen, E.C., Kiessling, L.L., 2007. Selective tumor cell targeting using low-affinity, multivalent interactions. ACS Chem. Biol. 2, 119–127; Owen, R.M., Carlson, C.B., Xu, J., Mowery, P., Fasella, E., Kiessling, L.L., 2007. Bifunctional ligands that target cells displaying the αVβ3 integrin. ChemBioChem. 8, 68–82.*

IgM and IgG molecules to α-gal bifunctional molecules on the tumor cell surface. Because of the relative low affinity of anti-Gal to individual α-gal epitopes (see Chapter 1), this antibody binding via its two binding sites of IgG or 10 binding sites of IgM to multiple α-gal epitopes presented on the cell membrane is much more effective than the one-on-one hapten/antibody interaction in solution. Anti-Gal IgM and IgG molecules bind to α-gal bifunctional molecules on the tumor cells, activate complement, and thus induce complement-mediated cytolysis of the tumor cells, as well as generation of complement cleavage chemotactic peptides C5a and C3a. The multiple macrophages recruited by C5a and C3a to the site of the tumor cells bind via their Fcγ receptors to the Fc portion of anti-Gal IgG molecules on the tumor cells and kill the cells by antibody-dependent cell-mediated cytolysis (ADCC). If the α-gal bifunctional molecule is bound to a bacterium or to a virus, the activated complement may bore holes in these pathogens, and the recruited macrophages are likely to phagocytize the opsonized pathogen, resulting in intracellular destruction by the content of lysosomal vesicles.

In studies on incubation of tumor cells with bifunctional α-gal-RGD mimetic peptide molecule, it was found that tumor cells expressing a high number of αVβ3 integrins were readily killed by the activated complement following anti-Gal binding to the α-gal bifunctional molecules (Owen et al., 2007; Carlson et al., 2007). However, viability of tumor cells presenting low number of RGD receptors was not affected because the number of anti-Gal antibodies binding to the bifunctional molecules was too low for induction of effective complement activation (Carlson et al., 2007). It would be of interest to determine whether infusion of bifunctional α-gal-RGD mimetic peptide molecules into the circulation of anti-Gal producing GT-KO mice carrying tumors that express high numbers of αVβ3 integrins will result in effective destruction of such tumor lesions. It should be stressed that to perform such studies, the tumor cells should lack α-gal epitopes. Tumor cells expressing these epitopes (that are the large majority of mouse tumor cell lines) are likely to be destroyed by the same mechanism as that mediating hyperacute rejection of xenografts by anti-Gal (see Chapter 6).

INTERACTION OF α-GAL BIFUNCTIONAL MOLECULES WITH VIRUSES

The principle of targeting anti-Gal to pathogens by an α-gal bifunctional molecule was demonstrated by Li et al. (1999) who described the synthesis of such a molecule with a polymannose that binds to a mannose receptor on *Escherichia coli* and which is linked by a spacer to the α-gal epitope. This bifunctional molecule readily bound to mannose receptors on *E. coli* and enabled the subsequent binding of anti-Gal to the bacteria.

Studies on α-gal bifunctional molecules binding to gp120 of HIV via a second arm with the binding region of the CD4 receptor (Perdomo et al., 2008) suggested that α-gal bifunctional molecules may be instrumental in assisting in protection against some viral infections. Perdomo et al. (2008) tagged a synthetic α-gal epitope disaccharide (Galα1-3Gal) to several partially overlapping 15aa peptides corresponding to the region of CD4 that is the docking receptor for HIV via gp120 binding to this receptor on T cells. The best bifunctional molecule was determined by ELISA with gp120 as solid-phase antigen and measuring human serum anti-Gal interaction with the α-gal bifunctional molecule that is bound to gp120. The α-gal bifunctional molecule with the peptide displaying best gp120 binding was validated for its activity, by demonstrating binding to cells infected with HIV. Incubation of HIV with this α-gal bifunctional molecule in human serum lacking or containing complement further indicated that anti-Gal binding to the virus via the α-gal epitope of the bifunctional molecule resulted in neutralization of the virus. This neutralization was even higher in the presence of complement, suggesting that complement-mediated lysis of the virus by the bound anti-Gal increased the neutralizing effect of the bifunctional molecule. Binding of anti-Gal to the bifunctional molecule on HIV infected cells also induced NK cell-mediated ADCC of the infected cells (Perdomo et al., 2008). Similar effects on HIV by an α-gal bifunctional molecule were observed when the α-gal epitope was linked to the T20 peptide (36aa) that recognizes the ectodomain portion of HIV envelope glycoprotein gp41(Naicker et al., 2004). These studies suggest that α-gal bifunctional molecules may be capable of neutralizing or destroying a variety of viruses, in the presence of anti-Gal, provided that virus-binding region ligand binds with high affinity to the virus.

α-GAL APTAMERS AS BIFUNCTIONAL MOLECULES TARGETING ANTI-GAL TO PATHOGENS

A versatile technique for the preparation of α-gal bifunctional molecules that harness anti-Gal against various pathogens is targeting by α-gal aptamers. The interest in developing α-gal aptamers as tools for overcoming bacterial infections by agents that are alternative to classical antibiotics therapies stems from the appearance of highly infective multidrug resistant bacteria, which cannot be killed by antibiotics (Kristian et al., 2015). Aptamers are short oligonucleotides of ss DNA or RNA (60–100 bases in length) selected *in vitro* to bind to a target molecule with high specificity and high affinity (Ellington and Szostak, 1990; Tuerk and Gold, 1990). The ssDNA aptamers have at their 5′ and 3′ ends common sequences that are 15–18 bases, whereas the remaining bases between these two ends have random sequence. The aptamers are available as libraries containing ~10^{15} different random sequences. This large variety of sequences results in "folding" of the aptamer in a very high number of different spatial structures that may bind specifically and with high affinity to various target proteins and other molecules. To isolate an aptamer that binds specifically to a target molecule (e.g., a protein on bacterial wall or on a virus), the library undergoes a selection process designated "Systematic Evolution of Ligands by Exponential Enrichment" (SELEX) (Keefe et al., 2010; Maier and Levy, 2016). An anti-VEGF aptamer "Macugen" preventing macular degeneration is already in use in the clinic, and several aptamers are being studied in clinical trials, including aptamers preventing blood clotting and aptamers targeted to various tumor cells (cf. Sundaram et al., 2013; Parashar, 2016; Pastor, 2016).

The process for selection of the appropriate aptamer and its conversion into an α-gal aptamer (called here alphamer) against target bacteria was described by Kristian et al. (2015) and is illustrated in Fig. 2. A specific protein, purified from the target bacteria, serves as the target for aptamer binding by linking it to a solid surface such as the surface of magnetic beads or of a plate. The target protein is exposed to the library of aptamers allowing few of them to bind to it. After removing the unbound aptamers, the aptamers attached to the target protein are detached by using urea and EDTA at high temperature. The detached aptamers are amplified by a polymerase chain reaction, and the resulting dsDNA oligonucleotides are converted by high temperature into ssDNA, which on cooling undergo their specific folding.

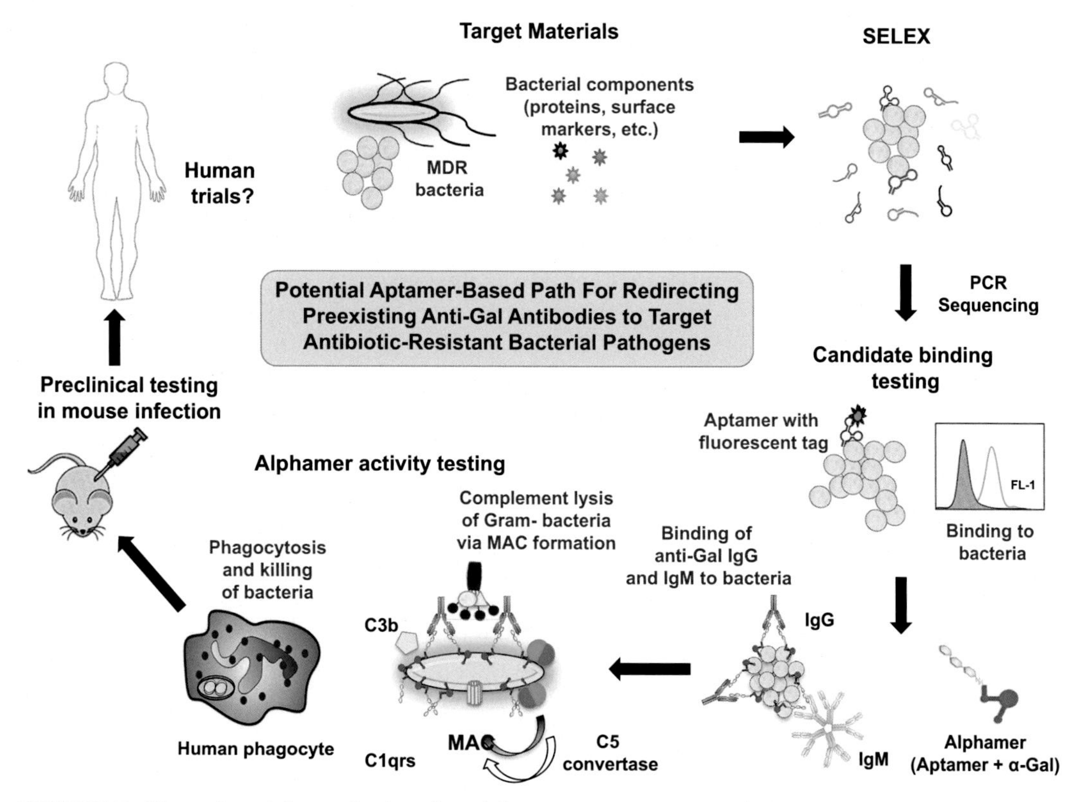

FIGURE 2 Illustration of the production of an alphamer reactive against multidrug resistant (MDR) bacteria. An aptamer is selected from a library of aptamers based on its binding to a protein or a marker specific to the target MDR bacteria, using the Systematic Evolution of Ligands by Exponential Enrichment (SELEX) process. This includes 10–15 rounds of isolation of the specifically binding aptamer and its amplification by PCR. The selected aptamer is sequenced, synthesized, tagged with fluorescein (FITC) and studied by flow cytometry for effective binding to the target bacteria. Following demonstration of effective binding, the aptamer is converted into an alphamer by tagging it with an α-gal epitope trisaccharide. The alphamer is confirmed to bind effectively to the target bacteria and to mediate binding of anti-Gal to the bacteria. The alphamer interaction with anti-Gal in the serum is further evaluated for mediating *in vitro* immunological functions including complement mediated lysis of the bacteria by C_{5b}-C_9 membrane attack complex (MAC) of activated complement, and phagocytosis by neutrophils and macrophages of anti-Gal opsonized bacteria. Alphamers found satisfactory in mediating these immunological functions are further studied for *in vivo* protective effects and safety in anti-Gal producing experimental animal models, prior to performing clinical trials in humans. *Reprinted from Kristian, S.A., Hwang J.H., Hall, B., Leire, E., Iacomini, J., Old, R., et al., 2015. Retargeting pre-existing human antibodies to a bacterial pathogen with an α-Gal conjugated aptamer. J. Mol. Med. (Berl.) 93, 619–631, with permission.*

These aptamers are subjected to another round of selection through the SELEX process. In a successful selection process, 10–15 rounds result in the isolation of an aptamer that binds specifically and with high affinity to the target protein.

The quality of the selected aptamer for binding to bacteria expressing the target protein is confirmed by demonstrating in flow cytometry effective binding of the fluorescein tagged aptamer to the bacteria. Aptamers displaying high-binding affinity are converted into alphamers by linking the synthetic α-gal epitope trisaccharide Galα1-3Galβ1-3GlcNAc to the aptamer via a spacer. The ability of the alphamer to bind anti-Gal to the target bacteria is determined by flow cytometry with anti-Gal IgG, by induction of bacterial lysis with anti-Gal IgM and complement and by opsonization of the bacteria with anti-Gal IgG for phagocytosis by macrophages and neutrophils. Success in these *in vitro* assays is followed by *in vivo* studies evaluating protection by the alphamer against infections with the target bacteria in anti-Gal producing experimental animal models such as α1,3galactosyltransferase knockout (GT-KO) mice and GT-KO pigs. Under certain circumstances, health regulatory agencies may expect demonstration of safety and efficacy also in a primate model prior to planning clinical trials in humans.

Kristian et al. (2015) studied alphamer activity against group A *Streptococcus* (GAS) bacteria. GAS are highly pathogenic bacteria that in some cases may be lethal. The aptamer chosen for conversion into an alphamer was a DNA aptamer (20A24P) that was found to bind specifically with high affinity to a conserved region of the surface-anchored M protein in multiple GAS strains (Hamula et al., 2011). Affinity of the aptamer to the M protein of GAS was increased by forming several truncation aptamers and identifying the resulting aptamer with the highest affinity to the target protein. Subsequently, an α-gal epitope was chemically linked to the 5′ end of the aptamer, converting it into an alphamer. The linking of an α-gal epitope to the aptamer did not alter the binding of the alphamer to GAS. Anti-Gal binding to this alphamer on GAS further induced neurtrophil-mediated phagocytosis of the anti-Gal opsonized bacteria. Moreover, incubation of GAS with the alphamer in human blood resulted in killing of the bacteria (Kristian et al., 2015). In view of these *in vitro* studies, it would be of interest to determine whether administration of this alphamer into anti-Gal producing GT-KO mice can help in protection against GAS infections.

CONCLUSIONS

Bifunctional molecules in which one arm is the α-gal epitope and the other arm is a ligand that enables binding to specific receptor on pathogens or on cells provide a unique tool for *in vivo* harnessing of the natural anti-Gal antibody for destruction of pathogens and tumor cells. Anti-Gal interaction with the α-gal epitope on the bifunctional molecule bound to target cells or to pathogens can destroy the target by complement-mediated cytolysis, ADCC, or opsonization for Fc-mediated and C3b-mediated phagocytosis followed by intracellular killing within neutrophils and macrophages. *In vitro* binding of α-gal bifunctional molecules was demonstrated with *E. coli*, HIV, and with tumor cells, all of which displayed subsequent binding of anti-Gal. Following anti-Gal binding, the bacteria, viruses, and tumor cells were destroyed or neutralized by the antibody. A recent development, which may provide limitless versatility to this technology, is the use of alphamers in which the antibody binding arm of the bifunctional molecule is the α-gal epitope, and the target binding arm is an aptamer selected from a library, based on its binding to the target. The combination of alphamers that have high affinity to the target molecule and a high density (i.e., concentration) of the target molecule on the pathogen may enable future development of new type of antibiotics that use anti-Gal as an antibody which kills pathogens. In view of the rapid ~100-fold increase in anti-Gal titers in humans immunized with xenoglycoproteins carrying α-gal epitopes (discussed in Chapter 1), it is possible that such an increase in anti-Gal activity may further amplify the *in vivo* efficacy of α-gal bifunctional molecules and that of alphamers.

References

Carlson, C.B., Mowery, P., Owen, R.M., Dykhuizen, E.C., Kiessling, L.L., 2007. Selective tumor cell targeting using low-affinity, multivalent interactions. ACS Chem. Biol. 2, 119–127.

Ellington, A.D., Szostak, J.W., 1990. In vitro selection of RNA molecules that bind specific ligands. Nature 346, 818–822.

Hamula, C.L., Le, X.C., Li, X.F., 2011. DNA aptamers binding to multiple prevalent M-types of *Streptococcus pyogenes*. Anal. Chem. 83, 3640–3647.

Keefe, A.D., Pai, S., Ellington, A., 2010. Aptamers as therapeutics. Nat. Rev. Drug Discov. 9, 537–550.

Kristian, S.A., Hwang, J.H., Hall, B., Leire, E., Iacomini, J., Old, R., et al., 2015. Retargeting pre-existing human antibodies to a bacterial pathogen with an α-Gal conjugated aptamer. J. Mol. Med. (Berl.) 93, 619–631.

Li, J., Zacharek, S., Chen, X., Wang, J., Zhang, W., Janczuk, A., Wang, P.G., 1999. Bacteria targeted by human natural antibodies using α-gal conjugated receptor-specific glycopolymers. Bioorg. Med. Chem. 7, 1549–1558.

Maier, K.E., Levy, M., 2016. From selection hits to clinical leads: progress in aptamer discovery. Mol. Ther. Methods Clin. Dev. 5, 16014.

McEnaney, P.J., Parker, C.G., Zhang, A.X., Spiegel, D.A., 2012. Antibody-recruiting molecules and emerging paradigm for engaging immune function in treating human disease. ACS Chem. Biol. 7, 1139–1151.

Naicker, K.P., Li, H., Heredia, A., Song, H., Wang, L.X., 2004. Design and synthesis of α-gal-conjugated peptide T20 as novel antiviral agent for HIV-immunotargeting. Org. Biomol. Chem. 2, 660–664.

Owen, R.M., Carlson, C.B., Xu, J., Mowery, P., Fasella, E., Kiessling, L.L., 2007. Bifunctional ligands that target cells displaying the αVβ3 integrin. ChemBioChem. 8, 68–82.

Parashar, A., 2016. Aptamers in therapeutics. J. Clin. Diagn. Res. 10, BE01–6.

Pastor, F., 2016. Aptamers: a new technological platform in cancer immunotherapy. Pharmaceuticals 9, 64.

Perdomo, M.F., Levi, M., Saellberg, M., Vahlne, A., 2008. Neutralization of HIV-1 by redirection of natural antibodies. Proc. Natl. Acad. Sci. U.S.A. 105, 12515–12520.

Sundaram, P., Kurniawan, H., Byrne, M.E., Wower, J., 2013. Therapeutic RNA aptamers in clinical trials. Eur. J. Pharm. Sci. 48, 259–271.

Tuerk, C., Gold, L., 1990. Systematic evolution of ligands by exponential enrichment: RNA ligands to bacteriophage T4 DNA polymerase. Science 249, 505–510.

CHAPTER

12

Acceleration of Wound and Burn Healing by Anti-Gal/α-Gal Nanoparticles Interaction

OUTLINE

INTRODUCTION

One of the main areas in medicine in which anti-Gal can be harnessed for a range of beneficial effects is the rapid recruitment of macrophages and their activation into "pro-healing" M2 macrophages that induce accelerated tissue repair and regeneration of external and internal injured tissues. Harnessing the immunological potential of anti-Gal was found to be particularly effective in acceleration of wound and burn healing in the skin. Macrophages are the major type of cells, which mediate early stages of wound repair and regeneration, (Martin, 1997; Singer and Clark, 1999; Gurtner et al., 2008; DiPietro and Koh, 2016). Macrophages

The Natural Anti-Gal Antibody as Foe Turned Friend in Medicine
http://dx.doi.org/10.1016/B978-0-12-813362-0.00012-9

usually arrive to the wound area as monocytes in capillaries, extravasate and convert into macrophages that migrate into the injured area along chemotactic gradients of substances such as monocyte chemoattractant protein-1, macrophages inflammatory protein-1, and regulated on activation, normal T cell expressed and secreted (RANTES) released from cells within and around injury sites (Wood et al., 1997; Piccolo et al., 1999; Shukaliak and Dorovini-Zis, 2000; Low et al., 2001; Heinrich et al., 2003; Shallo et al., 2003). The macrophages that are at a differentiation stage of "pro-inflammatory" M1 macrophages debride the wound of dead tissue, killed bacteria, and neutrophils. Subsequently, the macrophages reaching the wound become "anti-inflammatory" called also "pro-healing" M2 macrophages. These macrophages secrete a wide range of cytokines such as vascular endothelial growth factor (VEGF) inducing vascularization of the wound for generating the capillary network of the skin, epidermal growth factor (EGF) inducing regrowth of the injured epidermis for covering the wound and isolating it from the external environment, and fibroblasts growth factor (FGF) inducing regeneration of the dermis (Sica and Mantovani, 2012; DiPietro and Koh, 2016). Overall, cytokines secreted by M2 macrophages orchestrate healing and regeneration processes within the wound. In small injuries, the healing process may take only few days and result in complete restoration of the original structure of the skin. However, in large wounds and burns, the healing time is much longer and the risk of infection greatly increases, resulting in increased morbidity. Prolonged healing processes in wounds are usually mediated by the default repair mechanism of fibrosis of the injured tissue and scar formation. Thus, it is reasonable to assume that acceleration of macrophages recruitment into wounds and their activation into pro-healing macrophages may result in accelerated healing of large wounds and burns, thereby decreasing the risk of infections, scar formation, and morbidity in patients suffering of such injuries. Extensive research is being performed on the administering recombinant cytokines (e.g., platelets-derived growth factor (PDGF)) for accelerating wound healing (Meier and Nanney, 2006; Kiwanuka et al., 2012). This chapter describes an alternative therapeutic method in which macrophages are recruited rapidly to wounds following interaction between the natural anti-Gal antibody and α-gal nanoparticles applied to the wound. These nanoparticles present an abundance of α-gal epitopes, and their interaction with anti-Gal activates the complement system to generate chemotactic complement cleavage peptides, which induce rapid recruitment of macrophages into the treated wound (Fig. 1). The recruited macrophages bind anti-Gal-coated α-gal nanoparticles and are further activated into pro-healing macrophages that accelerate wound and burn healing.

ANTI-GAL INTERACTION WITH α-GAL NANOPARTICLES

α-Gal nanoparticles made of rabbit red blood cell membranes

The natural anti-Gal antibody is highly effective in activating the complement system as indicated by the *in vitro* cytolysis of cells presenting α-gal epitopes and incubated in fresh human serum (Chapter 10) and by the hyperacute rejection of porcine organs transplanted in nonhuman primates (see Chapter 6). Therefore, it is reasonable to assume that application to injury sites of nanoparticles that present multiple α-gal epitopes (designated α-gal nanoparticles) will result in extensive interaction with the natural anti-Gal antibody and thus activation

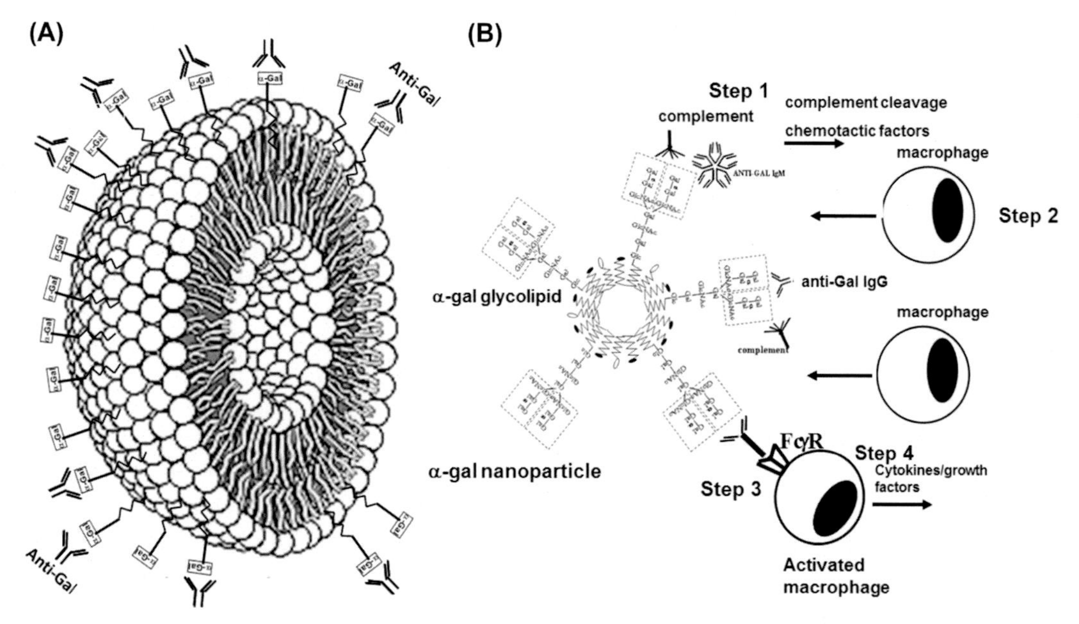

FIGURE 1 Schematic illustration of an α-gal nanoparticle (A) and suggested immunological processes occurring following application of α-gal nanoparticles to wounds (B). (A) The α-gal nanoparticles comprise phospholipids forming a lipid bilayer in which multiple α-gal glycolipids with α-gal epitopes (*rectangles*) are anchored. The natural anti-Gal antibody readily binds to these α-gal epitopes. (B) Application of α-gal nanoparticles to wounds results in recruitment and activation of macrophages according to the following steps: (1) Binding of the natural anti-Gal antibody to α-gal nanoparticles activates the complement system. (2) The chemotactic factors C5a and C3a, generated as complement cleavage peptides, induce rapid recruitment of macrophages to the α-gal nanoparticles. (3) The recruited macrophages interact via their Fcγ receptors (FcγR) with the Fc portion of anti-Gal coating the α-gal nanoparticles. (4) The Fc/FcγR interaction activates the macrophages to secrete a wide range of cytokines and growth factors that accelerate healing of the treated wound. *(A) Modified from Wikipedia, with permission. (B) Adapted from Hurwitz, Z., Ignotz, R., Lalikos, J., Galili, U., 2012. Accelerated porcine wound healing with α-gal nanoparticles. Plast. Reconstr. Surg. 129, 242–251, with permission.*

of the complement system. Such complement activation may not damage cells and not induce complement-mediated cytolysis because anti-Gal binds to α-gal epitopes on the surface of the α-gal nanoparticles and not to cells. However, this interaction is likely to generate the complement cleavage chemotactic peptides C5a and C3a, which may effectively accelerate the recruitment of macrophages into injury sites. Both anti-Gal and complement proteins are likely to reach the fluid in wounds together with other serum proteins that are released from ruptured capillaries. These assumptions were studied with α-gal nanoparticles consisting of α-gal glycolipids, phospholipids, and cholesterol that were readily produced from rabbit red blood cell (RBC) membranes (Galili et al., 2010; Wigglesworth et al., 2011; Hurwitz et al., 2012).

As discussed in Chapter 10, rabbit RBC membranes are the richest known source of α-gal epitopes in mammals, as they present ~2×10^6 α-gal epitopes/RBC (Galili et al., 1998). Many of these α-gal epitopes are carried by glycolipids with carbohydrate chains having 5–40 carbohydrates on 1–8 branches (antennae) (Eto et al., 1968; Stellner et al., 1973; Honma et al., 1981; Dabrowski et al., 1984; Egge et al., 1985; Hanfland et al., 1988). The α-gal glycolipids with 5–20 carbohydrates are illustrated in Fig. 3B of Chapter 10. The initial steps of the process for preparation of α-gal glycolipids (discussed in Chapter 10) are also used for preparation of α-gal nanoparticles. Rabbit RBC membranes are prepared by hypotonic shock of the RBC. Rabbit RBC membranes obtained from 1 L packed rabbit RBC are subjected to extraction of glycolipids, phospholipids, and cholesterol by 2 h mixing in chloroform:methanol 1:1 solution, followed by overnight extraction in 1:2 solution of chloroform:methanol with constant stirring (Fig. 3A in Chapter 10) (Galili et al., 2007, 2010; Wigglesworth et al., 2011). Rabbit RBC proteins are denatured in the chloroform:methanol solution, precipitate, and are removed by filtration. All neutral glycolipids except for the glycolipid ceramide trihexoside with the structure Galα1-4Galβ1-4Glc-Cer (which is also abundant in human RBC membranes, shown in Fig. 1 of Chapter 1), have α-gal epitopes and thus readily bind anti-Gal (Chapters 1 and 10).

The mixture of glycolipids, phospholipids, and cholesterol in the extract is dried, and then resuspended in saline in a sonication bath to generate α-gal liposomes (size of ~1 to 10 μm). The initial studies on anti-Gal acceleration of wound healing were performed with α-gal liposomes (Galili et al., 2010; Wigglesworth et al., 2011). However, in subsequent studies the α-gal liposomes were converted into submicroscopic α-gal liposomes, referred to as α-gal nanoparticles, by extensive sonication. The sonication of these liposomes with a sonication probe breaks the liposomes into smaller "spheres", these are the α-gal nanoparticles with size 10–300 nm (Wigglesworth et al., 2011; Hurwitz et al., 2012). The reduced size of α-gal nanoparticles enabled their sterilization by passing through a filter of 0.2 or 0.45 μm. The α-gal nanoparticles formed in this procedure have a phospholipid bilayer wall in which multiple α-gal glycolipids are anchored via their ceramide tail (Fig. 1A). Both α-gal liposomes and α-gal nanoparticles have the same structure as that illustrated in Fig. 1A and they differ only in their size (α-gal nanoparticles are ~10–100-fold smaller than the liposomes). This chapter describes studies on wound and burn healing performed with α-gal nanoparticles and those with α-gal liposomes that were performed prior to the conversion of these liposomes into nanoparticles by extensive sonication. Quantification of α-gal epitopes on the α-gal nanoparticles indicated the presentation of ~10^{15} epitopes per mg nanoparticles (Wigglesworth et al., 2011).

Hypothesis on the effects of α-gal nanoparticles in wounds

The extensive interaction of anti-Gal with α-gal nanoparticles applied to wounds is likely to initiate a series of localized processes in sequential steps, which ultimately may result in accelerated healing of the treated wound (Fig. 1B). It is hypothesized that these steps are as follows: Step 1. Binding of the natural anti-Gal antibody to the administered α-gal nanoparticles activates the complement system. Step 2. The chemotactic factors C5a and C3a generated as complement cleavage peptides induce rapid recruitment of macrophages to the site of applied α-gal nanoparticles. Step 3. The recruited macrophages interact via their Fcγ receptors (FcγR) with the Fc portion of anti-Gal coating the α-gal nanoparticles. It is possible (but not proven as yet) that C3b deposited on the α-gal nanoparticles may mediate similar interaction with C3b receptors on the macrophages. Step 4. Fc/FcγR interaction of anti-Gal opsonizing the α-gal nanoparticles with the macrophages activates these cells into "pro-healing" macrophages that secrete a wide range of cytokines and growth factors, which promote accelerated repair and regeneration of the wound. As discussed in Chapters 13 and 14, it is also hypothesized that in internal injuries, or in bio-implants containing α-gal nanoparticles, the cytokines/growth factors secreted by the recruited and activated macrophages further recruit stem cells into the injury sites. The recruited stem cells are thought to receive cues from the extracellular matrix (ECM) to differentiate into mature cells that restore the original structure and function of the injured tissue or of the bio-implant.

Anti-Gal/α-gal nanoparticles interaction in the skin induces rapid recruitment and activation of macrophages

The hypothesis illustrated in Fig. 1B was studied in α1,3galactosyltransferase knockout mice (GT-KO mice) (Galili et al., 2010; Wigglesworth et al., 2011) and in GT-KO pigs (Hurwitz et al., 2012). These knockout mice and pigs are the only known nonprimate mammals, which lack α-gal epitopes and can produce the anti-Gal antibody (Chapter 1). GT-KO mice do not naturally produce the anti-Gal antibody because they live in a sterile environment and receive sterile food, thus they do not have a natural flora that immunizes them to produce anti-Gal. Therefore, these mice are immunized with pig kidney membranes homogenate to induce production of the anti-Gal antibody (Tanemura et al., 2000; Galili et al., 2007). However, GT-KO pigs have a natural flora that immunizes them to produce the anti-Gal antibody in titers comparable to those found in humans (Dor et al., 2004; Fang et al., 2012; Galili, 2013a).

The step of complement activation and generation of chemotactic complement cleavage peptides C5a and C3a, following anti-Gal/α-gal nanoparticles interaction (Step 1 in Fig. 1B), was studied *in vivo* by evaluating the ability of these peptides to induce vasodilation (Vogt, 1986). Intradermal injection sites of α-gal nanoparticles in anti-Gal producing GT-KO mice were viewed from the basal aspect of the skin, 48 h post injection. As shown in Fig. 2, the areas adjacent to the injection sites of α-gal nanoparticles display redness, which represents vasodilation that is likely to be induced by C5a and C3a. Control nanoparticles lacking α-gal epitopes (nanoparticles produced from GT-KO pig RBC membranes) induced no such redness near the injection sites (Fig. 2).

The chemotactic effect of these complement cleavage peptides on macrophage recruitment (Step 2 in Fig. 1B) could be demonstrated by histological analysis of the α-gal nanoparticles injection site. Many neutrophils were observed at the injection site within 12h post injection (Wigglesworth et al., 2011). However, after 24h, most neutrophils disappeared and multiple mononuclear cells were observed migrating to the injection site (Fig. 3A). Almost all migrating cells were macrophages, as implied from the immunostaining of the cells with the macrophage-specific antibody F4/80 (Fig. 3B). The number of macrophages continued to increase for the next 6days. These cells were characterized by increased size, and ample cytoplasm suggested activation of the cells (Fig. 3C). These cells could be found in large numbers at the injection site, even on Day 14. However, by Day 21 post injection of the α-gal nanoparticles, all macrophages disappeared from the injection site. The normal histological features of the skin were restored without any indication of a granuloma, or any other chronic inflammatory response. In addition, the injection site displayed no keloid formation, fibrosis, or any other changes to the skin structure (Wigglesworth et al., 2011).

The recruitment of macrophages was dependent both on the presence of α-gal nanoparticles and on their interaction with the anti-Gal antibody. This could be inferred from observations indicating that intradermal injection of saline resulted in no recruitment of macrophages.

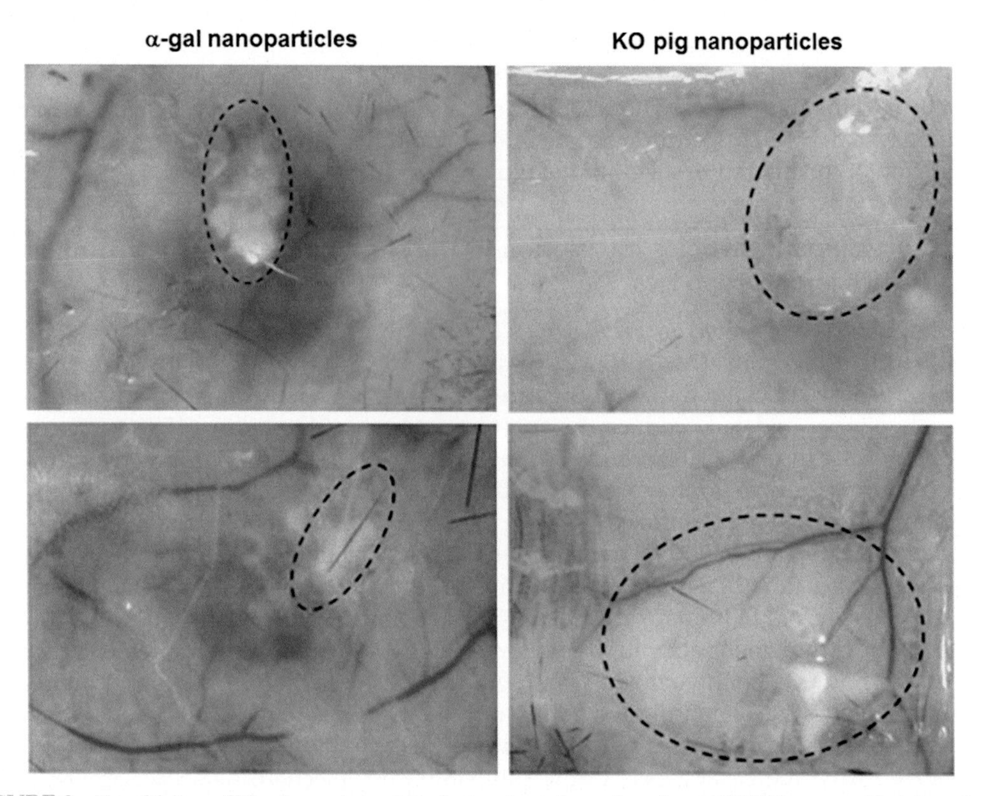

FIGURE 2 Vasodilation of blood vessels in α1,3galactosyltransferase knockout (GT-KO) mouse skin injected with α-gal nanoparticles or with nanoparticles lacking α-gal epitopes (produced from GT-KO pig red blood cell [KO pig nanoparticles]). *Dashed lines* indicate the border area of the injected nanoparticles. Complement cleavage peptides induce vasodilation of the blood vessels adjacent to the injection site of α-gal nanoparticles. In the absence of α-gal epitopes on the nanoparticles, no vasodilation is observed.

Moreover, intradermal injection of these nanoparticles into wild type (WT) mice (i.e., mice synthesizing α-gal epitopes and lacking the anti-Gal antibody) resulted in no induction of macrophage migration (Wigglesworth et al., 2011). The α-gal nanoparticles induced recruitment of macrophages also required activation of the complement system, as suggested in Step 1 in Fig. 1B. Inactivation of the complement system by injection of α-gal nanoparticles mixed with cobra venom factor (a complement activation inhibitor) resulted in no migration of macrophages to the injection site in the skin of GT-KO mice (Wigglesworth et al., 2011).

The identity of cells recruited by α-gal nanoparticles, as macrophages, was further confirmed by subcutaneous implantation of biologically inert sponge discs (made of polyvinyl alcohol—PVA, 10 mm diameter, 3 mm thickness) that contained 10 mg α-gal nanoparticles. Most of the cells (>90%) retrieved from the PVA sponge discs explanted after 6–9 days had the morphology of the cells in Fig. 3D. Immunostaining and analysis by flow cytometry revealed that the cells were stained positively with antibodies specific to the macrophage markers CD11b and CD14 (Fig. 4A) and not with antibodies to other cell populations, including B cells and T cells (Galili et al., 2010). Additional antibody staining indicated that most of the recruited macrophages are

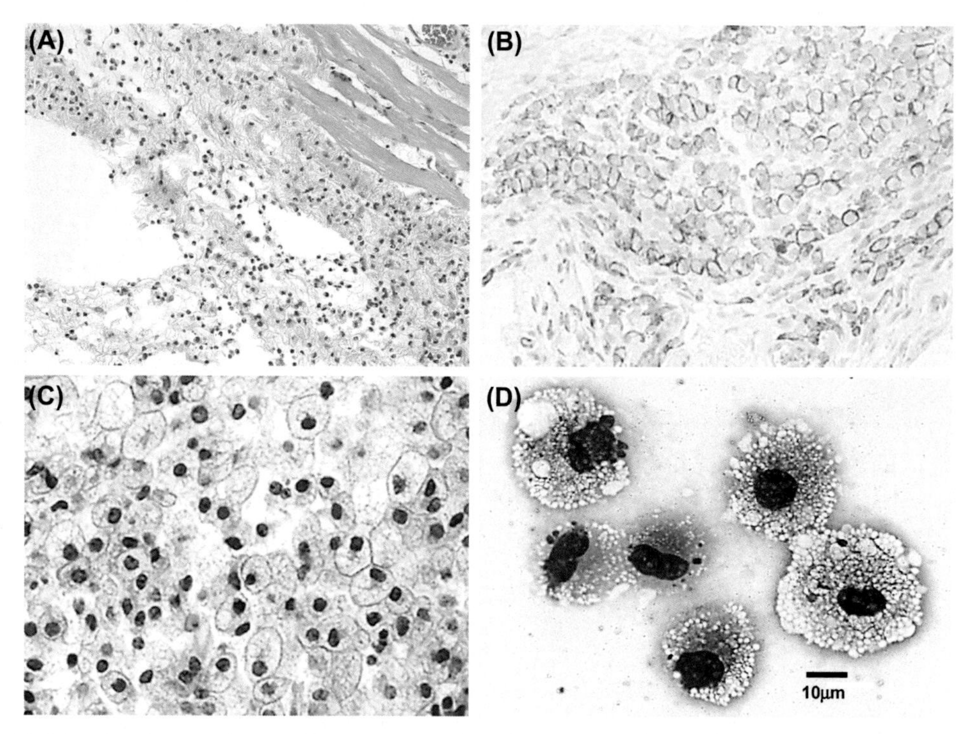

FIGURE 3 Recruitment of macrophages into the skin of α1,3galactosyltransferase knockout (GT-KO) mice, by α-gal nanoparticles. (A) Macrophage recruitment 24h after intradermal injection of 10mg α-gal nanoparticles. The empty area represents the injection site in which the nanoparticles were dissolved by alcohol during the fixation process (H&E ×100). (B) Immunostaining the α-gal nanoparticles injection site, 4 days postinjection, with macrophage-specific anti-F4/80 antibody coupled to peroxidase (HRP) (×200). (C) Staining of the intradermal injection site 7 days post injection of α-gal nanoparticles. Macrophages are large with ample cytoplasm (H&E ×400). (D) Individual macrophages recruited into a polyvinyl alcohol sponge disc containing 10mg α-gal nanoparticles and implanted subcutaneously into GT-KO mouse. The cells were obtained 7 days post implantation (Wright staining, ×1000). *(A–C) Reprinted with permission from Galili, U., 2015a. Acceleration of wound healing by α-gal nanoparticles interacting with the natural anti-Gal antibody. J. Immunol. Res. 2015. Article ID 589648. (D) Adapted with permission from Wigglesworth, K.M., Racki, W.J., Mishra, R., Szomolanyi-Tsuda, E., Greiner, D.L., Galili, U., 2011. Rapid recruitment and activation of macrophages by anti-Gal/α-gal liposome interaction accelerates wound healing. J. Immunol. 186, 4422–4432.*

M2 macrophages, as they are positive for IL10 and arginase staining and negative for IL12 staining (in preparation). The large size (20–30 μm) of the macrophages suggests that they were activated. In addition, they displayed multiple vacuoles in the cytoplasm (Fig. 3D). These vacuoles are likely to represent the extensive uptake of anti-Gal opsonized α-gal nanoparticles following the Fc/FcγR interaction with macrophages. Nanoparticles within the vacuoles were dissolved and removed by the alcohol during the fixation step.

The binding of anti-Gal opsonized α-gal nanoparticles to macrophages via Fc/FcγR interaction (Step 3 in Fig. 1B) is illustrated by scanning electron microscopy in Fig. 5. Anti-Gal-coated α-gal nanoparticles were incubated *in vitro* with macrophages grown from monocytes of GT-KO pigs. Within 2h of incubation, the α-gal nanoparticles bound extensively to

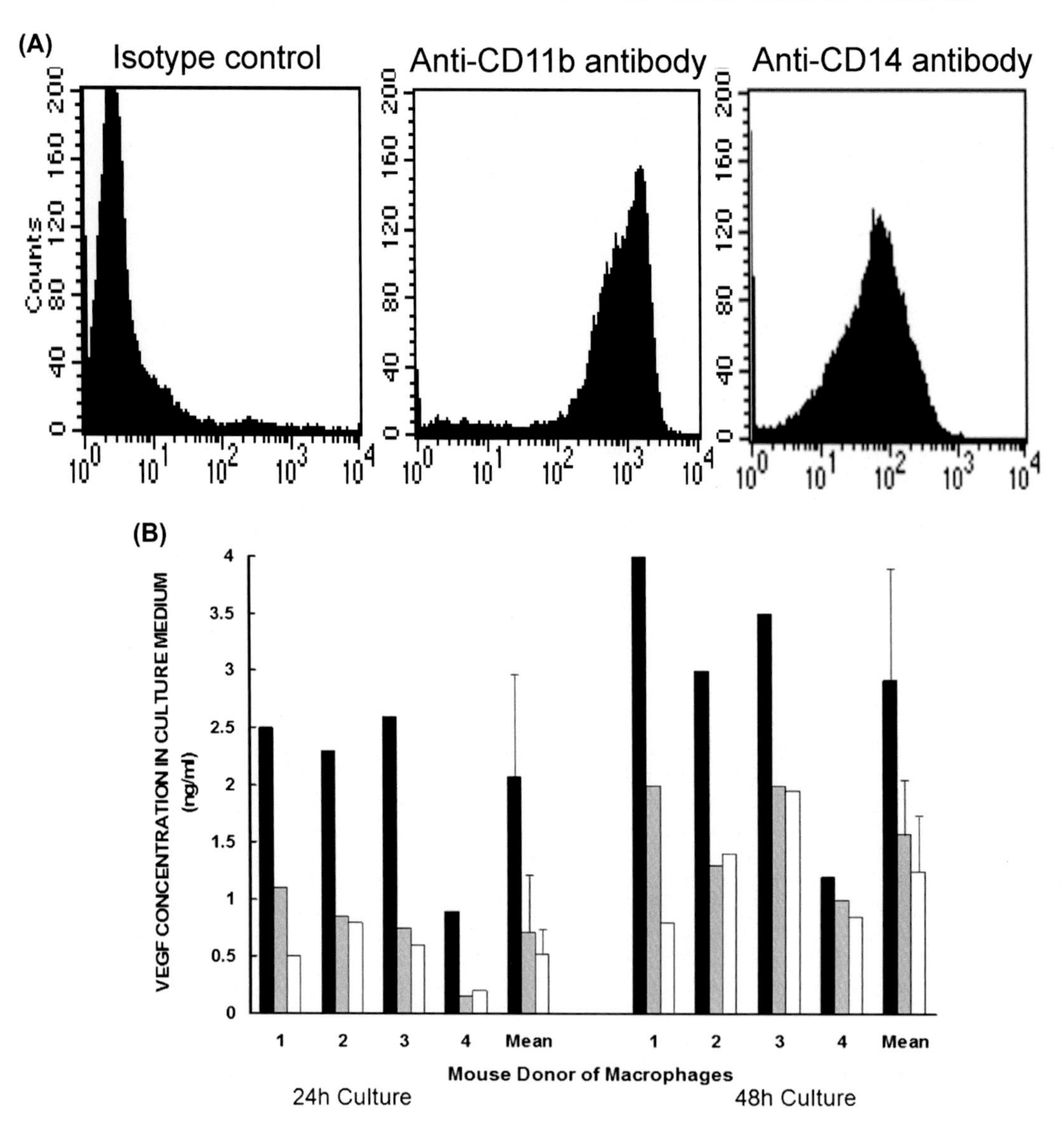

FIGURE 4 Characterization of α1,3galactosyltransferase knockout (GT-KO) mouse macrophages recruited by α-gal nanoparticles (A) and secretion of VEGF by the macrophages binding anti-Gal-coated α-gal nanoparticles (B). (A) A polyvinyl alcohol (PVA) sponge disc containing 10mg α-gal nanoparticles was implanted subcutaneously in GT-KO mice for 7days. Subsequently, the recruited cells were obtained from explanted PVA sponge discs and subjected to immunostaining and flow cytometry analysis. More than 90% of the cells were stained with the macrophage-specific anti-CD11b and anti-CD14 antibodies. (B) Peritoneal macrophages were cultured with anti-Gal-coated α-gal nanoparticles (*closed columns*), α-gal nanoparticles without anti-Gal (*gray columns*), or as macrophages alone (*open columns*). Vascular endothelial growth factor (VEGF) secretion by the macrophages was measured in culture media after 24 or 48h. Data with macrophages from 4 GT-KO mice and mean+SD. *(A) Adapted with permission from Galili, U., Wigglesworth, K., Abdel-Motal, U.M., 2010. Accelerated healing of skin burns by anti-Gal/α-gal liposomes interaction. Burns 36, 239–251. (B) Reprinted with permission from Galili, U., 2015b. Avoiding detrimental human immune response against mammalian extracellular matrix implants. Tissue Eng. Part B 21, 231–241.*

FIGURE 5 Scanning electron microscopy (SEM) describing binding of anti-Gal-coated α-gal nanoparticles to adherent α1,3galactosyltransferase knockout (GT-KO) pig macrophage. (A–E) The α-gal nanoparticles coated with natural GT-KO pig anti-Gal antibody were incubated with adherent GT-KO pig macrophages for 2h at room temp. The macrophages were then extensively washed to remove nonadherent nanoparticles and subjected to SEM analysis. The surface of representative macrophages is covered with α-gal nanoparticles as a result of Fc/FcγR interactions. The insets in (A) and (C) are enlarged in (B) and (D). In (A–D) the size of α-gal nanoparticles is ~100–300nm, in (E) the size is ~10–30nm. (F) A macrophage incubated with α-gal nanoparticles that were not coated with anti-Gal. *Adapted with permission from Galili, U., 2013b. Macrophages recruitment and activation by α-gal nanoparticles accelerate regeneration and can improve biomaterials efficacy in tissue engineering. Open Tissue Eng. Regen. Med. J. 6, 1–11; Galili, U., 2015b. Avoiding detrimental human immune response against mammalian extracellular matrix implants. Tissue Eng. Part B 21, 231–241.*

macrophages covering much of their surface. In addition to binding of α-gal nanoparticles with a size of 100–300 nm (Fig. 5A–D), binding of smaller nanoparticles of 10–30 nm could be demonstrated at higher magnifications (Fig. 5E). In the absence of anti-Gal, α-gal nanoparticles did not bind to macrophages (Fig. 5F).

It is hypothesized that the Fc/FcγR interaction between α-gal nanoparticles and macrophages signals activation of a variety of cytokine producing genes in the recruited macrophages (Step 4 in Fig. 1B). Such activation was studied with GT-KO mouse macrophages incubated for 24–48 h at 37°C, alone or with α-gal nanoparticles coated with anti-Gal or lacking the antibody. In the absence of anti-Gal, the macrophages incubated with α-gal nanoparticles secreted only background level of VEGF, as determined by ELISA measuring this cytokine (Fig. 4B). However, coincubation of the macrophages with anti-Gal-coated α-gal nanoparticles for 24 and 48 h resulted in secretion of VEGF by the activated macrophages, at levels that were significantly higher than the background levels (Fig. 4B). Skin specimens from anti-Gal producing GT-KO mice injected with α-gal nanoparticles or with nanoparticles lacking α-gal epitopes (i.e., made of GT-KO pig RBC membranes) were further assayed after 48 h for activation of various cytokine genes, by performing quantitative real-time PCR. Skin receiving intradermal injection of α-gal nanoparticles was found to display increased activation of the VEGF, FGF, IL1, PDGF, and CSF genes (Wigglesworth et al., 2011). This increased activity of the cytokines' genes is likely to be associated with activation of the recruited macrophages because control nanoparticles lacking α-gal epitopes did not induce recruitment of macrophages and production of these cytokines was significantly lower.

ACCELERATED HEALING OF α1,3GALACTOSYLTRANSFERASE KNOCKOUT MOUSE WOUNDS TREATED WITH α-GAL LIPOSOMES AND α-GAL NANOPARTICLES

To determine whether recruitment and activation of macrophages by α-gal nanoparticles applied to wounds, can affect wound healing, anesthetized anti-Gal producing GT-KO mice were subjected to full-thickness skin oval wound formation (~6 mm × 9 mm) in the dorsal region of the mouse flank. Wound treatment consisted of application of a spot bandage dressing, in which the pad was coated with 10 mg α-gal nanoparticles in saline, 10 mg α-gal liposomes (i.e., particles with the same structure as α-gal nanoparticles, however, with size of 1–10 μm), liposomes lacking α-gal epitopes (from GT-KO pig RBC membranes), or with saline. Wounds were evaluated for healing every 3 days. Wounds treated with saline or with control liposomes lacking α-gal epitopes were found to heal (i.e., become covered with regenerating epidermis) 12–14 days after wounding. However, wounds treated with α-gal nanoparticles displayed 95%–100% healing already on Day 6 (Fig. 6). At that time point, wounds treated with saline or with liposomes lacking α-gal epitopes displayed <20% healing, whereas wounds treated with α-gal liposomes displayed ~60% healing (Fig. 6). The wounds treated with α-gal nanoparticles further displayed on Day 6 increased vascularization, fibroblast migration, and collagen deposition in the dermis, in comparison to saline-treated wounds (Wigglesworth et al., 2011). These observations suggest that the multiple cytokines secreted by macrophages that are activated because of interaction with anti-Gal-coated α-gal nanoparticles induce simultaneous repair and regenerations processes in the injury site, which may decrease the healing time in GT-KO mice by ~50%.

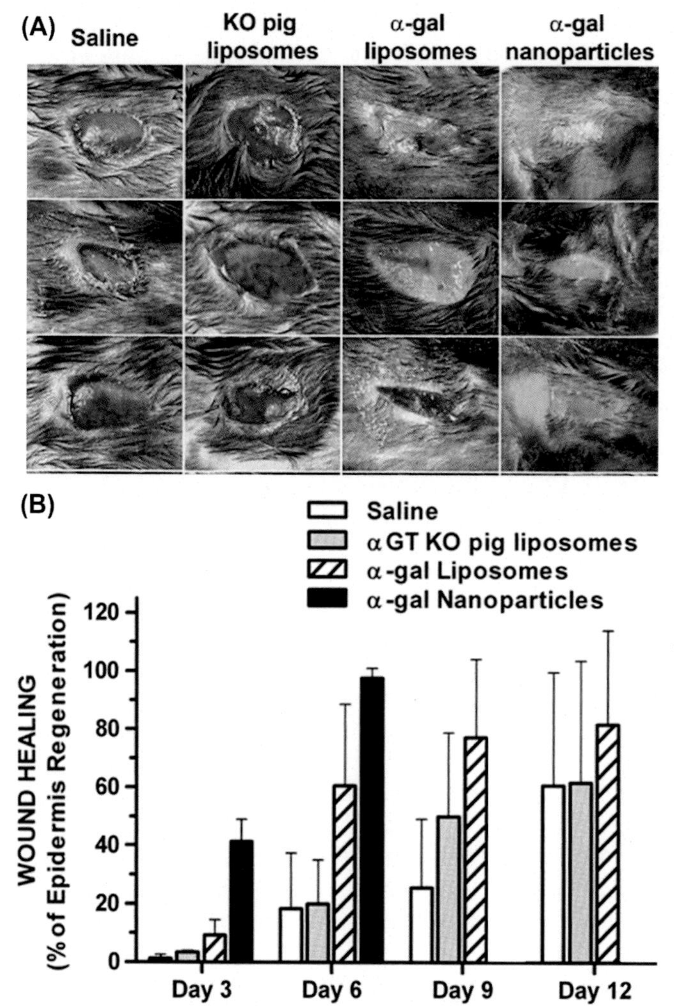

FIGURE 6 Accelerated healing of representative wounds in α1,3galactosyltransferase knockout (GT-KO) mice treated with α-gal liposomes and α-gal nanoparticles. (A) Morphology on Day 6 of oval 6mm×9mm full thickness excisional wounds treated with: saline, 10mg liposomes from GT-KO (KO) pig red blood cell membranes lacking α-gal epitopes, α-gal liposomes, or with α-gal nanoparticles. (B) Healing of wounds at various days post wounding, as measured by percentage of wound area covered by regenerating epidermis. Wounds (as in Fig. 6A) were treated with 10mg α-gal nanoparticles (*closed columns*), α-gal liposomes, which are large (1–10μm) α-gal nanoparticles (*hatched columns*), 10mg αGT-KO pig liposomes lacking α-gal epitopes (*gray columns*), or saline (*open columns*). Data presented as Mean+SD from ≥5 mice/group ($p<.05$). On day 6, there were n=20 mice/group with wounds treated with α-gal nanoparticles, α-gal liposomes, or with saline. *Adapted with permission from Wigglesworth, K.M., Racki, W.J., Mishra, R., Szomolanyi-Tsuda, E., Greiner, D.L., Galili, U., 2011. Rapid recruitment and activation of macrophages by anti-Gal/α-gal liposome interaction accelerates wound healing. J. Immunol. 186, 4422–4432.*

The accelerated healing was also observed with α-gal liposomes, however, this healing was somewhat slower than that with the α-gal nanoparticles, although both liposomes and nanoparticles are comprised of the same materials and differ only in size. The "breaking" of α-gal liposomes into nanoparticles by a sonication probe results in many more particles presenting α-gal epitopes in 10mg α-gal nanoparticles than in 10mg α-gal liposomes. It is possible that because of the higher number of α-gal nanoparticles, they disperse faster and uniformly throughout the wound than the liposomes, thus wound healing is faster with the nanoparticles.

Healing of chronic wounds in diabetes by α-gal nanoparticles

α-Gal nanoparticles may also be studied as treatment for inducing wound healing in patients with impaired ability to heal wounds because of low migration of macrophages into wounds, such as chronic wounds in diabetic patients and in the elderly population. Both populations were found to produce the natural anti-Gal antibody within the normal range (Galili et al., 1995; Wang et al., 1995). The effective recruitment of macrophages by α-gal nanoparticles may enable overcoming the deficient physiologic migration of macrophages into wounds in such populations. Preliminary studies were performed in mice that were chemically induced to become diabetic by intraperitoneal injections of streptozotocin (50 mg/kg). One month after the mice became hyperglycemic, they underwent wounding as in Fig. 6. The wounds were treated with 10 mg α-gal nanoparticles on spot bandage or with saline. The wounds treated with α-gal nanoparticles healed within 12 days, whereas wounds treated with saline became chronic wounds and displayed no healing even after 30 days (Galili, 2017). Thus, it would be of interest to determine whether administration of α-gal nanoparticles into chronic wounds of diabetic patients by application of the nanoparticles on dressing, or by injection into and around the wound, can induce healing.

Treatment of α1,3galactosyltransferase knockout mouse wounds with α-gal liposomes decreases scar formation

Scar formation is often observed in healing of large wounds and in deep wounds. Scars are generated by the default physiologic mechanism of fibrosis, which is aimed to form, in relatively short time, effective barrier between pathogens in the environment, and the internal tissues of the body. The fibrosis process consists of multiple fibroblasts migrating into the injury site and secreting dense collagen ECM. The regenerating epidermis covering the fibrosis area is usually thicker than that of uninjured skin because of hyperplasia of the cells. The combination of dense connective tissue in the dermis and hyperplasia of the epidermis forms the scar tissue. Scar tissues do not enable regrowth of skin appendages such as hair, sebaceous glands, smooth muscles, and fat tissue. In wound healing, the process of fibrosis initiates within several days post injury. If the injury is small and restoration of the normal skin structure is completed in relatively short time, no scar formation is observed. However, in large injuries, the fibrosis of the wound occurs before restoration of normal tissue structure can be completed, thus an irreversible scar tissue is formed.

Studies on healed wounds indicated that, in addition to the acceleration of wound repair and regeneration, α-gal liposomes (and likely α-gal nanoparticles) also have long-term effects. Wounds treated with dressing covered with saline and inspected histologically 4 weeks post injury, demonstrated distinct fibrosis and scar formation, as usually observed in large injuries that heal. The dermis of these healed wounds contained dense fibrotic tissue (collagen stained deep blue in Trichrome staining), no skin appendages regrowth, and hyperplasia of the epidermis (Fig. 7). However, in wounds treated with α-gal liposomes, the healed skin had normal histological structure, which includes loose connective tissue in the dermis, regrowth of skin appendages such as hair shafts and sebaceous glands, fat tissue and smooth muscle in the hypodermis, and normal thin epidermis (Fig. 7) (Wigglesworth et al., 2011). It is possible that the prevention of fibrosis and scar formation in wounds treated with α-gal liposomes is associated with the accelerated healing of these wounds. This accelerated healing, induced by rapidly recruited and activated macrophages, restores the normal structure of the injured skin, prior to onset of the fibrosis process and scar formation. Thus, α-gal liposomes and nanoparticles may be viewed as helping in restoration of the normal structure and function

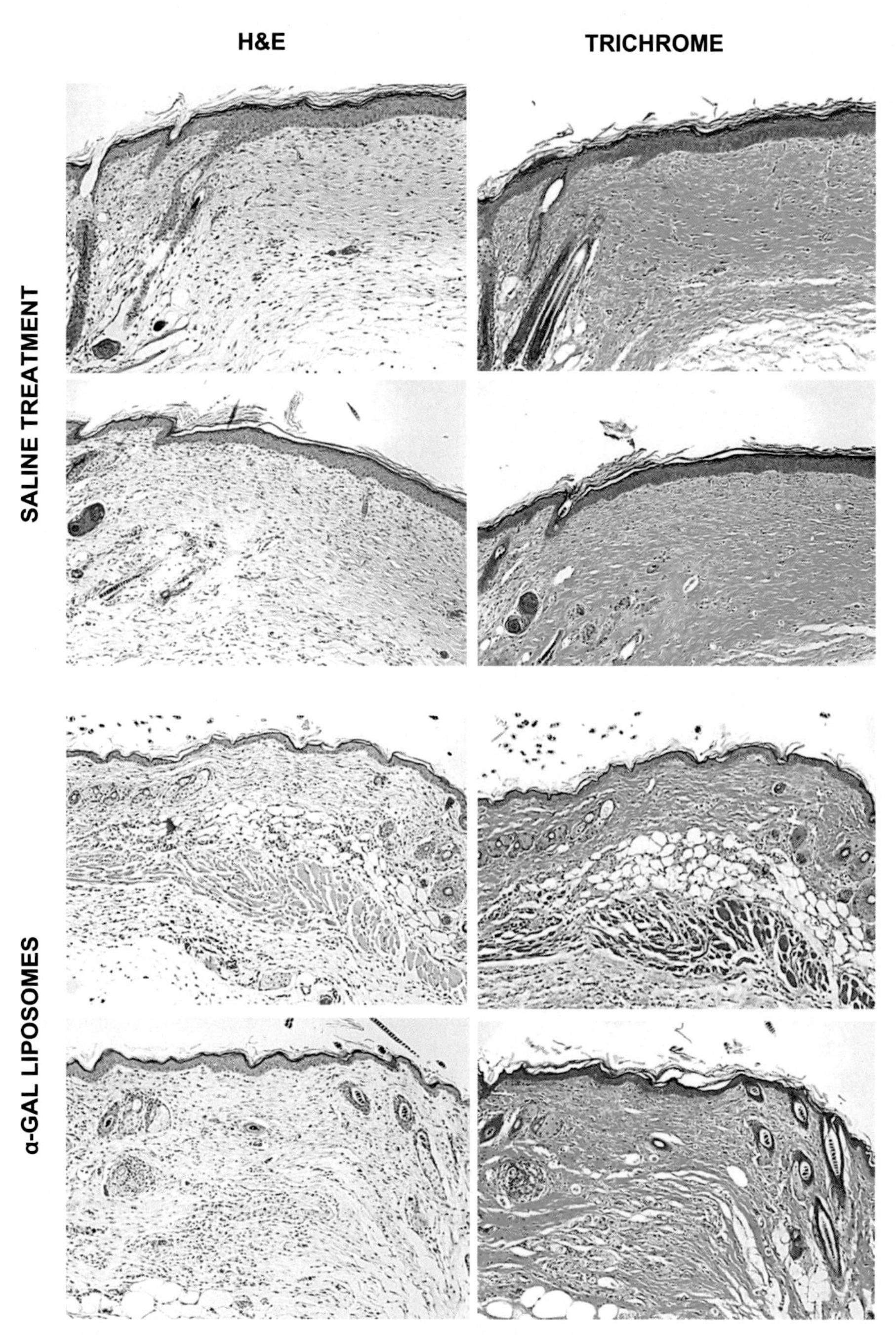

FIGURE 7 Treatment of α1,3galactosyltransferase knockout (GT-KO) mouse wounds with α-gal liposomes prevents scar formation. Wounds treated with 10 mg α-gal liposomes or with saline were excised after 28 days and stained with H&E and with Trichrome (staining collagen blue). Saline-treated wounds developed fibrosis, resulting in scar formation, characterized by dense connective tissue, no skin appendages, and hypertrophic epidermis. The hair shafts in the left are part of the uninjured tissue. In contrast, α-gal liposomes treated wounds display restoration of normal histology of the skin. The epidermis is thin as in normal mouse skin in Fig. 8A, the dermis comprises loose connective tissue and skin appendages, including hair and sebaceous glands. In addition, fat cells and muscle cells are observed in the hypodermis (×100). Specimens are representative of five mice per group. In each pair, the figure on the right is a trichrome staining of the figure on the left (stained by H&E). *Adapted with permission from Wigglesworth, K.M., Racki, W.J., Mishra, R., Szomolanyi-Tsuda, E., Greiner, D.L., Galili, U., 2011. Rapid recruitment and activation of macrophages by anti-Gal/α-gal liposome interaction accelerates wound healing. J. Immunol. 186, 4422–4432.*

of the injured tissue during the "race" between this pathway of repair and regeneration and the default pathway resulting in scar formation.

Fibrosis and scar formation processes also occur in internal injuries such as surgical incisions, injured myocardium following ischemia due to occlusion of coronary arteries, and in nerve injuries. The observed prevention of scar formation in wounds treated with α-gal nanoparticles raises the possibility that this treatment may be effective in prevention of scar formation in these internal injuries, as well. Discussions on the hypothetical possibility of using α-gal nanoparticles in inducing regeneration of ischemic myocardium following myocardial infarction and regeneration of injured nerves are included in Chapters 14 and 15, respectively.

ACCELERATING BURN HEALING IN α1,3GALACTOSYLTRANSFERASE KNOCKOUT MICE BY α-GAL LIPOSOMES

The early studies on anti-Gal-mediated healing of injuries were performed with α-gal liposomes applied to skin burns in anti-Gal producing GT-KO mice (Galili et al., 2010). Thermal injuries were performed under anesthesia by brief touch on shaven skin of mice with the heated end of a metal spatula. This brief touch caused a ~2 mm × 3 mm burn in the skin, which is comparable to a second-degree burn in humans, in which the epidermis and approximately half of the dermis are injured (Fig. 8B) (Galili et al., 2010). The burns were treated with spot bandage covered with saline, or with 10 mg α-gal liposomes, and the proportion of the wound surface covered by regenerating epidermis was determined at various time points.

The effect of α-gal liposomes on recruitment of cells into the burns was observed already on Day 3. The number of macrophages and neutrophils in burns treated with these liposomes was several-fold higher than that observed in saline-treated wounds (Fig. 8D and C, respectively). By Day 6, a large proportion of burns treated with α-gal liposome displayed complete regeneration of the epidermis, including formation of *stratum corneum*, whereas no significant epidermis regeneration was observed in saline-treated burns (Fig. 8F and E, respectively). Healing of burns treated with saline was also slower on Day 9 than that of α-gal liposomes treated burns, whereas on Day 12 the wounds treated saline or with α-gal liposomes displayed 100% regeneration of the epidermis (Galili et al., 2010). Burn healing studies in WT mice demonstrated no significant differences between wounds treated with α-gal liposomes and those treated with saline. Healing in both groups took 12–14 days (Galili et al., 2010). These observations suggest that accelerated healing of burns, as that of wounds, depends both on presence of anti-Gal (which is absent in WT mice) and of the α-gal liposomes.

ACCELERATED WOUND HEALING BY α-GAL NANOPARTICLES IN α1,3GALACTOSYLTRANSFERASE KNOCKOUT PIGS

The histology of mouse skin is different from that of humans. Whereas mouse skin is very thin and the epidermis consists only 2–3 layers of cells (Fig. 8A), human skin is much thicker, with 5–10 layers of cells in the epidermis and extensive interdigitation with the underlying

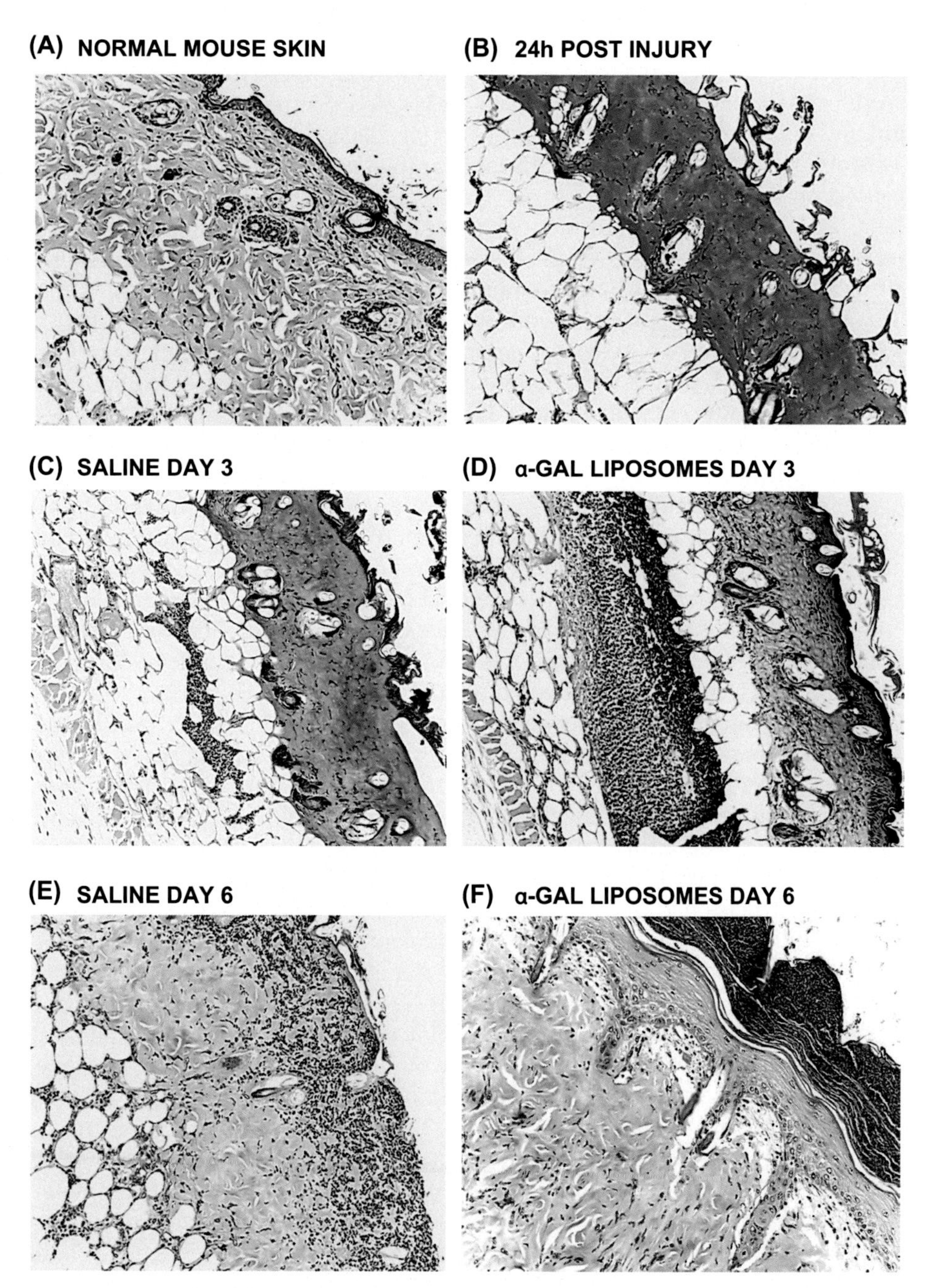

FIGURE 8 Accelerated healing of burns in α1,3galactosyltransferase knockout (GT-KO) mice treated with α-gal liposomes. (A) Histology of normal GT-KO mouse skin. (B) Histology of a burn 24 h post injury, displaying damage to the skin similar to second-degree burn in humans. (C) Saline-treated wound on Day 3. (D) α-Gal liposomes treated wound on Day 3 displaying extensive recruitment of macrophages and neutrophils in the dermis. (E) Saline-treated wound on Day 6. Macrophages and neutrophils migrate toward the surface of the wound. (F) α-Gal liposomes treated wound on Day 6 displaying regeneration of the epidermis, including *stratum corneum*. Recruited macrophages and neutrophils are found above the intact epidermis. Each burns pair on Days 3 and 6 is from the same representative mouse of five mice per time point. *Reprinted with permission from Galili, U., Wigglesworth, K., Abdel-Motal, U.M., 2010. Accelerated healing of skin burns by anti-Gal/α-gal liposomes interaction. Burns 36, 239–251.*

dermis. Pig skin has a structure, which is histologically like that of human skin. GT-KO pigs lack α-gal epitopes (Lai et al., 2002; Phelps et al., 2003; Kolber-Simonds et al., 2004) and produce the natural anti-Gal antibody like humans (Dor et al., 2004; Fang et al., 2012; Galili, 2013a). Thus, it was of interest to determine in this large animal model whether α-gal nanoparticles have any effect on wound healing.

Excisional 20 mm × 20 mm square wounds (~3 mm deep) were formed under anesthesia on the back of 3-month old GT-KO pigs. Borders of the wounds were marked by tattooed dots prior to wounding. Suspensions containing ~100 mg α-gal nanoparticles, 10 mg α-gal nanoparticles, or saline were placed in the wound cavities. Because each pig had several wounds, the effects of the two doses of nanoparticles versus that of saline could be compared in every pig. The wounds were covered with dressing that prevented spilling of the suspensions from wounds (Hurwitz et al., 2012). The dressings were changed every 3–4 days, morphology of the wound recorded and α-gal nanoparticles or saline were applied to the wounds with each change of wound dressing. Pigs were euthanized on Days 7, 13, and 60, wounds were excised and subjected to histological analysis. Wounds evaluated on Day 3 displayed no changes in comparison to wound morphology right after wounding. However, all wounds studied on Day 7 were filled with granulation tissue. The size of saline-treated wounds (defined as area of the wound not covered by regenerating epidermis) was not significantly different from that of α-gal nanoparticles treated wounds (Fig. 9E). However, the latter wounds contained many more macrophages and displayed initial deposits of collagen, which could not be detected in saline-treated wounds (Hurwitz et al., 2012).

Day 10 wounds treated with 100 mg and 10 mg α-gal nanoparticles were ~60% and ~35% smaller than saline-treated wounds, respectively (Fig. 9E). Major differences in wound healing were observed on Day 13. Many of the wounds treated with 100 mg α-gal nanoparticles were completely covered by regenerating epidermis, whereas saline-treated wounds displayed physiologic healing in which an area of ~25 mm^2 (i.e., ~0.5 cm × 0.5 cm) remained exposed without regenerating epidermis (Figs. 9E and 10). Wounds treated with 100 and 10 mg α-gal nanoparticles had on Day 13 ~90% and 80% smaller noncovered areas, respectively, than saline-treated wounds (Fig. 9E). The accelerating effects of α-gal nanoparticles on wound healing seemed to be dose dependent as wounds treated with 10 mg α-gal nanoparticles healed faster than saline-treated wounds but slower than wounds treated with 100 mg α-gal nanoparticles (Fig. 10). Healing wounds display wound contraction (marked by stretching of tattooed dots), however, no differences in wound contraction were observed between α-gal nanoparticles treated and saline-treated wounds (Fig. 10). Full healing of saline-treated wounds was observed on Days 18–22 (not shown). Thus, healing time of wounds treated with 100 mg α-gal nanoparticles was shortened by ~40%.

Wounds treated with α-gal nanoparticles for 13 days and which were not completely covered with epidermis were subjected to histological evaluation and compared with saline-treated wounds in the same pig (Fig. 9A and C vs. Fig. 9B and D, respectively). There were many more blood vessels and cells in α-gal nanoparticles treated wounds than in saline-treated wounds. The observed increase in vascularization of the wounds treated with α-gal nanoparticles may be associated with elevated concentration of VEGF within the granulation tissue. As discussed above, this local elevation in VEGF concentration is likely to be the product of the macrophages recruited by anti-Gal-coated α-gal nanoparticles and activated by the Fc/FcγR interaction with anti-Gal coating the nanoparticles (Figs. 1B and 4B).

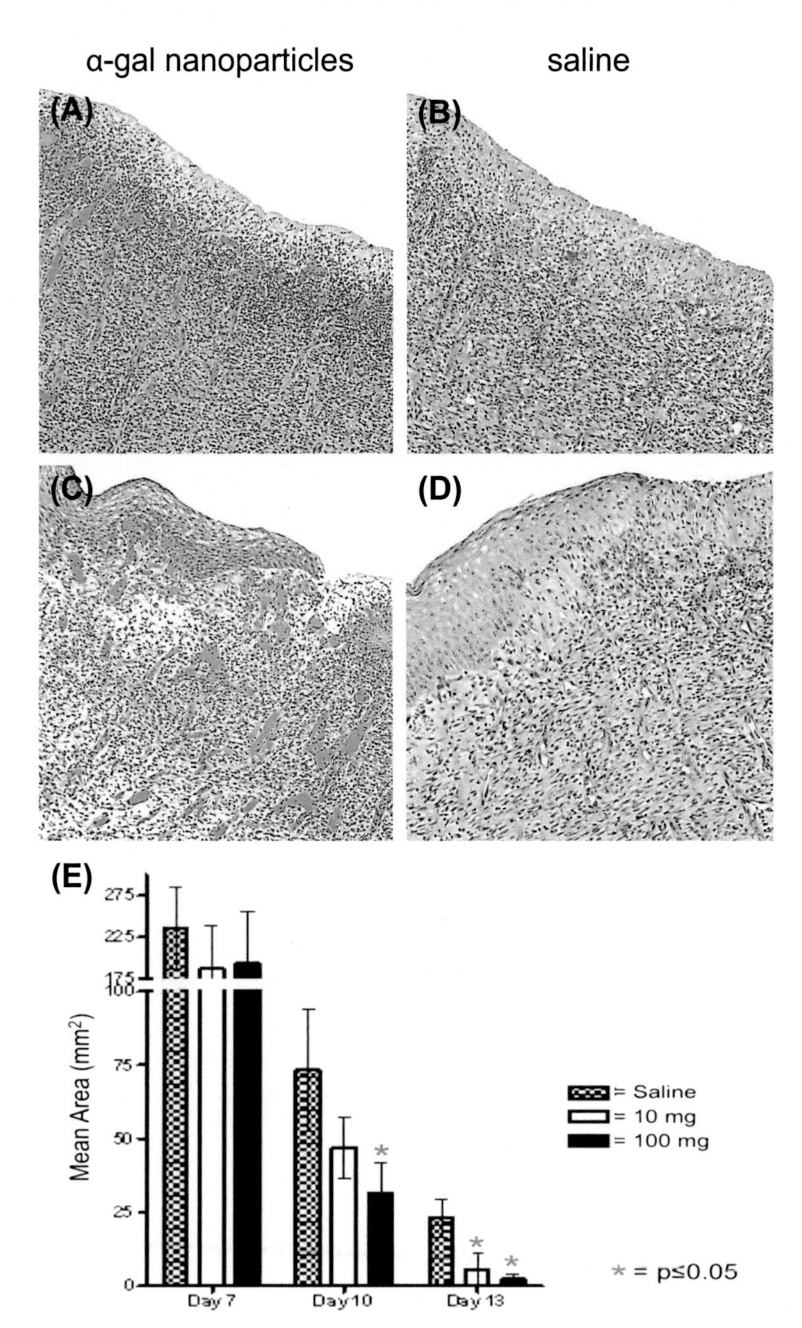

FIGURE 9 Vascularization and healing of wounds treated with α-gal nanoparticles (A and C) or saline (B and D) and wound size on Days 7–13 in α1,3galactosyltransferase knockout pigs (E). Excisional wounds (20 mm × 20 mm, 3 mm deep) were treated with 100, or 10 mg α-gal nanoparticles, or with saline, monitored for healing (area not covered by regenerating epidermis), excised and subjected to histological analysis. (A–D) Wounds from one representative pig describing area not covered with epidermis (A and B) or area under epidermis leading front (C and D). Note the much higher vascularization and many more cells in the granulation tissue in the wound treated with 100 mg α-gal nanoparticles (A and C) than in saline-treated wound (B and D). (E) Healing of wounds as measured by the surface area not covered by regenerating epidermis on Days 7, 10, and 13 post wounding and treatment. Mean ± SD from nine pigs on day 7 and eight pigs on days 10 and 13. *(A–D) Reprinted with permission from Galili, U., 2015a. Acceleration of wound healing by α-gal nanoparticles interacting with the natural anti-Gal antibody. J. Immunol. Res. 2015, 589648; (E) Reprinted with permission from Hurwitz, Z., Ignotz, R., Lalikos, J., Galili, U., 2012. Accelerated porcine wound healing with α-gal nanoparticles. Plast. Reconstr. Surg. 129, 242–251.*

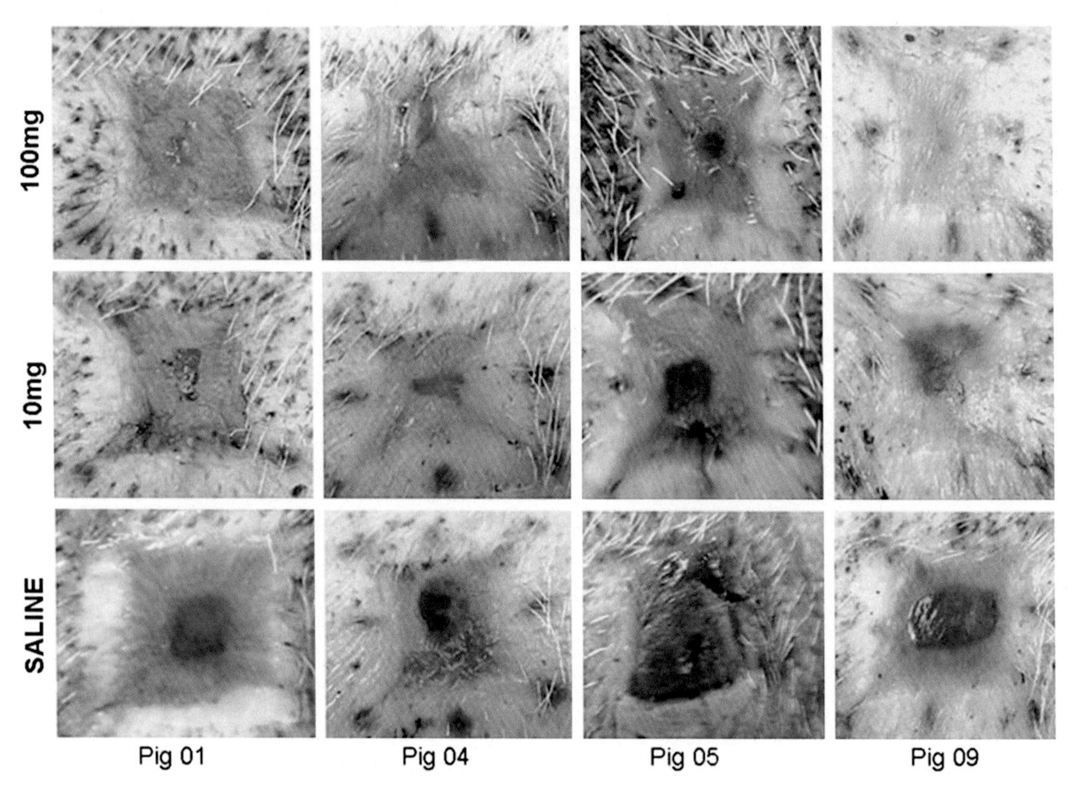

FIGURE 10 Healing of excisional wounds in the skin of α1,3galactosyltransferase knockout pigs treated with α-gal nanoparticles or with saline and viewed on Day 13 (gross morphology). Wounds as in Fig. 9 treated with 100, 10 mg α-gal nanoparticles, or with saline in each column, are from the same pig. Border of wounds was marked by tattooed dots to determine contraction during healing. Note that the physiologic healing in saline-treated wounds is slower than that in wounds treated with 10 mg α-gal nanoparticles. Wounds treated with 100 mg of the nanoparticles are the fastest healing wounds, with three of the four representative wounds completely covered by regenerating epidermis. *Adapted with permission from Hurwitz, Z., Ignotz, R., Lalikos, J., Galili, U., 2012. Accelerated porcine wound healing with α-gal nanoparticles. Plast. Reconstr. Surg. 129, 242–251; Galili, U., 2015a. Acceleration of wound healing by α-gal nanoparticles interacting with the natural anti-Gal antibody. J. Immunol. Res. 2015, 589648.*

Monitoring health of the pigs and studying their kidney and heart histology on Days 7, 13, and 60 posttreatment revealed no local or systemic toxicity responses in any of the treated pigs. GT-KO pigs with wounds treated with α-gal nanoparticles and monitored for 60 days displayed no keloid formation in both α-gal nanoparticles and saline-treated wounds. Furthermore, regenerating dermis in α-gal nanoparticles treated wounds displayed the presence of skin appendages such as hair. However, because pigs tend not to form distinct scars following physiologic wound healing, it was impossible to determine whether α-gal nanoparticles prevent fibrosis and scar formation in this experimental model. Overall, the study in GT-KO pigs (Hurwitz et al., 2012) demonstrated accelerated wound healing in α-gal nanoparticles treated wounds, like that observed in GT-KO mice (Wigglesworth et al., 2011).

METHODS FOR APPLICATION OF α-GAL NANOPARTICLES TO INJURIES

The α-gal nanoparticles are highly stable particles. The activity of these nanoparticles does not decrease even after 4 years of storage as suspension at 4°C or frozen at −20°C. Furthermore, their activity (i.e., anti-Gal binding) does not diminished for at least a year as nanoparticles dried on wound dressing and stored at room temp. The reason for this high stability is that carbohydrate chains are highly stable molecules because, in contrast to proteins, they do not have tertiary structures and they do not tend to undergo oxidation. The stability of α-gal epitopes on the α-gal nanoparticles can be further inferred from studies on blood group A and B antigens, which have structure similar to that of the α-gal epitope (Chapter 3). Stability of these blood group antigens has enabled their detection in Egyptian mummies that are >2000 years old (Crainic et al., 1989). Stability of liposomes made of phospholipids is high, as well. The combined stabilities of carbohydrate chains and phospholipids result in an overall high stability of α-gal nanoparticles, which makes them amenable to application in various forms.

In addition to application of α-gal nanoparticles as suspension in saline (as used in the experimental studies described above), these nanoparticles may be applied in a dried form on wound dressings, or as aerosol sprayed on wounds and burns. To prevent their spread beyond the injury area (e.g., when applied into internal injuries such as surgical incisions), the nanoparticles may be applied in a semisolid medium as water base ointment, hydrogel, plasma clot, or incorporated into biodegradable scaffold materials, such as collagen sheets. Diffusion of anti-Gal and complement proteins into the semisolid medium containing the α-gal nanoparticles will result in anti-Gal/α-gal nanoparticles interaction, complement activation, and recruitment of macrophages into the biodegradable material to initiate the accelerated healing process illustrated in Fig. 1B. An example of α-gal nanoparticles application to wound in a semisolid state is within a plasma clot described in Fig. 11. α-Gal nanoparticles were introduced into human plasma prior to induction of clot formation. The clot containing α-gal nanoparticles was applied onto wounds of GT-KO mice. Macrophages were observed to be recruited within 3 days (Fig. 11A). However, the plasma clots were filled with recruited macrophages after 6 days, and the epidermis grew in that period to cover the clot (Fig. 11B). This rapid epidermis growth may reflect the cytokines and growth factor secreted within the clot by activated macrophages, as suggested in Fig. 1B.

CONCLUSIONS

Healing of wounds and burns may be accelerated by applying α-gal liposomes or α-gal nanoparticles to the injured tissue. These liposomes and nanoparticles are made of phospholipids, glycolipids, and cholesterol extracted from rabbit RBC membranes and present multiple α-gal epitopes (~10^{15} α-gal epitopes/mg). α-Gal nanoparticles are α-gal liposomes that were fractured into much smaller liposomes (<300 nm) by extensive probe sonication. The α-gal nanoparticles also may be prepared of synthetic glycolipids and phospholipids. Both α-gal liposomes and α-gal nanoparticles bind the natural anti-Gal antibody and activate the complement system, resulting in localized production of complement cleavage chemotactic peptides, which induce rapid migration of macrophages into the injury. Anti-Gal bound to

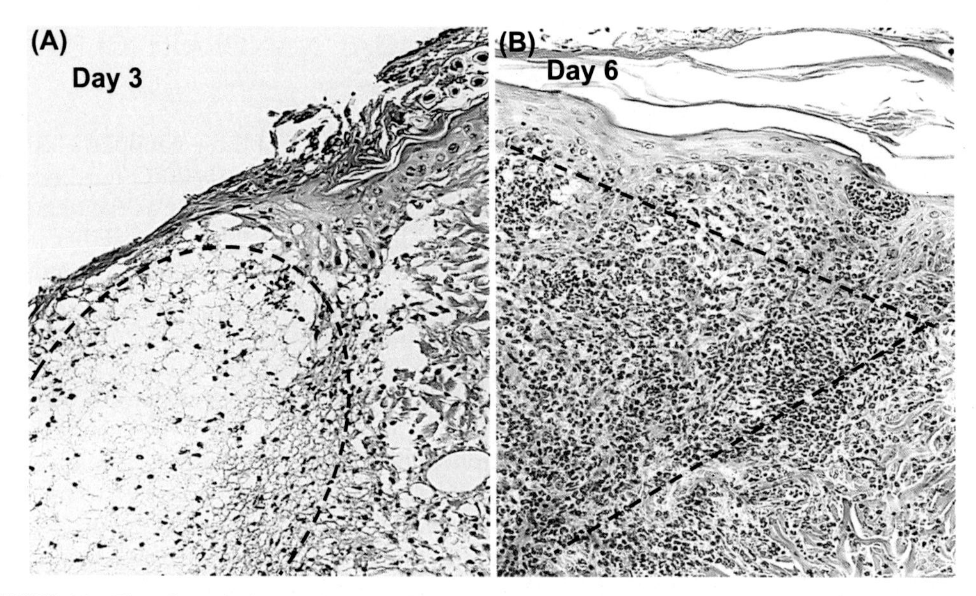

FIGURE 11 The effect of plasma clots containing 10 mg α-gal nanoparticles on macrophages recruitment and wound healing in α1,3galactosyltransferase knockout mice. (A) Representative clot on Day 3 post wounding demonstrating the beginning of macrophage recruitment into the clot and in the area surrounding it. (B) Day 6 post wounding. The clot contains multiple recruited macrophages that migrated into it, and its surface is covered by regenerating epidermis. The dashed lines mark the edge of the clot. Representative specimens from five mice at each time point. *Reprinted with permission from Galili, U. 2013b. Macrophages recruitment and activation by α-gal nanoparticles accelerate regeneration and can improve biomaterials efficacy in tissue engineering. Open Tissue Eng. Regen. Med. J. 6, 1–11.*

α-gal nanoparticles or liposomes further binds to macrophages via Fc/FcγR interaction. This interaction activates macrophages to secrete various pro-healing cytokines and growth factors. Topical application of α-gal nanoparticles or liposomes onto injuries in anti-Gal producing GT-KO mice or pigs accelerates healing of wounds and burns so that the regeneration time of the epidermis is shortened by 40%–60%. This acceleration of wound healing in mice occurs prior to the onset of fibrosis. Therefore, wounds do not develop scars in anti-Gal producing mice treated with α-gal liposomes and nanoparticles. Preliminary studies in mice with chemically induced diabetes impaired wound healing suggest that α-gal nanoparticles can induce healing of wounds, which are chronic and fail to heal without treatment. These nanoparticles and liposomes are highly stable for years in suspension or in a dried state on wound dressings. Studies in GT-KO mice and pigs demonstrated no toxic effects of α-gal nanoparticles on the treated animals. Thus, it would be of interest study the feasibility and efficacy of α-gal nanoparticles in inducing accelerated healing of burns and acute or chronic wounds in humans.

References

Crainic, K., Durigon, M., Oriol, R., 1989. ABO tissue antigens of Egyptian mummies. Forensic Sci. Int. 43, 113–124.

Dabrowski, U., Hanfland, P., Egge, H., Kuhn, S., Dabrowski, J., 1984. Immunochemistry of I/i-active oligo- and polyglycosylceramides from rabbit erythrocyte membranes. Determination of branching patterns of a ceramide pentadecasaccharide by 1H nuclear magnetic resonance. J. Biol. Chem. 259, 7648–7651.

DiPietro, L.A., Koh, T.J., 2016. Macrophages and wound healing. Adv. Wound Care 2, 71–75.

Dor, F.J., Tseng, Y.L., Cheng, J., Moran, K., Sanderson, T.M., Lancos, C.J., et al., 2004. α1,3-Galactosyltransferase gene-knockout miniature swine produce natural cytotoxic anti-Gal antibodies. Transplantation 78, 15–20.

Egge, H., Kordowicz, M., Peter-Katalinic, J., Hanfland, P., 1985. Immunochemistry of I/i-active oligo- and polyglycosylceramides from rabbit erythrocyte membranes. J. Biol. Chem. 260, 4927–4935.

Eto, T., Ichikawa, Y., Nishimura, K., Ando, S., Yamakawa, T., 1968. Chemistry of lipids of the posthemolytic residue or stroma of erythrocytes. XVI. Occurrence of ceramide pentasaccharide in the membrane of erythrocytes and reticulocytes in rabbit. J. Biochem. (Tokyo) 64, 205–213.

Fang, J., Walters, A., Hara, H., Long, C., Yeh, P., Ayares, D., et al., 2012. Anti-gal antibodies in α1,3-galactosyltransferase gene-knockout pigs. Xenotransplantation 19, 305–310.

Galili, U., 2013a. α1,3galactosyltransferase knockout pigs produce the natural anti-Gal antibody and simulate the evolutionary appearance of this antibody in primates. Xenotransplantation 20, 267–276.

Galili, U., 2013b. Macrophages recruitment and activation by α-gal nanoparticles accelerate regeneration and can improve biomaterials efficacy in tissue engineering. Open Tissue Eng. Regen. Med. J. 6, 1–11.

Galili, U., 2015a. Acceleration of wound healing by α-gal nanoparticles interacting with the natural anti-Gal antibody. J. Immunol. Res. 2015:Article ID 589648.

Galili, U., 2015b. Avoiding detrimental human immune response against mammalian extracellular matrix implants. Tissue Eng. Part B 21, 231–241.

Galili, U., 2017. α-Gal nanoparticles in wound and burn healing acceleration. Adv. Wound Care 6, 81–91.

Galili, U., Tibell, A., Samuelsson, B., Rydberg, L., Groth, C.G., 1995. Increased anti-Gal activity in diabetic patients transplanted with fetal porcine islet cell clusters. Transplantation 59, 1549–1556.

Galili, U., LaTemple, D.C., Radic, M.Z., 1998. A sensitive assay for measuring α-gal epitope expression on cells by a monoclonal anti-Gal antibody. Transplantation 65, 1129–1132.

Galili, U., Wigglesworth, K., Abdel-Motal, U.M., 2007. Intratumoral injection of α-gal glycolipids induces xenograft-like destruction and conversion of lesions into endogenous vaccines. J. Immunol. 178, 4676–4687.

Galili, U., Wigglesworth, K., Abdel-Motal, U.M., 2010. Accelerated healing of skin burns by anti-Gal/α-gal liposomes interaction. Burns 36, 239–251.

Gurtner, G.C., Werner, S., Barrandon, Y., Longaker, M.T., 2008. Wound repair and regeneration. Nature 453, 314–321.

Hanfland, P., Kordowicz, M., Peter-Katalinic, J., Egge, H., Dabrowski, J., Dabrowski, U., 1988. Structure elucidation of blood group B-like and I-active ceramide eicosa- and pentacosasaccharides from rabbit erythrocyte membranes by combined gas chromatography-mass spectrometry; electron-impact and fast atom-bombardment mass spectrometry; and two-dimensional correlated, relayed-coherence transfer, and nuclear Overhauser effect 500-MHz 1H-n.m.r. spectroscopy. Carbohydr. Res. 178, 1–21.

Heinrich, S.A., Messingham, K.A., Gregory, M.S., Colantoni, A., Ferreira, A.M., Dipietro, L.A., et al., 2003. Elevated monocyte chemoattractant protein-1 levels following thermal injury precede monocyte recruitment to the wound site and are controlled, in part, by tumor necrosis factor-α. Wound Repair Regen. 11, 110–119.

Honma, K., Manabe, H., Tomita, M., Hamada, A., 1981. Isolation and partial structural characterization of macroglycolipid from rabbit erythrocyte membranes. J. Biochem. 90, 1187–1196.

Hurwitz, Z., Ignotz, R., Lalikos, J., Galili, U., 2012. Accelerated porcine wound healing with α-gal nanoparticles. Plast. Reconstr. Surg. 129, 242–251.

Kiwanuka, E., Junker, J., Eriksson, E., 2012. Harnessing growth factors to influence wound healing. Clin. Plast. Surg. 39, 239–248.

Kolber-Simonds, D., Lai, L., Watt, S.R., Denaro, M., Arn, S., Augenstein, M.L., et al., 2004. Production of α1,3-galactosyltransferase null pigs by means of nuclear transfer with fibroblasts bearing loss of heterozygosity mutations. Proc. Natl. Acad. Sci. U.S.A. 101, 7335–7340.

Lai, L., Kolber-Simonds, D., Park, K.W., Cheong, H.T., Greenstein, J.L., Im, G.S., et al., 2002. Production of α-1,3-galactosyltransferase knockout pigs by nuclear transfer cloning. Science 295, 1089–1092.

Low, Q.E., Drugea, I.A., Duffner, L.A., Quinn, D.G., Cook, D.N., Rollins, B.J., et al., 2001. Wound healing in MIP-1α(-/-) and MCP-1(-/-) mice. Am. J. Pathol. 159, 457–463.

Martin, P., 1997. Wound healing-aiming for perfect skin regeneration. Science 276, 75–81.

Meier, K., Nanney, L.B., 2006. Emerging new drugs for wound repair. Expert Opin. Emerg. Drugs 11, 23–37.

Phelps, C.J., Koike, C., Vaught, T.D., Boone, J., Wells, K.D., Chen, S.H., et al., 2003. Production of α1,3-galactosyltransferase-deficient pigs. Science 299, 411–414.

Piccolo, M.T., Wang, Y., Sannomiya, P., Piccolo, N.S., Piccolo, M.S., Hugli, T.E., et al., 1999. Chemotactic mediator requirements in lung injury following skin burns in rats. Exp. Mol. Pathol. 66, 220–226.

Shallo, H., Plackett, T.P., Heinrich, S.A., Kovacs, E.J., 2003. Monocyte chemoattractant protein-1 (MCP-1) and macrophage infiltration into the skin after burn injury in aged mice. Burns 29, 641–647.

Shukaliak, J., Dorovini-Zis, K., 2000. Expression of the [β]-chemokines RANTES and MIP-1β by human brain microvessel endothelial cells in primary culture. J. Neuropathol. Exp. Neurol. 59, 339–352.

Sica, A., Mantovani, A., 2012. Macrophage plasticity and polarization: in vivo veritas. J. Clin. Investig. 122, 787–795.

Singer, A.J., Clark, R.A., 1999. Cutaneous wound healing. N. Engl. J. Med. 341, 738–746.

Stellner, K., Saito, H., Hakomori, S., 1973. Determination of aminosugar linkage in glycolipids by methylation. Aminosugar linkage of ceramide pentasaccharides of rabbit erythrocytes and of Forssman antigen. Arch. Biochem. Biophys. 133, 464–472.

Tanemura, M., Yin, D., Chong, A.S., Galili, U., 2000. Differential immune response to α-gal epitopes on xenografts and allografts: implications for accommodation in xenotransplantation. J. Clin. Investig. 105, 301–310.

Vogt, W., 1986. Anaphylatoxins: possible roles in disease. Complement 3, 177–188.

Wang, L., Anaraki, F., Henion, T.R., Galili, U., 1995. Variations in activity of the human natural anti-Gal antibody in young and elderly populations. J. Gerontol. Med. Sci. 50A, M227–M233.

Wigglesworth, K.M., Racki, W.J., Mishra, R., Szomolanyi-Tsuda, E., Greiner, D.L., Galili, U., 2011. Rapid recruitment and activation of macrophages by anti-Gal/α-gal liposome interaction accelerates wound healing. J. Immunol. 186, 4422–4432.

Wood, G.W., Hausmann, E., Choudhuri, R., 1997. Relative role of CSF-1, MCP-1/JE, and RANTES in macrophage recruitment during successful pregnancy. Mol. Reprod. Dev. 46, 62–69.

SECTION 4

FUTURE DIRECTIONS

(Hypothetical Therapies Requiring Experimental Validation)

CHAPTER

13

Anti-Gal and Anti-non Gal Antibodies in Regeneration of Extracellular Matrix Bio-Implants

OUTLINE

INTRODUCTION

Much of the research on tissue and organ engineering has been focused on the use of bio-implants of mammalian origin for the repair or replacement of injured or impaired human tissues and organs. Because of similarity in size between pigs and humans, porcine tissues and organs have been used as a major source of implants in bio-engineering research. Chapter 6 discusses the use of live organ or tissue xenografts and the immunologic barriers in xenotransplantation, including the natural anti-Gal antibody barrier. The present chapter describes the experience associated with these immunologic barriers, which has been obtained by grafting

The Natural Anti-Gal Antibody as Foe Turned Friend in Medicine
http://dx.doi.org/10.1016/B978-0-12-813362-0.00013-0

porcine bio-implants into humans and nonhuman primates. Bio-implants are defined here as processed tissues or organs lacking live cells in contrast to xenografts that contain live cells. Extensive research is being conducted on developing methods for using "patches" of extracellular matrix (ECM) bio-implants of various "specialized" tissues (i.e., tissues other than connective tissue) such as myocardium (Hirt et al., 2014; Feric and Radisic, 2016) and urinary bladder (Song et al., 2014) for replacement of injured or impaired tissues (Atala, 2009; Badylak et al., 2012; Peloso et al., 2015). The ECM consists of a highly complex, tissue-specific network of proteins, proteoglycans and glycosaminoglycans, which help regulate many cellular functions. The ECM further provides the scaffold that specifically organizes each tissue. This chapter describes the immunological challenges, including that of the anti-Gal barrier, in inducing regeneration of specialized tissues by using ECM bio-implants, in comparison with regeneration of connective tissues. The chapter further proposes hypothetical uses of α-gal nanoparticles for harnessing the immunologic potential of the natural anti-Gal antibody in increasing the likelihood of ECM implant regeneration into a functional autologous tissue or organ.

EXTRACELLULAR MATRIX BIO-IMPLANTS AND TISSUE REGENERATION

As described in Chapter 12, repair and regeneration of epidermis in external wounds is the result of regrowth over the granulation tissue of the uninjured epidermis bordering the wound area. Much of the regeneration of other epithelial tissues (e.g., epithelium lining the intestine and urinary bladder) also is mediated by cell division at the border of the injured areas and "resurfacing" of the injury with healthy epithelial cells (Coletti et al., 2013). Nevertheless, stem cells were reported to be found in the skin (Fuchs, 2008) and in intestinal epithelium (Marshman et al., 2002). Resurfacing of blood vessels with endothelial cells was found to be a combination of continuous division of these cells in the leading edge of the regenerating endothelium and activity of stem cells that attach to the surface of blood vessels lacking endothelial cells (Asahara et al., 1997; Garmy-Susini and Varner, 2005). However, in internal injuries involving mesenchymal tissues, there seem to be two types of regeneration: (1) Stem cell-mediated regeneration of specialized tissues (e.g., muscle, cartilage, and bone) and (2) Fibroblast-mediated regeneration of various types of connective tissues. In both types of regeneration, ECM bio-implant can be instrumental in the regeneration process because it provides the scaffold for directing the regeneration process, as well as for providing instructions for the stem cells to differentiate into the specialized cells that restore the structure and function of the treated tissue or organ. Both human and porcine ECM bio-implants are generated from tissues that have been decellularized by treatment with various detergents (Bader et al., 1998; Booth et al., 2002; Atala, 2009; Liu et al., 2009; Wang et al., 2010; Badylak et al., 2012). Incubation of tissues with these detergents (e.g., sodium dodecyl sulfate [SDS] and Triton X-100) dissolves the lipid part of cell membranes. Subsequently, the cytoplasmic content of cells destroyed by the detergents is removed by washes. DNase is used in many protocols for eliminating the DNA released from nuclei following the destruction of nuclear membranes by the detergents. Whole organs, such as heart or kidney, can also be decellularized by injection of various detergents into their vascular system (Atala, 2009; Crapo et al., 2011; Badylak et al., 2012; Arenas-Herrera et al., 2013; Orlando et al., 2013; Katari et al., 2014).

Human ECM bio-implants may be obtained from human tissue banks. Alternatively, a mammalian ECM, mostly of porcine origin, may be obtained from various sources providing pig tissues. Human ECM implants elicit a relatively low immune response in human recipients, as suggested by the wide use of human allogeneic tendons for replacement of ruptured anterior cruciate ligament (ACL). In contrast, porcine ECM implants elicit an extensive anti-porcine immune response, which might be detrimental to the interaction of stem cells with the ECM bio-implant. However, the quality of human ECM implants is likely to vary from one cadaveric donor to the other because of age differences, as well as individual differences between donors. In contrast, porcine ECM implants can be produced with standardized high quality because they may be obtained from a well-monitored herd and at an optimal age of the donor. The issues associated with the use of a porcine ECM implants are discussed below, whereas the possible uses of human ECM implants are discussed at the end of this chapter. It is suggested that the probability for success of both types of ECM implants may hypothetically be improved, if they are processed to contain α-gal nanoparticles.

Stem cell-mediated regeneration

Regeneration of injured mesenchymal tissues, other than connective tissue, is usually mediated by mesenchymal stem cells that arrive to the injury site and receive "ques" from the ECM to differentiate into cells, like the original cells of the tissue (Bader et al., 1998; Atala, 2009; Liu et al., 2009; Badylak et al., 2012). Thus, stem cells reaching an ischemic myocardium, post myocardial infarction, are thought to be "instructed" by the myocardium ECM to differentiate into cardiomyocytes, whereas the same stem cells reaching an injured urinary bladder wall will be instructed by the ECM in the urinary bladder to differentiate into smooth muscle cells. Additional cells with regenerative ability are progenitor cells (Laflamme and Murry, 2011) and pericytes (Caplan, 2009; Lolmede et al., 2009; Wong et al., 2015; Murray et al., 2017), which reside within the healthy tissue, and are activated to differentiate into functional cells to replace adjacent injured parts of the tissue.

As discussed in Chapter 12, if the regeneration process of wounds does not occur within a given period post injury, the wound is healed by migrating fibroblasts via the default mechanism of fibrosis and scar formation. The same default process of fibrosis occurs in internal injuries. One example is ischemia post myocardial infarction. If the injury of the ischemic myocardium is relatively large, the injured tissue will undergo fibrosis resulting in irreversible formation of a scar tissue in the myocardium. The hypothetical approach for induction of myocardium regeneration into functional tissue, by accelerated recruitment of stem cells with α-gal nanoparticles injected into the injured myocardium, is discussed in the next chapter. However, as discussed below, the need for accelerated recruitment of stem cells, to avoid fibrosis, is also applicable to myocardial patches and to other ECM bio-implants. If stem cells do not reach the bio-implant within few weeks post implantation, fibrosis is likely to occur, as in wounds and in internal injuries.

One approach to achieve regeneration without fibrosis has been to seed ECM bio-implants with stem cells *in vitro*, prior to implantation (Vashi et al., 2015). This approach raises the *in vivo* challenge of providing vascularization to the bio-implant prior to the death of differentiated cells in the implant because of ischemia. Another approach has been to load the bio-implant with cytokines that recruit stem cells *in vivo* (Ko et al., 2013). Fast depletion of such cytokines because of their diffusion from the bio-implant may decrease the efficacy of this approach. An alternative hypothetical approach, suggested is this chapter, is the recruitment of macrophages into the bio-implant by α-gal nanoparticles, and activation of these macrophages to secrete cytokines that recruit stem cells and progenitor cells into the bio-implant.

Regeneration of connective tissue

Regeneration of connective tissue does not seem to require interaction between stem cells and ECM because the default mechanism for injury repair is fibrosis mediated by fibroblasts migrating into external or internal injury sites (Wynn and Ramalingam, 2012; Johnson and DiPietro, 2013). Thus, if the ECM bio-implant is constructed of porcine collagen fibers, human fibroblasts migrating into such implants repopulate them, secrete their own ECM, and regenerate the ECM into an autologous connective tissue of the recipient. Such regeneration of a porcine connective tissue ECM implants has been observed in patients receiving porcine loose connective tissue implants of small intestine submucosa (SIS) ECM (Badylak et al., 1999; McPherson et al., 2000; Hiles et al., 2009) and in patients receiving porcine dermis ECM (Reing et al., 2010). These loose connective tissues regenerating from the ECM implants, "hold" organs in place and attach epithelial tissues to other underlying tissues. In contrast to specialized tissues, the loose connective tissue ECMs are usually not subjected to substantial bio-mechanical stress and do not perform any specific function, such as contraction, as in muscle cells or shock absorbing as in cartilage. Because connective tissue formation is the default regeneration mechanism, ECM bio-implants that require stem cells for regeneration but fail to regenerate due to lack of sufficient stem cells or because of a potent anti-implant immune response (see below), may ultimately be replaced by fibrotic tissue (Galili, 2015).

Antibody response against porcine extracellular matrix

The immune response in mammals against porcine implants is a complex process that involves multiple cells and mechanisms. The reader is directed to a recent book (Badylak, 2015) and to Badylak and Gilbert (2008) that review various aspects of this immune response. Because the ECM bio-implants are decellularized and thus are devoid of live cells, no cytolysis mechanisms, including complement mediated cytolysis, cytotoxic T cells, NK cells, or antibody dependent cell-mediated cytolysis, take place in immune rejection of bio-implants. Much of the destruction of porcine bio-implants in humans and in nonhuman primate recipients is mediated by natural and elicited antibodies (discussed below). These antibodies bind to porcine bio-implants, activate the complement system via these antigen/antibody interactions, thereby generating complement cleavage peptides such as C5a and C3a (Galili, 2015). These peptides are potent chemotactic factors (chemoattractants) that induce recruitment of macrophages into bio-implants (Klos et al., 2013). Recruited macrophages bind via their Fcγ receptors to the Fc "tail" of antibodies immunocomplexed with the porcine ECM (Ravetch and Bolland, 2001) and via their C3b receptors (C3bR, also known as CR1 and CD35) to C3b complement deposits on the ECM (van Lookeren Campagne et al., 2007; Murray and Wynn, 2011; Mantovani et al., 2013). Because the macrophages that adhere to the ECM cannot internalize the large network of ECM molecules linked to each other, they secrete their granules-containing proteases and glycosidases, which degrade the ECM and cause the destruction of the bio-implant.

A second mechanism by which antibodies can prevent regeneration of ECM bio-implants is prevention of the appropriate "cues," which "instruct" stem cells to differentiate into cells restoring the structure and function of the implant. The cues provided to the stem cells for differentiation are the result of interaction between stem cells and a variety of ECM proteins, glycoproteins, proteoglycans, and glycosaminoglycans, which comprise the ECM unique to each tissue (Atala, 2009; Zhang et al., 2009; Guilak et al., 2009; Hidalgo-Bastida and Cartmell, 2010; Badylak et al., 2012; Peloso et al., 2015). Interaction of antibodies with the implanted ECM may mask the ECM molecules and prevent their interaction with stem cells. This assumption is supported by observations on inhibition of stem cells adhesion to ECM with antibodies to ECM or with antibodies to cell surface molecules on stem cells (Williams et al., 1991; Shakibaei, 1998; Jung et al., 2005). Thus, reconstruction of the bio-implant into the target tissue fails following degradation of the ECM by macrophages and in the absence of appropriate interaction of stem cells and the porcine ECM bio-implant. Under these circumstances, the default fibrosis process is likely to convert the ECM bio-implant into connective tissue. There are two types of antibodies that can hinder the appropriate regeneration of porcine ECM bio-implants: (1) The anti-Gal antibody and (2) Anti-non-gal antibodies (Galili, 2015).

ANTI-GAL AS THE FIRST ANTIBODY BARRIER TO PORCINE EXTRACELLULAR MATRIX BIO-IMPLANTS IN MONKEYS AND HUMANS

As discussed in Chapters 1 and 6, anti-Gal is continuously produced in humans as ~1% of immunoglobulins and it binds to α-gal epitopes on glycolipids, glycoproteins and proteoglycans. Most porcine tissues (and those of other nonprimate mammals), as well as their ECM, contain high concentrations of α-gal epitopes (Galili et al., 1988; Oriol et al., 1999; Tanemura et al., 2000; Maruyama et al., 2000). Quantification of α-gal epitopes in porcine tissues and cells demonstrated in meniscus cartilage as many as 2×10^{11} α-gal epitopes/mg (Stone et al., 1998), in tendon 1×10^{11} α-gal epitopes/mg (Stone et al., 2007a), and on porcine endothelial and epithelial cells, at least 1×10^{7} and 3×10^{7} α-gal epitopes/cell, respectively (Galili et al., 1988). If a porcine ECM (or any other mammalian tissue) presenting α-gal epitopes is implanted into monkeys or humans, anti-Gal readily binds to these epitopes and is likely to induce rapid destruction of the bio-implant.

Anti-Gal-mediated destruction of porcine ECM bio-implant is further exacerbated by the activity of elicited anti-Gal antibodies. Introduction of xenoglycoproteins with multiple α-gal epitopes into monkeys and humans results in activation of the ~1% quiescent anti-Gal B cells to produce and secrete anti-Gal IgG antibodies, which increase the titer of this antibody by 30- to >100-fold (Galili et al., 1995, 1997, 2001; Stone et al., 2007b). This extensive anti-Gal antibody response to porcine implants is demonstrated in Fig. 1 in which a rhesus monkey was implanted with porcine tendon across the knee to replace ruptured ACL. The porcine tendon was implanted within tunnels drilled in the femur and tibia, anchored in the femur end by the bone block, and sutured at the free end to the tibia (Stone et al., 2007b). The activity of anti-Gal was determined by ELISA with synthetic α-gal epitopes linked to bovine serum albumin (α-gal BSA) as solid-phase antigen. The titer of anti-Gal increased by ~300-fold within a period of 2 weeks post implantation (Fig. 1) (Stone et al., 2007b).

Extensive elevation in anti-Gal production was observed also in humans exposed to xenograft tissues presenting α-gal epitopes. A marked increase in anti-Gal response was observed following grafting of porcine fetal islet cells presenting α-gal epitopes. This grafting resulted in 30–80-fold increase in anti-Gal titer in diabetic patients receiving a kidney allograft and the porcine fetal pancreatic islet cells (Galili et al., 1995). This elicited anti-Gal response occurred despite immunosuppression treatment that was potent enough to prevent rejection of the kidney allograft (Groth et al., 1994). Patients with impaired liver function, who were treated by temporary extracorporeal perfusion of their blood through a pig liver, displayed a similar marked increase in anti-Gal titers (Cotterell et al., 1995; Yu et al., 1999). This implies that the release of xenoglycoproteins from the pig liver, perfused for only several hours, was sufficient to induce activation of the many quiescent anti-Gal B cells for the increased production of the anti-Gal antibody. Elevation in anti-Gal activity was also observed in patients implanted with glutaraldehyde cross-linked porcine heart valve to replace impaired autologous heart valve (Konakci et al., 2005). These marked increases in the titer of anti-Gal following exposure to α-gal epitopes on porcine glycoproteins is the combination of increased concentration of the antibody in the serum and increased affinity of the antibody. This increased affinity is the result of an affinity maturation process in which anti-Gal B cell clones with high affinity B cell receptor undergo preferential expansion in comparison with low affinity anti-Gal B cell clones (Galili and Matta, 1996; Galili et al., 2001).

Both natural and elicited anti-Gal antibodies binding to porcine ECM implants are likely to have a synergistic detrimental effect, which can cause accelerated degradation the implant. An example for such a destructive antibody activity is the rhesus monkey in Fig. 1, which was implanted with an unprocessed porcine tendon. The implant completely degraded and disappeared within 2 months. Thus, successful implantation of porcine ECM implants in humans requires the elimination of the anti-Gal/α-gal epitopes interaction barrier. As discussed below, this barrier can be eliminated by either enzymatic treatment with α-galactosidase or by using ECM from α1,3galactosyltransferase knockout (GT-KO) pigs. This need for removal of α-gal epitopes is further demonstrated in a higher efficacy of porcine dermis ECM implanted in monkeys when the ECM is devoid of α-gal epitopes in comparison with porcine dermis ECM expressing α-gal epitopes (Xu et al., 2009).

An exception to the detrimental effects of anti-Gal/α-gal epitope interaction on porcine ECM bio-implants is porcine SIS ECM bio-implants. Such implants in monkeys and in GT-KO mice demonstrated no detrimental effects of anti-Gal on SIS bio-implants that were not treated for the elimination of α-gal epitopes (Raeder et al., 2002; Daly et al., 2009). This lack of antibody effect is likely to be a result of the very low concentration of α-gal epitopes on SIS because this bio-implant is comprised mostly of porcine collagen fibers. There are only few α-gal epitopes on porcine collagen or on collagen of other species. Because there is a relatively "great distance" between the α-gal epitopes on SIS, complement activation by anti-Gal binding to these epitopes is very low (McPherson et al., 2000). Accordingly, SIS was found to be effective as ECM bio-implant in hernia repair in many patients (Hiles et al., 2009). In contrast, porcine cartilage and tendon ECM (both containing multiple α-gal epitopes) implanted into Old World monkeys are destroyed by natural and elicited anti-Gal antibodies within a period of 2 months post implantation (Stone et al., 1998, 2007b).

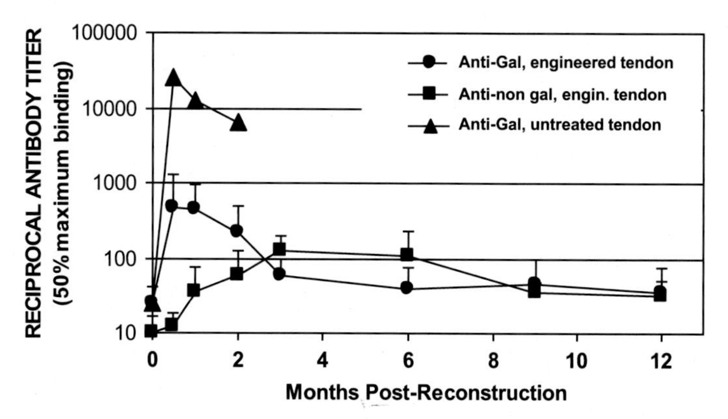

FIGURE 1 Anti-Gal and anti-non-gal antibody production in rhesus monkeys implanted with untreated or processed (engineered) porcine tendon for anterior cruciate ligament reconstruction. One monkey was implanted with untreated tendon and five monkeys with tendons treated with α-galactosidase and partial cross-linking with glutaraldehyde. Anti-Gal antibody activity was determined by ELISA with synthetic α-gal linked to bovine serum albumin as solid-phase antigen. Anti-non-gal antibody activity was measured in sera depleted of anti-Gal (adsorption on glutaraldehyde fixed rabbit red blood cells), by ELISA with a homogenate of fragmented pig tendon as solid-phase antigen. Data presented as mean + standard deviation of antibody titer, i.e., reciprocal of serum dilution displaying half of maximum antibody binding. *Adapted with permission from Stone, K.R., Walgenbach, A.W., Turek, T.J., Somers, D.L., Wicomb, W., Galili, U., 2007b. Anterior cruciate ligament reconstruction with a porcine xenograft: a serologic, histologic, and bio-mechanical study in primates. Arthroscopy 23, 411–419.*

The effective binding of anti-Gal to the multiple α-gal epitopes in most porcine ECM and the extensive elevation in anti-Gal titers in human or nonhuman primate recipients of ECM containing α-gal epitopes both suggest that successful regeneration of porcine ECM into human functional tissue requires the use of ECM that is devoid of α-gal epitopes. Two methods have been developed for this purpose: (1) The use of decellularized porcine tissues or organs that are treated with α-galactosidase and (2) Preparation of ECM from tissues or organs obtained from α1,3galactosyltransferase knockout (GT-KO) pigs.

Elimination of α-gal epitopes by recombinant α-galactosidase

α-Galactosidase is an enzyme that cleaves the terminal α-galactosyl unit of the α-gal epitopes and converts this epitope into a disaccharide called *N*-acetyllactosamine (i.e., Galα1-3Galβ1-4GlcNAc-R converted into Galβ1-4GlcNAc-R) as illustrated in Fig. 2 in Chapter 7. The *N*-acetyllactosamine epitope does not bind anti-Gal and is nonimmunogenic in humans because it is naturally present on human cells (Ogawa and Galili, 2006). Enzymatic elimination of α-gal epitopes from porcine tissues was performed using recombinant α-galactosidase that was cloned from coffee beans and produced in a yeast expression system (Zhu et al., 1995, 1996). Initial studies on porcine meniscus and articular cartilage indicated that incubation of these tissues in a solution of 100 units/mL of recombinant α-galactosidase resulted in complete elimination of α-gal epitopes (Stone et al., 1998). Such elimination could be demonstrated by the lack of binding of a monoclonal anti-Gal antibody called M86 to the homogenate of the cartilage tissue. The elimination of α-gal epitopes by α-galactosidase was further confirmed by demonstration of no elicited anti-Gal antibody production in monkeys implanted with porcine cartilage treated with this enzyme (Stone et al., 1998).

The contribution of anti-Gal/α-gal epitope interaction to immune-mediated rejection of porcine ECM in monkeys could be evaluated by comparing the rejection of untreated porcine meniscus and articular cartilage implanted subcutaneously in the suprapatellar pouch of cynomolgus monkey versus that of the same cartilage specimens incubated in α-galactosidase, i.e., cartilage completely devoid of α-gal epitopes. The cartilage implants were explanted after 2 months and inspected histologically. Implants of untreated porcine cartilage displayed an extensive inflammatory response of mononuclear cells that filled the implant and were comprised of equal numbers of macrophages and T cells (Stone et al., 1998). Less than 5% of the infiltrating cells were B cells. However, the inflammatory response in α-galactosidase treated cartilage implants was ~95% lower than that in the untreated implants. These observations implied that most of the immune response against the cartilage bio-implant, in the first two months post implantation is mediated by the natural and elicited anti-Gal antibodies. Treatment of the bio-implant with α-galactosidase eliminated completely α-gal epitopes and avoided the detrimental effects of anti-Gal. The mild immune response observed in the absence of α-gal epitopes is likely to be associated with the anti-non-gal antibody response that is discussed below.

Effective elimination of α-gal epitopes by α-galactosidase was further observed in porcine tendons (Stone et al., 2007a,b). These tendons, attached to bone blocks, also were subjected to partial cross-linking with glutaraldehyde, implanted in rhesus monkeys (Stone et al., 2007b), and subsequently in humans to replace ruptured ACL (Stone et al., 2007a). In the absence of α-gal epitopes, the elicited anti-Gal response to such implants was ~98% lower than that observed with untreated tendon implants (Fig. 1). The residual production of elicited anti-Gal antibodies was the result of the immune response to α-gal epitopes on red blood cells (RBC) and bone marrow cells encased within cavities in the porcine bone blocks. The enzyme cannot reach these cells, and thus α-gal epitopes on glycoproteins remaining in the bone cavities leach out post implantation in course of the bone remodeling. These glycoproteins with α-gal epitopes induce some stimulation of anti-Gal B cells for a period of 2–4 months, until the porcine bone tissue is remodeled into bone of the implant recipient (Stone et al., 2007a,b, 2017). Recombinant α-galactosidase also was reported to eliminate α-gal epitopes in pig heart valves implants (Park et al., 2009), which contain high concentration of these epitopes (Kasimir et al., 2005; Naso et al., 2013). Indeed, no anti-Gal binding to pig valves was observed following incubation of these valves with α-galactosidase, whereas untreated valves displayed extensive binding of the antibody (Park et al., 2009).

α1,3Galactosyltransferase knockout pigs

Anti-Gal/α-gal epitope barrier in xenotransplantation (Chapter 6) and the success in knocking out the α1,3GT (*GGTA1*) gene in mice (Thall et al., 1995; Tearle et al., 1996) motivated researchers to inactivate this gene in pigs. Disruption of the α1,3GT gene by the knockout technology was initially achieved in pigs 12–15 years ago (Lai et al., 2002; Phelps et al., 2003; Kolber-Simonds et al., 2004; Takahagi et al., 2005). These pigs are completely devoid of α-gal epitopes as indicated by their ability to produce the natural anti-Gal antibody at titer that are at least as high as those found in humans (Dor et al., 2004; Fang et al., 2012; Galili,

2013). Accordingly, monkeys transplanted with GT-KO pig xenografts do not display any production of elicited anti-Gal antibodies (Chen et al., 2005; Kuwaki et al., 2005; Tseng et al., 2005; Yamada et al., 2005; Ezzelarab et al., 2006; Hisashi et al., 2008; Yeh et al., 2010), and SIS implants from GT-KO pig elicit no anti-Gal response (Daly et al., 2009). These observations correlate with the much lower immune rejection of GT-KO pigs heart valve implants in monkey recipients in comparison with immune rejection and calcification of wild-type pig valve implants (McGregor et al., 2013).

Overall, the observations above imply that porcine ECM bio-implants treated with recombinant α-galactosidase, or those obtained from GT-KO pigs, lack α-gal epitopes and thus, may not be affected by the anti-Gal/α-gal epitope barrier. The decellularization process, which is usually part of the processing that converts tissues into ECM, provides the ECM with the porosity that is likely to enable subsequent penetration of α-galactosidase throughout the ECM, resulting in enzymatic elimination of α-gal epitopes. Thus, both approaches are likely to be suitable for obtaining porcine ECM devoid of α-gal epitopes.

ANTI-NON-GAL ANTIBODIES AS THE SECOND ANTIBODY BARRIER TO PORCINE EXTRACELLULAR MATRIX BIO-IMPLANTS

Grafting of porcine ECM into humans results in exposing the immune system of the recipient to many immunogenic proteins even in the absence of the α-gal epitopes. Whereas the immune response to allogeneic graft is primarily against few transplantation antigens, the immune response against porcine tissue xenografts or ECM is directed against most porcine proteins in the graft. Porcine immunogenic proteins induce production of a wide range of antibodies called anti-non-gal antibodies (also mentioned in Chapter 6 as part of the immune response to xenografts). The immunogenicity of porcine proteins in humans is the result of different mutations, which have accumulated during the separate evolution of the various mammalian lineages (Wilson, 1985). As many as 0.5% of nucleotides in the noncoding DNA of most mammalian lineages mutate in a period of a million years (Britten, 1986). Thus, on average, the homologous sequences of noncoding DNA regions in species of lineages that separated >75 million years in the "great mammalian radiation" (e.g., mice vs. rabbits or humans vs. pigs) differ by 35%–40% from each other. In apes these differences are smaller, possibly because of the longer life span of these species. The homologous coding regions of genes have lower sequence differences, which depend on the functional constraints of each protein, or of specific regions within a protein, e.g., catalytic domain of an enzyme or the ligand binding domain of a receptor vs. the tether of membrane bound enzymes or receptors. In few proteins, such as histones or collagen, most mutations may be detrimental to the function of the protein and thus cannot be tolerated. Therefore, porcine and bovine collagen have relatively low immunogenicity in humans and can be used in cosmetic surgery as "fillers" that are sustained *in vivo* for months. However, in most proteins, mutations are tolerated to different extents in various regions of the protein. Majority of homologous proteins in pigs and humans differ from each other in 3%–40% of their amino acid sequences.

Because many of the amino acid sequences present in porcine proteins (or any other nonprimate mammal) of various ECMs are absent in homologous human proteins, most porcine

ECM proteins are immunogenic in human recipients and induce production of anti-non-gal antibodies (Galili, 2012). The gradual release of these immunogenic proteins or peptides from the ECM implants results in activation of T helper cells and B cells, which produce anti-non-gal antibodies (Galili, 2012). Such antibodies were reported to be produced against xenogeneic ECM molecules such as laminin (Lissitzky et al., 1988) and fibronectin (Sabari et al., 2011) and are likely to be produced by the human immune system against many porcine ECM proteins. Anti-non-gal antibodies bind to their corresponding antigens in the ECM, hinder stem cells and progenitor cells interaction with the ECM molecules, activate the complement system, and generate chemotactic factors that recruit macrophages and other mononuclear cells mediating a chronic inflammatory response within the bio-implant. Binding of the recruited macrophages, to the ECM by Fc/Fc receptor and C3b/C3b receptor interactions, is followed by secretion of proteases and glycosidases that gradually degrade the ECM. This inflammatory process may ultimately result in replacement of ECM bio-implants with fibrotic tissue, as observed in young recipients of decellularized porcine heart valves (Cicha et al., 2011).

Long-term human anti-non-gal antibody response against porcine extracellular matrix

Production of anti-non-gal antibodies in human recipients of porcine ECM bio-implants may continue for prolonged periods, as long as the porcine proteins of the implant are present in the recipient. This could be demonstrated in a study on implantation of processed porcine patellar tendon to replace ruptured human ACL (Stone et al., 2007a, 2017). The tendons had a patellar and tibia bone blocks, which could anchor the ACL in the femur and tibia. As described above, the porcine tendons were incubated for 12h in a recombinant α-galactosidase solution to remove α-gal epitopes. Subsequently, the tendons were washed and underwent partial cross-linking for 12h in 0.1% glutaraldehyde. Glutaraldehyde groups that remained free were blocked with 0.1M glycine. The processed tendons were irradiated for final sterilization and stored frozen until use. These porcine tendons were implanted across the knee joint through femoral and tibia tunnels and their bone blocks were anchored to the tunnel walls with interference screws. It was hypothesized that the mild cross-linking by glutaraldehyde slows the penetration of macrophages that were recruited by anti-non-gal antibodies produced against the porcine ligament proteins. This slow penetration further enables fibroblasts of the recipient, which follow the infiltrating macrophages, to align with the porcine collagen fibers scaffold and secrete their own collagen and other ECM proteins. This process of slow destruction of the porcine ECM implant and infiltration of fibroblasts ultimately results in the remodeling of the porcine tendon implant into an autologous human ACL. Indeed, five evaluable patients implanted with processed pig tendon displayed regenerated ACL that continues to function for >15years (Stone et al., 2007a,2017).

Production of anti-non-gal antibodies against the porcine tendon could be monitored in the implanted patients by ELISA in which the solid-phase antigen was a homogenate of porcine tendon consisting of tissue fragments with the size of ~20 to 200 μm. The sera of the monitored patients were depleted of anti-Gal prior to the assay by adsorption on glutaraldehyde fixed rabbit RBC. As shown in Fig. 2, sera tested prior to implantation displayed only background anti-non-gal antibody activity. An increase in activity of these antibodies, because of the immune response to the porcine tendon, was observed 2months post implantation and

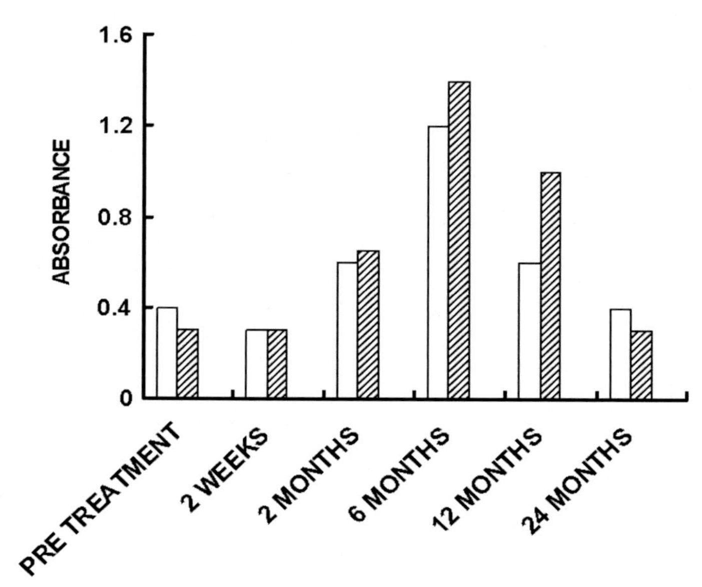

FIGURE 2 Anti-non-gal IgG antibody response in two representative orthopedic patients who were grafted with porcine patellar tendon bio-implants for the replacement of ruptured anterior cruciate ligament. Anti-non-gal antibody activity was determined by ELISA at various time points and presented as optical density (O.D.) units at serum dilution of 1:640. ELISA was performed as described in Fig. 1. Patient 1—open columns; Patient 2—hatched columns. *Reprinted with permission from Galili, U., 2012. Induced anti-non gal antibodies in human xenograft recipients. Transplantation 93, 11–16.*

peaked 6 months post implantation. Subsequently, the activity of these antibodies decreased because of the gradual replacement of the porcine ECM with human tissue in the remodeling process. After 2–2.5 years, the activity of anti-non-gal antibodies returned to the pre-implantation level, suggesting that all the porcine tissue was eliminated, thus, there was no more antigenic stimulation for production of these antibodies. A similar kinetics of anti-non-gal antibody production was observed in monkeys implanted with processed porcine tendon. Anti-non-gal antibodies in these monkeys were detected in the serum only 4 weeks post implantation, peaked at 3–6 months and subsequently decreased in their activity as the porcine tissue was reconstructed into a monkey ACL with autologous fibroblasts and collagen fibers (Fig. 1) (Stone et al., 2007b).

The wide diversity in specificities of anti-non-gal antibodies produced in humans implanted with porcine tendon ECM could be demonstrated by Western blot analysis of pig tendon proteins with sera depleted of anti-Gal antibodies by adsorption on glutaraldehyde fixed rabbit RBC (Fig. 1 in Chapter 6) (Stone et al., 2007a). Porcine tendon proteins separated by SDS polyacrylamide gel electrophoresis (PAGE) and blotted on nitrocellulose paper did not bind any IgG antibodies from pre-implantation sera, implying that these sera do not contain antibodies that bind to porcine tendon proteins. However, sera obtained 6 months post implantation displayed extensive production of anti-non-gal antibodies, which bound to multiple pig tendon proteins, implying that many pig tendon proteins are immunogenic in humans (Fig. 1 in Chapter 6) (Stone et al., 2007a). Because of the high number of proteins binding anti-non-gal antibodies, the binding pattern is of a "smear" in which the bands partially overlap rather than display a pattern of individual protein bands. Some of the anti-non-gal binding proteins were also found in porcine kidney, suggesting that these are proteins common to different tissues. The specificity of these anti-non-gal antibodies to porcine proteins is indicated by the observation that none of these antibodies bound to human ligament proteins

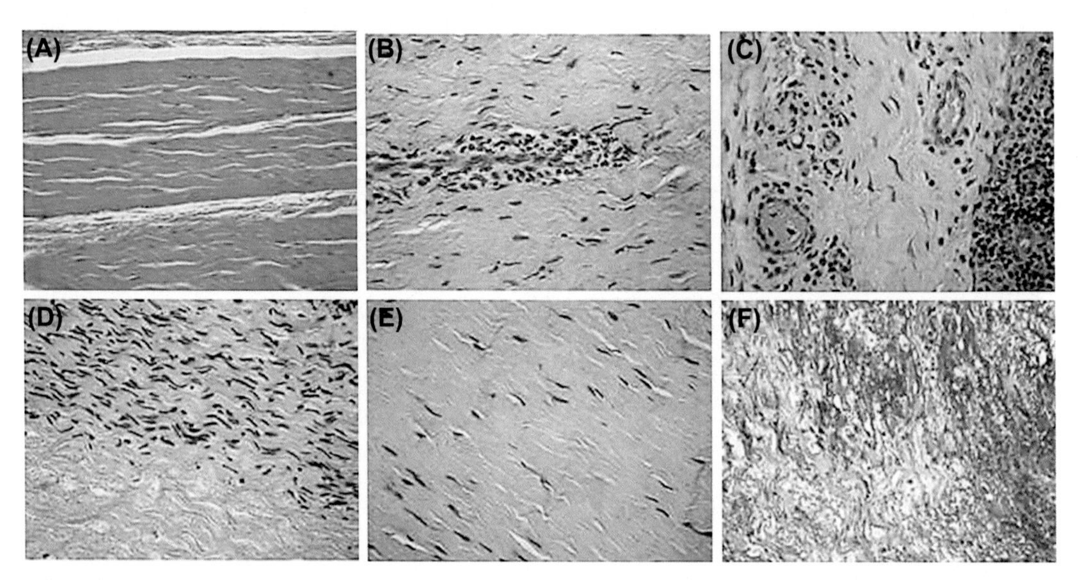

FIGURE 3 Reconstruction of processed porcine tendon bio-implants into autologous human anterior cruciate ligament, evaluated histologically in patients who ruptured their bio-implants in ski and biking accidents. (A) Pre-implantation porcine patellar tendon processed for removal of α-gal epitopes and partially cross-linked with glutaraldehyde. (B) Infiltration of macrophages and a capillary into the porcine tendon bio-implant. (C) Vascularization of the implant in a region near macrophage infiltrates and extravasation of the macrophages. (D) Fibroblasts infiltrating a region of the implant in which they align with the porcine collagen fibers scaffold. (E) An advanced stage in the bio-implant remodeling in which the repopulating fibroblasts are far apart from each other because of the collagen fibers and other extracellular matrix proteins, proteoglycans, and glycosaminoglycan that the recipient's fibroblasts have secreted. (F) Trichrome staining of regenerating porcine tendon bio-implant in which the collagen fibers are stained blue. (A–E): hematoxylin and eosin, (F): trichrome (×200). *Reprinted with permission from Stone, K.R., Abdel-Motal, U.M., Walgenbach, A.W., Turek, T.J., Galili, U., 2007a. Replacement of human anterior cruciate ligaments with pig ligaments: a model for anti-non-gal antibody response in long-term xenotransplantation. Transplantation 83, 211–219.*

(Fig. 1 in Chapter 6). The lack of antibody binding to human ligament proteins further suggests that effective immune tolerance mechanisms in human recipients of porcine ECM prevents production of autoantibodies to human proteins when the immune system is stimulated by homologous porcine immunogenic proteins.

Histological analysis of porcine ligament bio-implants in patients who ruptured them within the first-year post implantation was feasible because of skiing or biking accidents. Fig. 3 illustrates a composite of the remodeling process of the porcine bio-implant (Stone et al., 2007a). One of the first events observed in the bio-implant is the infiltration of macrophages which may be recruited by anti-non-gal antibodies binding to the ECM (Fig. 3B), followed by ingrowth of capillaries. These capillaries enable the extravasation of additional macrophages, which slowly destroy the porcine ECM fibers scaffold (Fig. 3C). Because of the partial cross-linking, the destruction of the ECM is gradual and slow, so that infiltrating fibroblasts that follow the macrophages align with the porcine ECM fibers scaffold and secrete their own ECM (Fig. 3D). Ultimately, the porcine ECM scaffold is eliminated and replaced by the ECM produced by fibroblasts of the patient to form an autologous ACL (Fig. 3E). A trichrome staining indicates the collagen fibers within the remodeled ligament (Fig. 3F). In the absence of cross-linking, the recruited macrophages are likely to destroy the porcine ECM collagen fibers scaffold prior to the penetration of fibroblasts. Thus, production of the collagen fibers of the autologous ECM would have been without the distinct parallel orientation that provides the strength required for the appropriate bio-mechanical function of the ligament.

Kinetics of anti-non-gal antibody production in monkeys and humans greatly differs from that of anti-Gal antibody (Figs. 1 and 2). Because quiescent anti-Gal B cells comprise ~1% of B cells (Galili et al., 1993), their activation in response to α-gal epitopes on xenoglycoproteins results in a peak response of anti-Gal within a short period of 2 weeks (Fig. 3 in Chapter 1) (Galili et al., 2001; Stone et al., 2007a). However, production of anti-non-gal antibodies is much slower because the initial number of B cells in each of the multiple anti-non-gal B cell clones is very small. Expansion of B cells in these clones to numbers that can produce antibodies at measurable titers takes several weeks to two months (Fig. 2) (Galili et al., 2001; Galili, 2012; Stone et al., 2007a). As discussed below, it is suggested that the period required for the expansion of anti-non-gal B cell clones may be exploited for avoiding the detrimental effects of anti-non-gal antibodies, by using α-gal nanoparticles to accelerate stem cells recruitment into the ECM bio-implant.

STEM CELLS MAY BE RECRUITED BY α-GAL NANOPARTICLES BINDING THE ANTI-GAL ANTIBODY

The α-gal epitope may be eliminated from porcine ECM by disruption (i.e., knockout) of the α1,3GT gene (Lai et al., 2002; Phelps et al., 2003; Kolber-Simonds et al., 2004; Takahagi et al., 2005), thereby avoiding the detrimental effects of natural and elicited anti-Gal antibodies. However, this genetic engineering approach cannot be used for the prevention of anti-non-gal antibody effects on porcine ECM bio-implants because of the large number of immunogenic porcine proteins that induce anti-non-gal antibody response. In addition, immune suppression for preventing anti-non-gal antibody production is likely to require the use doses of immunosuppressive drugs that cause complete prevention of antibody production (Galili et al., 1995; Galili, 2012). Studies with diabetic patients who received a kidney allograft and porcine fetal pancreatic islet cell clusters indicated that immunosuppressive regimens that were potent enough to prevent rejection of kidney allografts, did not prevent production of anti-Gal and anti-non-gal antibodies against the porcine islet cells (Groth et al., 1994; Galili et al., 1995).

Rapid recruitment of colony-forming cells by α-gal nanoparticles

A suggested approach that may avoid the detrimental effects of anti-non-gal antibodies is acceleration of the regenerative process within the ECM by early recruitment of stem cells with α-gal nanoparticles that are introduced into the ECM implants (Galili, 2015). It is hypothesized that recruitment of stem cells into ECM bio-implants may be accelerated to occur before anti-non-gal antibodies reach detrimental titers. Stem cells reaching the ECM bio-implant prior to production of high titers of anti-non-gal antibodies may interact with the ECM without hindrance of such antibodies, which can mask the ECM, and before degradation of the ECM by macrophages binding to anti-non-gal antibodies immunocomplexed with the ECM.

As detailed in Chapter 12, α-gal nanoparticles are biodegradable submicroscopic liposomes that present multiple α-gal epitopes and are prepared from rabbit RBC membranes. α-Gal nanoparticles are comprised of phospholipids, cholesterol, and α-gal glycolipids. Their size is ~10 to 300nm, and they present ~10^{15} α-gal epitopes/mg (Wigglesworth et al., 2011), i.e., ~10,000-fold higher concentration of α-gal epitopes than that in pig cartilage or in pig tendon (Stone et al., 1998, 2007a). These α-gal nanoparticles, which are highly stable, readily bind anti-Gal and activate the complement system because of the extensive anti-Gal/α-gal epitopes interaction. When applied to wounds or burns, this interaction generates chemotactic complement cleavage peptides that induce rapid recruitment of macrophages and the subsequent activation of these cells via Fc/Fc receptor interaction, with anti-Gal coating the nanoparticles. The activated macrophages secrete a wide range of cytokines that accelerate wound and burn healing (Chapter 12). It was hypothesized that among the cytokines secreted by the activated macrophages, there are cytokines that induce the recruitment of stem cells (Galili, 2015).

The hypothesized recruitment of stem cells by α-gal nanoparticles was suggested by studies with biologically inert polyvinyl alcohol (PVA) sponge discs containing 1–10mg α-gal nanoparticles that were implanted subcutaneously in anti-Gal producing GT-KO mice. The α-gal nanoparticles within these sponge discs induce rapid recruitment of macrophages (Galili et al., 2010). As indicated in Fig. 4 of Chapter 12, the PVA sponge discs explanted after 6–8days contained ~0.5 to 1×10^6 macrophages (CD11b+, CD14+), which could be retrieved by repeatedly squeezing these sponge discs in PBS (Galili et al., 2010). The morphology of these macrophages is shown in Fig. 3D in Chapter 12. Most of the recruited macrophages are nondividing cells, which display in flow cytometry characteristics of M2 macrophages, including positive staining with the macrophage specific anti-F4/80 antibody, production of arginase and IL10, lack of IL12, and negative staining for the neutrophil marker Ly6G/C (in preparation). These M2 macrophages, referred to as "anti-inflammatory" or "pro-healing" macrophages, were shown to orchestrate the regeneration of ECM into skeletal muscle tissue (Sicari et al., 2014). Culturing the cells retrieved from the PVA sponge discs on coverslips for 5days demonstrated the presence of very few cells (approximately one in 1×10^5 recruited cells), which have the ability of extensive cell division and formation of colonies (Fig. 4). Such colonies contain ~300 to 1000 cells, raising the possibility that among the recruited macrophages there are also few stem cells, which display fast proliferation with a cell cycle time as short as 12h. Incubation of these cultures for 7days resulted in expansion of the colonies to a size of 3–7mm (not shown).

Studies on PVA sponge discs kept subcutaneously in GT-KO mice for 6weeks demonstrated the production of nerve bundles (Fig. 5A) and skeletal (striated) muscle myotubes (Fig. 5B), suggesting migration of pluripotent stem cells capable of differentiating into various types of tissue in these sponge discs. Migration of pluripotential stem cells capable of proliferation and differentiation into various cell types was also observed in other synthetic scaffolds that were inspected several weeks post implantation (Ko et al., 2013). The observation of proliferating cells reaching PVA sponge discs containing α-gal nanoparticles within 6days suggests that the recruited macrophages that are activated by the interaction with anti-Gal-coated nanoparticles, secrete cytokines that induce accelerated recruitment of stem cells. Similar fast recruitment of stem cells into ECM bio-implants containing α-gal nanoparticles

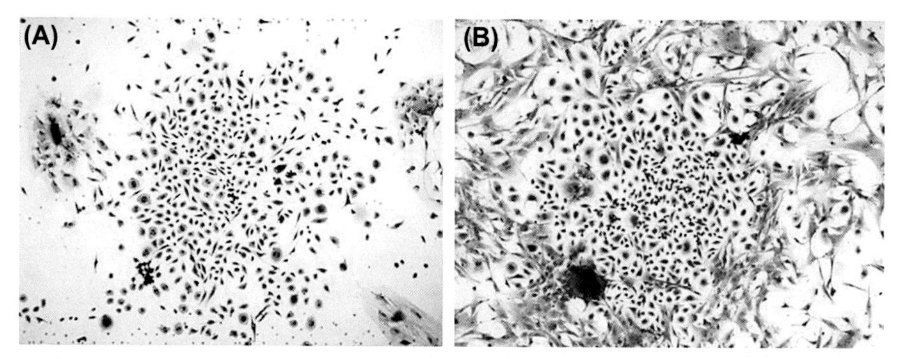

FIGURE 4 Two colonies (A and B) formed by proliferating cells among the macrophages obtained from polyvinyl alcohol sponge discs containing 10 mg/mL α-gal nanoparticles and harvested 6 days post subcutaneous implantation. The harvested cells were plated on round coverslip discs in 24 well plates, as 2×10^5 per well, cultured for 5 days and then washed, fixed, and stained with Wright staining. Approximately 1 in 1×10^5 recruited cells display extensive proliferating activity, which results in formation of colonies, each including ~300 to 1000 cells.

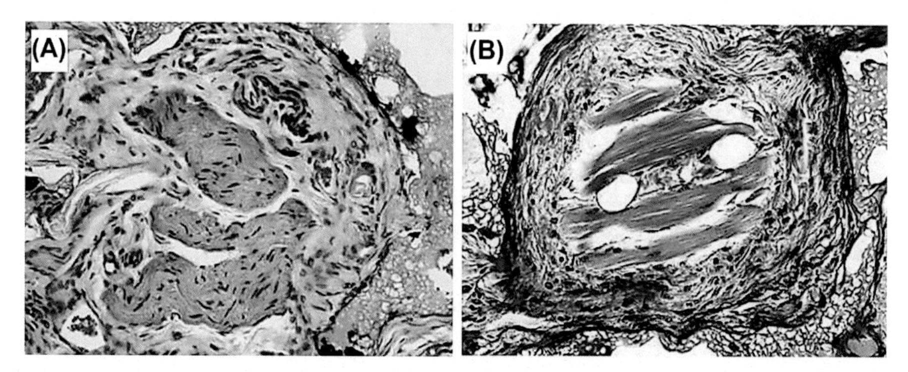

FIGURE 5 Differentiation of stem cells into nerve cells (A) or into skeletal muscle myotubes (B) in polyvinyl alcohol (PVA) sponge discs implanted subcutaneously for 6 weeks in anti-Gal producing α1,3galactosyltransferase knockout mice. The specimen in (A) is stained hematoxylin and eosin (H&E) (PVA sponge stained purple in the right region). The specimen in (B) stained with trichrome, staining collagen blue, and myotubes red (PVA sponge stained blue in the right region). Three nerve bundles are observed in (A) and several myotubes surrounded by connective tissue in (B) (×200).

may be crucial for induction of appropriate regeneration of such implants because it may occur prior to the detrimental anti-non-gal antibody production.

The notion that rapid recruitment of stem cells into ECM implants is essential for the success of implant regeneration into a functioning tissue has prompted researchers to study the possible effects of various cytokines such as granulocyte-colony-stimulating factor, stem cell factor, hepatocyte growth factor, and monocyte chemotactic proteins introduced into the ECM implant prior to implantation, for recruitment of stem cells (Ko et al., 2013). The window of time during which stem cells may be recruited by soluble cytokines within an ECM bio-implant may be very limited because of the diffusion of such cytokines out of the implant. However, the rapid recruitment of macrophages within the implant by anti-Gal/α-gal nanoparticles interaction is followed by Fc/Fc receptors interaction which activate the recruited macrophages. The resulting local secretion of a wide range of cytokines by these macrophages may provide for a prolonged activity of stem cell-recruiting macrophages within the ECM, as well as generation of a microenvironment within the ECM that is conducive to stem cell proliferation and differentiation.

Cartilage formation in polyvinyl alcohol sponge discs containing α-gal nanoparticles and porcine cartilage extracellular matrix

The ability of the putative stem cells recruited by α-gal nanoparticles to differentiate into mature specialized cells was further determined in PVA sponge discs containing porcine meniscus fibrocartilage ECM (Galili, 2015). PVA sponge discs containing a mixture of microscopic pig meniscus cartilage homogenate (50 mg/mL), and 10 mg/mL α-gal nanoparticles were implanted subcutaneously in anti-Gal producing GT-KO mice. α-Gal epitopes were removed from the porcine meniscus cartilage by incubation in recombinant α-galactosidase, to prevent formation of immune complexes between anti-Gal and these cartilage ECM fragments. The sponge disc implants were explanted after 5 weeks, sectioned, and stained. As shown in Fig. 6, the PVA sponge discs contained areas, which included relatively few cells and large amounts of extracellular material stained red by hematoxylin and eosin (H&E) (Fig. 6A and B). The blue color of this extracellular material in trichrome staining indicates that it is comprised primarily of the collagen fibers (Fig. 6C and D). The fibrocartilage produced, is similar in structure to porcine meniscus cartilage, characterized by relatively few fibrochondrocytes (identified by their elongated nuclei) and large amounts of ECM comprised of the fibrocartilage secreted by these cells (Fig. 6F). The cellular organization of the fibrochondrocytes in the PVA sponge discs differs from that in the meniscus tissue, possibly because the limited space in the sponge cavities does not enable for the parallel organization of the cells as in the meniscus. Nevertheless, the *de novo* formed fibrocartilage tissue "fills" these spaces and thus it assumed the shape of the sponge cavity border (Fig. 6B–D). The empty areas between the fibrocartilage and the sponge material are caused by tissue contracting because of dehydration due to alcohol fixation during the staining procedure. Implantation of PVA sponge discs containing fibrocartilage ECM but no α-gal nanoparticles resulted in no formation of fibrocartilage, and the cavities were filled primarily with fat tissue (Fig. 6E). These findings suggest that the proliferative cells recruited with macrophages by α-gal nanoparticles are stem cells that differentiate into fibrochondroblasts producing cartilage after receiving cues from the fragmented porcine fibrocartilage ECM.

As discussed below, the studies on formation of cartilage in PVA sponge discs containing α-gal nanoparticles and porcine cartilage homogenate raise the possibility that α-gal nanoparticles introduced into porcine ECM "patch" bio-implants may induce rapid regeneration of such implants, prior to formation of anti-non-gal antibodies. In addition, these studies may suggest a possible novel therapy to be studied for reconstruction of cartilage in defects within articular cartilage of the joints. Because cartilage does not undergo spontaneous regeneration, such defects are difficult to treat. It would be of interest to determine whether a "paste" containing a mixture of α-gal nanoparticles and homogenate of porcine articular cartilage, devoid of α-gal epitopes, may induce articular cartilage regeneration in these defects. The cartilage defects may undergo abrasion prior to the application of such a paste to remove any fibrotic tissue and to release anti-Gal from ruptured capillaries. This antibody interacts with the α-gal nanoparticles, activates complement, and recruits macrophages and stem cells. The recruited stem cells may receive cues from the cartilage homogenate to differentiate into chondroblasts that secrete collagen and other ECM molecules which comprise the regenerating articular cartilage. Because macrophages and stem cells need to

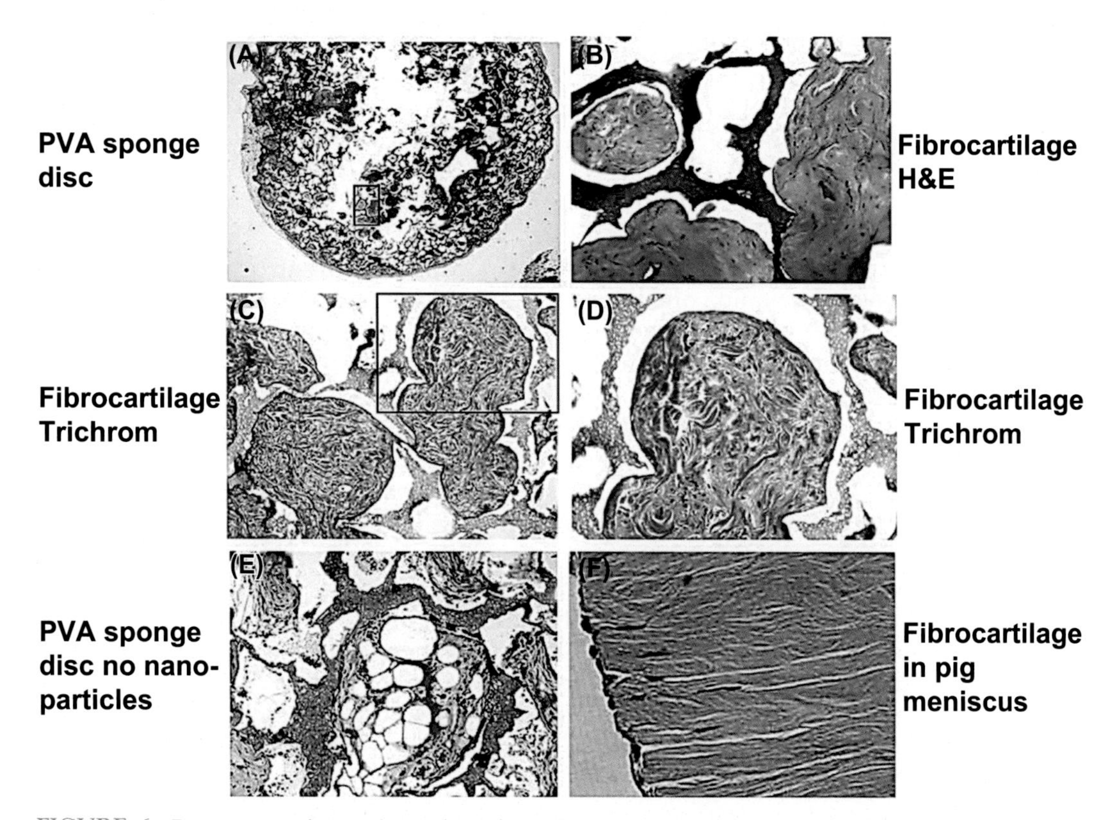

FIGURE 6 Recruitment of α1,3galactosyltransferase knockout (GT-KO) mouse stem cells by 10 mg/mL α-gal nanoparticles into polyvinyl alcohol (PVA) sponge discs implanted subcutaneously. These stem cells differentiate into fibrocartilage producing cells in the presence of fragmented porcine meniscus cartilage extracellular matrix (ECM) (50 mg/mL). (A) Section within the PVA sponge disc, 5 weeks post implantation (hematoxylin and eosin (H&E) ×10). Areas of fibrocartilage growth are stained red. (B) Magnification of the inset in (A), demonstrating the fibrocartilage growth (red). The PVA sponge material is stained red purple. (C) As in (B), however, the fibrocartilage is dyed with trichrome that stains the collagen fibers blue (×100). (D) The inset in (C) magnified (×200). The fibrocartilage is characterized by multiple collagen fibers and relatively few cells (nuclei stained purple). (E) H&E staining of the tissue formed in PVA sponge disc implant containing only meniscus cartilage ECM fragments (×100). In the absence of α-gal nanoparticles, the formed cells are mostly fat cells, and no fibrocartilage formation is observed. (F) Porcine meniscus displaying parallel organization of fibrochondrocytes with elongated nuclei and parallel orientation of the collagen fibers (H&E, ×200). Alcohol fixation of the histologic sections in Fig. 5B–D results in shrinking that forms space observed between the fibrocartilage tissue and the PVA sponge walls. Representative sections are from 5 GT-KO mice per group. *Reprinted with permission from Galili, U., 2015. Avoiding detrimental human immune response against mammalian extracellular matrix implants. Tissue Eng. B 21, 231–241.*

reach the α-gal nanoparticles and the cartilage fragments, the paste should be of a consistency that enables its retention in the defects, as well as allows for migration of these cells through it. A similar rationale may apply to a possible therapeutic approach for treatment of bone fractures. Application to such fractures of a paste containing microscopic porcine bone fragments, devoid of α-gal epitopes and mixed with α-gal nanoparticles, may result in accelerated healing of the fracture, thereby decreasing morbidity associated with this healing process. It would be of further interest to determine the regenerative ability of such pastes containing human cartilage or bone fragments and α-gal nanoparticles in inducing repair of injured cartilage or bone in humans.

SUGGESTED METHOD FOR ACCELERATING EXTRACELLULAR MATRIX REGENERATION WITH α-GAL NANOPARTICLES: MYOCARDIAL EXTRACELLULAR MATRIX AS AN EXAMPLE

Development of methods that will enable the use of ECM implant patches would be of significance in inducing regeneration of injured tissues that have undergone fibrosis. A specific hypothetical example is that of porcine myocardium ECM patch, which may be considered for regenerating ischemic myocardium that converted into scar tissue, following myocardial infarction (Fig. 7A). The ischemic myocardium in many patients post myocardial infarction heals within several weeks by the default repair mechanism of fibrosis and scar formation. The partial loss of left ventricle function is a risk factor associated with heart failure and early death of the patient. The use of porcine myocardium ECM patches for regeneration of ischemic myocardium is an attractive option because it provides the ECM, which may instruct stem cells, or progenitor cells of the adjacent uninjured myocardium, to differentiate into functioning cardiomyocytes that repopulate the ECM bio-implant. A major challenge in the use of a porcine myocardium ECM patch is the induction of stem cell recruitment into the patch. Hypothetically, porcine myocardium ECM implantation may be successful if stem cells are actively recruited into the bio-implant. Moreover, based on the considerations above, it may be assumed that rapid recruitment of stem cells into the ECM bio-implant, and vascularization of the ECM may significantly improve the probability of ECM regeneration prior to production of detrimental anti-non-gal antibodies.

Conversion of porcine myocardium into ECM involves the alternating incubation of the tissue in detergents such as SDS, Triton X-100, or deoxycholate (Atala, 2009; Badylak et al., 2012; Perea-Gil et al., 2015). These detergents solubilize the cell and nuclear membranes by removing the phospholipids, glycolipids, and cholesterol from the tissue. To further remove the DNA released from decellularized cells, the ECM is incubated with DNase and undergoes several washes. The optimal concentrations of detergents should be determined empirically as those that effectively remove lipids and cholesterol but cause only minimal disruption of the bonds between the various proteins, glycosaminoglycans, and proteoglycans of the ECM. This conversion of tissues into decellularized ECMs also makes them porous. Therefore, incubation of washed ECM for 12–24 h in a suspension of α-gal nanoparticles, with mild shaking, is likely to result in diffusion of these nanoparticles throughout the ECM. Alternatively, the α-gal nanoparticles or larger α-gal liposomes may be introduced into the ECM patch by multiple injections of small volumes throughout the patch. Because α-gal nanoparticles are stable

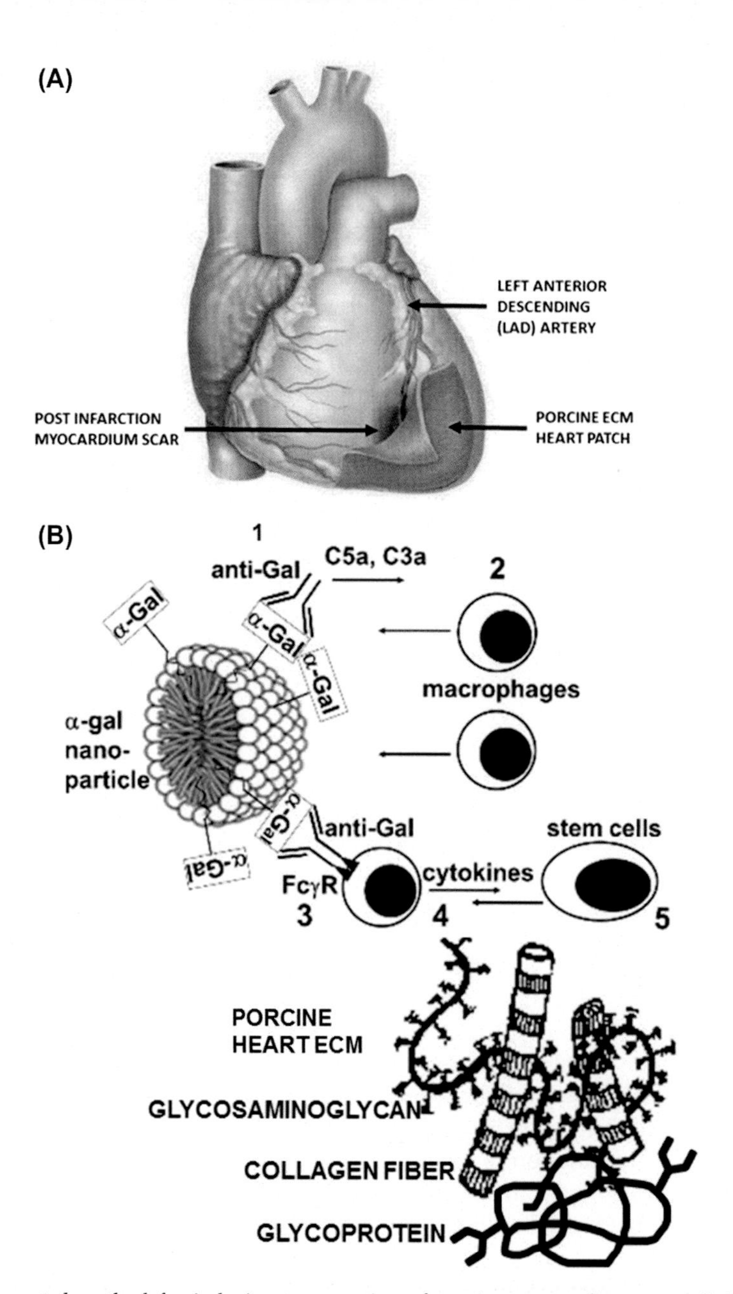

FIGURE 7 Suggested method for inducing regeneration of porcine myocardium extracellular matrix (ECM) "patch" bio-implant attached on or in place of part of the scar tissue that is formed following ischemia caused by myocardial infarction. (A) Illustration of a hypothetical porcine myocardium ECM patch placed on the scar tissue in the left ventricular wall of a human heart. (B) Schematic illustration of the processes that may induce accelerated ECM bio-implant regeneration by α-gal nanoparticles. The effects of α-gal nanoparticles administered within ECM bio-implants may be divided into several steps: (1) Anti-Gal binding to α-gal epitopes on the nanoparticles activates the complement system to generate complement cleavage chemotactic peptides, such as C5a and C3a. (2) These chemotactic peptides induce rapid recruitment of macrophages into the ECM implant. (3) The Fc "tails" of anti-Gal bound to α-gal nanoparticles interact with Fcγ receptors (FcγR) on the recruited macrophages. (4) Macrophages activated by the Fc/FcγR interaction are induced to produce a wide range of anti-inflammatory, "prohealing" cytokines/growth factors, including those inducing vascularization and stem cell/progenitor cell recruitment. (5) Recruited stem cells receive "cues" from the porcine myocardium ECM, which direct them to differentiate into functional cardiomyocytes that repopulate the ECM bio-implant, prior to production of detrimental anti-non-gal antibodies. PVA, polyvinyl alcohol. *(A) Adapted with permission from Topnews.ea 06/25/2013. (B) Adapted with permission from Galili, U., 2017. α-Gal nanoparticles in wound and burn healing acceleration. Adv. Wound Care 6, 81–91.*

for years when stored at 4 or −20°C, the ECM containing these nanoparticles may be stored frozen until use.

It is suggested that implantation of porcine myocardial patches over scars or partially replacing scars of post infarction ischemic myocardium may be followed by a series of steps resemble accelerated wound healing (discussed in Chapter 12) and which are illustrated in Fig. 7B: **Step #1**. Anti-Gal diffusing into the myocardial ECM binds to the multiple α-gal epitopes on α-gal nanoparticles within the ECM and activates the complement system. This activation of the complement system is similar to complement activation that initiates hyperacute rejection of the porcine xenograft presenting α-gal epitopes and transplanted in Old World monkeys (discussed in Chapter 6). However, unlike xenograft rejection, no cells are damaged by this anti-Gal/α-gal epitope interaction because it occurs on the nanoparticles rather than on cells. To ensure the rapid diffusion of anti-Gal and complement proteins into the bio-implant containing α-gal nanoparticles, abrasion of the implantation area may be needed to release serum proteins from the adjacent tissue. **Step #2**. The complement cleavage chemotactic peptides generated as a product of the complement system activation induce rapid migration of macrophages into the myocardial ECM implant. As shown in Fig. 3 in Chapter 12, recruitment of macrophages is observed already within 24h and continues for ~7days following intradermal injection of α-gal nanoparticles. This recruitment was found to be inhibited by cobra venom factor that inhibits complement activation (Wigglesworth et al., 2011). As described in Figs. 2 and 3 in Chapter 14, similar migration of macrophages by this mechanism is observed in GT-KO mouse hearts containing α-gal nanoparticles that are implanted subcutaneously and in GT-KO pig hearts injected via a catheter with these nanoparticles. **Step #3**. The recruited macrophages migrating into the ECM bio-implant bind α-gal nanoparticles by interaction between their Fc receptors and the Fc portion of anti-Gal coating these nanoparticles. This interaction is illustrated in Fig. 5 in Chapter 12. **Step #4**. The Fc/Fc receptor interaction between anti-Gal coating α-gal nanoparticles and macrophages activates the macrophages to internalize the nanoparticles, produce and secrete multiple pro-healing cytokines. One of these cytokines is VEGF, which induces angiogenesis within the implant (Fig. 4B in Chapter 12 and Fig. 3C above). It is hypothesized that among the secreted cytokines, there are also cytokines, which recruit stem cells and progenitor cells into the ECM bio-implant (Fig. 5). **Step #5**. The recruited stem cells receive cues from the porcine ECM, proliferate, and differentiate into human cardiomyocytes that repopulate the ECM and restore the structure and function of the myocardium implant. Ultimately, the regenerating ECM may become a functional part of the contracting left ventricular wall. It is possible that introduction of injectable biomaterial extracts of porcine heart ECM into the patch together with the α-gal nanoparticles may further add to the conducive microenvironment for the differentiation of the recruited stem cells into cardiomyocytes (Radisic and Christman, 2013; Gaetani et al., 2016).

It should be stressed that no studies demonstrating regenerative acceleration of myocardial ECM patch implants containing α-gal nanoparticles have been performed yet. Preliminary studies on this process may be feasible in GT-KO pigs, which naturally produce the anti-Gal antibody at levels as those in humans (Dor et al., 2004; Fang et al., 2012; Galili, 2013). It would also be of interest to determine whether α-gal nanoparticles administered within ECM implants may accelerate regeneration of ECM devoid of α-gal epitopes from other mammalian tissues, such as liver and urinary bladder. Whereas α-gal nanoparticles may be of significance in accelerating *in vivo* regeneration of ECM tissue patches, it is not clear whether these nanoparticles may be of use in whole decellularized organ implants. Decellularization of organs such as porcine heart, kidney, pancreas, and liver for generating their ECM has been shown to be feasible by infusion of detergent through their vascular system (Atala, 2009; Crapo et al., 2011; Badylak et al., 2012). Seeding such ECM with stem cells is being studied as a method for *in vitro* generation of such organ, based on the interaction of the stem cells with the various ECMs within these decellularized organs and their differentiation into the corresponding cell types. Because macrophages do not seem to participate in this *in vitro* differentiation processes, α-gal nanoparticles may not contribute to the *in vitro* regeneration processes. However, it would be of interest to determine in the future whether administration of these nanoparticles into the *in vitro* regenerated organs, prior to their transplantation, may contribute to processes for effective *in vivo* completion of the regeneration processes, by accelerating vascularization of the organ.

REGENERATION OF HUMAN EXTRACELLULAR MATRIX WITH α-GAL NANOPARTICLES

The elimination of the α-gal epitope from porcine ECM bio-implant leaves the natural anti-non-gal antibodies as a major immunologic obstacle for success of the implant. One theoretical approach that may be considered for overcoming the anti-non-gal immune obstacle is the use of ECM bio-implants prepared from human cadaveric tissue, because immunogenicity of human ECM is low. In the recent decades, there has been increased use of collagen-based human orthopedic tissues such as bone and bone fillers, ligament allografts, cartilage, and decellularized dermis, from cadaver donors, for grafting in patients with injured or impaired tissues. The effective reconstruction of human tendon allografts into ACL (Wydra et al., 2016) and that of decellularized dermis allografts (Ellis and Kulber, 2012) support the assumption that the immune response to ECM allografts is weak. Thus, in most recipients of ECM allografts, there is no significant immune response, which may interfere with reconstruction of the bio-implant into an autologous tissue. Whereas regeneration of collagen-based orthopedic tissue seems to be mediated primarily by fibroblasts infiltrating the ECM bio-implant, *in vivo* regeneration of specialized tissues, such as heart, urinary bladder, or liver, by ECM patches requires the infiltration of stem cells into the implant. In the absence of any mechanism for active *in vivo* recruitment of stem cells into human ECM bio-implants, migration of stem cells to implanted ECM may be poor and insufficient for effective reconstruction of the bio-implant into an autologous functioning tissue. The latter scenario may ultimately result in fibrosis of the human bio-implant.

It is hypothesized that α-gal nanoparticles introduced into human ECM bio-implants, such as decellularized human heart patches that are prepared from cadaver donors, induce rapid *in vivo* recruitment of macrophages. These macrophages are activated by interaction with the anti-Gal-coated α-gal nanoparticles and secrete cytokines/growth factors that induce vascularization of the bio-implant and recruit stem cells, as suggested in Fig. 7B for porcine myocardial ECM bio-implants. If the ECM is intact (i.e., not degraded because of extensive necrosis) then recruited stem cells may receive cues from the myocardial ECM to differentiate into cardiomyocytes that repopulate the bio-implant and convert it into a functional myocardium. Intactness of the ECM may be determined by assays indicating that components of the ECM are not degraded, e.g., SDS PAGE and Western blot identification of ECM components' size. Because human ECM is of very low immunogenicity or no immunogenicity, it is possible that the reconstruction of the ECM may occur without production of anti-non-gal antibodies. An experimental animal model that could be suitable for studying this hypothetical treatment is the GT-KO pig that produces natural anti-Gal antibody with characteristics like those of human anti-Gal (Galili, 2013). α-Gal nanoparticles were found in this model to effectively bind anti-Gal, recruit macrophages, and induce accelerated healing of wounds (Hurwitz et al., 2012; Chapter 12). Successful demonstration of reconstruction of GT-KO pig ECM patches of various tissues implanted on GT-KO pig injured myocardium, urinary bladder wall, and liver may provide initial information on the efficacy of α-gal nanoparticles in inducing regeneration of same species ECM bio-implants of specialized tissues.

No Conclusions section is included since the proposed therapy in this chapter is hypothetical.

References

Arenas-Herrera, J.E., Ko, I.K., Atala, A., Yoo, J.J., 2013. Decellularization for whole organ bioengineering. Biomed. Mater. 8 (1).

Asahara, T., Murohara, T., Sullivan, A., Silver, M., van der Zee, R., Li, T., et al., 1997. Isolation of putative progenitor endothelial cells for angiogenesis. Science 275, 964–967.

Atala, A., 2009. Engineering organs. Curr. Opin. Biotechnol. 575–592.

Bader, A., Schilling, T., Teebken, O.E., Brandes, G., Herden, T., Steinhoff, G., et al., 1998. Tissue engineering of heart valves-human endothelial cell seeding of detergent acellularized porcine valves. Eur. J. Cardiothorac. 14, 279–284.

Badylak, S.F., 2015. Host response to biomaterials. In: Badylak, S.F. (Ed.), The Impact of Host Response on Biomaterial Selection. Elsevier, Academic Press, Publishers.

Badylak, S.F., Gilbert, T.W., 2008. Immune response to biologic scaffold materials. Semin. Immunol. 20, 109–116.

Badylak, S., Liang, A., Record, R., Tullius, R., Hodde, J., 1999. Endothelial cell adherence to small intestinal submucosa: an acellular bioscaffold. Biomaterials 20, 2257–2263.

Badylak, S.F., Weiss, D.J., Caplan, A., Macchiarini, P., 2012. Engineered whole organs and complex tissues. Lancet 379, 943–952.

Booth, C., Korossis, S.A., Wilcox, H.E., Watterson, K.G., Kearney, J.N., Fisher, J., et al., 2002. Tissue engineering of cardiac valve prostheses I: development and histological characterization of an acellular porcine scaffold. J. Heart Valve Dis. 11, 457–462.

Britten, R.J., 1986. Rates of DNA sequence evolution differ between taxonomic groups. Science 231, 1393–1398.

Caplan, A.I., 2009. Why are MSCs therapeutic? New data: new insight. J. Pathol. 217, 318–324.

Chen, G., Qian, H., Starzl, T., Sun, H., Garcia, B., Wang, X., et al., 2005. Acute rejection is associated with antibodies to non-gal antigens in baboons using gal-knockout pig kidneys. Nat. Med. 11, 1295–1298.

Cicha, I., Rüffer, A., Cesnjevar, R., Glöckler, M., Agaimy, A., Daniel, W.G., et al., 2011. Early obstruction of decellularized xenogenic valves in pediatric patients: involvement of inflammatory and fibroproliferative processes. Cardiovasc. Pathol. 20, 222–231.

Coletti, D., Teodori, L., Lin, Z., Beranudin, J.F., Adamo, S., 2013. Restoration versus reconstruction: cellular mechanisms of skin, nerve and muscle regeneration compared. Regen. Med. Res. 1, 4.

Cotterell, A.H., Collins, B.H., Parker, W., Harland, R.C., Platt, J.L., 1995. The humoral immune response in humans following cross-perfusion of porcine organs. Transplantation 60, 861–868.

Crapo, P.M., Gilbert, T.W., Badylak, S.F., 2011. An overview of tissue and whole organ decellularization processes. Biomaterials 32, 3233–3243.

Daly, K.A., Stewart-Akers, A.M., Hara, H., Ezzelarab, M., Long, C., Cordero, K., et al., 2009. Effect of the α-gal epitope on the response to small intestinal submucosa extracellular matrix in a nonhuman primate model. Tissue Eng. A 15, 3877–3888.

Dor, F.J., Tseng, Y.L., Cheng, J., Moran, K., Sanderson, T.M., Lancos, C.J., et al., 2004. α1,3-galactosyltransferase gene-knockout miniature swine produce natural cytotoxic anti-gal antibodies. Transplantation 78, 15–20.

Ellis, C.V., Kulber, D., 2012. Acellular dermal matrices in hand reconstruction. Plast. Reconstr. Surg. 130 (5 Suppl. 2), 256S–269S.

Ezzelarab, M., Hara, H., Busch, J., Rood, P.P., Zhu, X., Ibrahim, Z., et al., 2006. Antibodies directed to pig non-gal antigens in naïve and sensitized baboons. Xenotransplantation 13, 400–407.

Fang, J., Walters, A., Hara, H., Long, C., Yeh, P., Ayares, D., et al., 2012. Anti-gal antibodies in α1,3-galactosyltransferase gene-knockout pigs. Xenotransplantation 19, 305–310.

Feric, N.T., Radisic, M., 2016. Strategies and challenges to myocardial replacement therapy. Stem Cells Transl. Med. 5, 410–416.

Fuchs, E., 2008. Skin stem cells: rising to the surface. J. Cell Biol. 180, 273–284.

Gaetani, R., Yin, C., Srikumar, N., Braden, R., Doevendans, P.A., Sluijter, J.P., et al., 2016. Cardiac-derived extracellular matrix enhances cardiogenic properties of human cardiac progenitor cells. Cell Transplant. 25, 1653–1663.

Galili, U., 2012. Induced anti-non gal antibodies in human xenograft recipients. Transplantation 93, 11–16.

Galili, U., 2013. α1,3galactosyltransferase knockout pigs produce the natural anti-Gal antibody and simulate the evolutionary appearance of this antibody in primates. Xenotransplantation 20, 267–276.

Galili, U., 2015. Avoiding detrimental human immune response against mammalian extracellular matrix implants. Tissue Eng. B 21, 231–241.

Galili, U., 2017. α-Gal nanoparticles in wound and burn healing acceleration. Adv. Wound Care 6, 81–91.

Galili, U., Matta, K.L., 1996. Inhibition of anti-Gal IgG binding to porcine endothelial cells by synthetic oligosaccharides. Transplantation 62, 256–262.

Galili, U., Shohet, S.B., Kobrin, E., Stults, C.L.M., Macher, B.A., 1988. Man, apes, and old world monkeys differ from other mammals in the expression of α-galactosyl epitopes on nucleated cells. J. Biol. Chem. 263, 17755–17762.

Galili, U., Anaraki, F., Thall, A., Hill-Black, C., Radic, M., 1993. One percent of circulating B lymphocytes are capable of producing the natural anti-Gal antibody. Blood 82, 2485–2493.

Galili, U., Tibell, A., Samuelsson, B., Rydberg, B., Groth, C.G., 1995. Increased anti-Gal activity in diabetic patients transplanted with fetal porcine islet cell clusters. Transplantation 59, 1549–1556.

Galili, U., LaTemple, D.C., Walgenbach, A.W., Stone, K.R., 1997. Porcine and bovine cartilage transplants in cynomolgus monkey: II. Changes in anti-Gal response during chronic rejection. Transplantation 63, 646–651.

Galili, U., Chen, Z.C., Tanemura, M., Seregina, T., Link, C.J., 2001. Induced antibody response (in xenograft recipients). Graft 4, 32–35.

Galili, U., Wigglesworth, K., Abdel-Motal, U.M., 2010. Accelerated healing of skin burns by anti-Gal/α-gal liposomes interaction. Burns 36, 239–251.

Garmy-Susini, B., Varner, J.A., 2005. Circulating endothelial progenitor cells. Br. J. Cancer 93, 855–858.

Groth, C.G., Korsgren, O., Tibell, A., Tollemar, J., Möller, E., Bolinder, J., et al., 1994. Transplantation of porcine fetal pancreas to diabetic patients. Lancet 344, 1402–1404.

Guilak, F., Cohen, D.M., Estes, B.T., Gimble, J.M., Liedtke, W., Chen, C.S., 2009. Control of stem cell fate by physical interactions with the extracellular matrix. Cell Stem Cell 5, 17–26.

Hidalgo-Bastida, L.A., Cartmell, S.H., 2010. Mesenchymal stem cells, osteoblasts and extracellular matrix proteins: enhancing cell adhesion and differentiation for bone tissue engineering. Tissue Eng. B 16, 405–412.

Hiles, M., Record Ritchie, R.D., Altizer, A.M., 2009. Are biologic grafts effective for hernia repair?: a systematic review of the literature. Surg. Innov. 16, 26–37.

Hirt, M.N., Hansen, A., Eschenhagen, T., 2014. Cardiac tissue engineering: state of the art. Circ. Res. 114, 354–367.

Hisashi, Y., Yamada, K., Kuwaki, K., Tseng, Y.L., Dor, F.J., Houser, S.L., et al., 2008. Rejection of cardiac xenografts transplanted from α1,3-galactosyltransferase gene-knockout (GalT-KO) pigs to baboons. Am. J. Transplant. 8, 2516–2526.

Hurwitz, Z., Ignotz, R., Lalikos, J., Galili, U., 2012. Accelerated porcine wound healing with α-gal nanoparticles. Plast. Reconstr. Surg. 129, 242–251.

Johnson, A., DiPietro, L.A., 2013. Apoptosis and angiogenesis: an evolving mechanism for fibrosis. FASEB J. 27, 3893–3901.

Jung, Y., Wang, J., Havens, A., Sun, Y., Wang, J., Jin, T., et al., 2005. Cell-to-cell contact is critical for the survival of hematopoietic progenitor cells on osteoblasts. Cytokine 32, 155–162.

Kasimir, M.T., Rieder, E., Seebacher, G., Wolner, E., Weigel, G., Simon, P., 2005. Presence and elimination of the xenoantigen gal (α1,3) gal in tissue-engineered heart valves. Tissue Eng. 11, 1274–1280.

Katari, R., Peloso, A., Zambon, J.P., Soker, S., Stratta, R.J., Atala, A., et al., 2014. Renal bioengineering with scaffolds generated from human kidneys. Nephron Exp. Nephrol. 126, 119.

Klos, A., Wende, E., Wareham, K.J., Monk, P.N., 2013. International Union of Pharmacology. LXXXVII. Complement peptide C5a, C4a, and C3a receptors. Pharmacol. Rev. 65, 500–543.

Ko, I.K., Lee, S.J., Atala, A., Yoo, J.J., 2013. In situ tissue regeneration through host stem cell recruitment. Exp. Mol. Med. 45, e57.

Kolber-Simonds, D., Lai, L., Watt, S.R., Denaro, M., Arn, S., Augenstein, M.L., et al., 2004. Production of α-1,3-galactosyltransferase null pigs by means of nuclear transfer with fibroblasts bearing loss of heterozygosity mutations. Proc. Natl. Acad. Sci. U.S.A. 101, 7335–7340.

Konakci, K.Z., Bohle, B., Blumer, R., Hoetzenecker, W., Roth, G., Moser, B., et al., 2005. α-Gal on bioprosthesis: xenograft immune response in cardiac surgery. Eur. J. Clin. Investig. 35, 17–23.

Kuwaki, K., Tseng, Y.L., Dor, F.J., Shimizu, A., Houser, S.L., Sanderson, T.M., et al., 2005. Heart transplantation in baboons using α1,3-galactosyltransferase gene-knockout pigs as donors: initial experience. Nat. Med. 11, 29–31.

Laflamme, M.A., Murry, C.E., 2011. Heart regeneration. Nature 473, 326–335.

Lai, L., Kolber-Simonds, D., Park, K.W., Cheong, H.T., Greenstein, J.L., Im, G.S., et al., 2002. Production of α-1,3-galactosyltransferase knockout pigs by nuclear transfer cloning. Science 295, 1089–1092.

Lissitzky, J.C., Charpin, C., Bignon, C., Bouzon, M., Kopp, F., Delori, P., et al., 1988. Laminin biosynthesis in the extracellular matrix-producing cell line PFHR9 studied with monoclonal and polyclonal antibodies. Biochem. J. 250, 843–852.

Liu, Y., Bharadwaj, S., Lee, S.J., Atala, A., Zhang, Y., 2009. Optimization of a natural collagen scaffold to aid cell-matrix penetration for urologic tissue engineering. Biomaterials 30, 3865–3873.

Lolmede, K., Campana, L., Vezzoli, M., Bosurgi, L., Tonlorenzi, R., Clementi, E., et al., 2009. Inflammatory and alternatively activated human macrophages attract vessel-associated stem cells, relying on separate HMGB1- and MMP-9-dependent pathways. J. Leukoc. Biol. 85, 779–787.

Mantovani, A., Biswas, S.K., Galdiero, M.R., Sica, A., Locati, M., 2013. Macrophage plasticity and polarization in tissue repair and remodeling. J. Pathol. 229, 176–185.

Marshman, E., Booth, C., Potten, C.S., 2002. The intestinal epithelial stem cell. Bioassays 24, 91–98.

Maruyama, S., Cantu 3rd, E., Galili, U., D'Agati, V., Godman, G., Stern, D.M., et al., 2000. α-Galactosyl epitopes on glycoproteins of porcine renal extracellular matrix. Kidney Int. 57, 655–663.

McGregor, C.G., Kogelberg, H., Vlasin, M., Byrne, G.W., 2013. Gal-knockout bioprostheses exhibit less immune stimulation compared to standard biological heart valves. J. Heart Valve Dis. 22, 383–390.

McPherson, T.B., Liang, H., Record, D., Badylak, S.F., 2000. Galα(1,3)Gal epitope in porcine small intestinal submucosa. Tissue Eng. 6, 233–239.

Murray, P.J., Wynn, T.A., 2011. Protective and pathogenic functions of macrophage subsets. Nat. Rev. Immunol. 11, 723–737.

Murray, I.R., Baily, J.E., Chen, W.C., Dar, A., Gonzalez, Z.N., Jensen, A.R., et al., 2017. Skeletal and cardiac muscle pericytes: functions and therapeutic potential. Pharmacol. Ther. 171, 65–74.

Naso, F., Gandaglia, A., Bottio, T., Tarzia, V., Nottle, M.B., d'Apice, A.J., et al., 2013. First quantification of α-gal epitope in current glutaraldehyde-fixed heart valve bioprostheses. Xenotransplantation 20, 252–261.

Ogawa, H., Galili, U., 2006. Profiling terminal *N*-acetyllactoamines of glycans on mammalian cells by an immunoenzymatic assay. Glycoconj. J. 23, 663–674.

Oriol, R., Candelier, J.J., Taniguchi, S., Balanzino, L., Peters, L., Niekrasz, M., et al., 1999. Major carbohydrate epitopes in tissues of domestic and African wild animals of potential interest for xenotransplantation research. Xenotransplantation 6, 79–89.

Orlando, G., Booth, C., Wang, Z., Totonelli, G., Ross, C.L., Moran, E., et al., 2013. Discarded human kidneys as a source of ECM scaffold for kidney regeneration technologies. Biomaterials 34, 5915–5925.

Park, S., Kim, W.H., Choi, S.Y., Kim, Y.J., 2009. Removal of α-gal epitopes from porcine aortic valve and pericardium using recombinant human α-galactosidase. J. Korean Med. Sci. 24, 1126–1131.

Peloso, A., Dhal, A., Zambon, J.P., Li, P., Orlando, G., Atala, A., 2015. Current achievements and future perspectives in whole-organ bioengineering. Stem Cell Res. Ther. 6, 107.

Perea-Gil, I., Uriarte, J.J., Prat-Vidal, C., Gálvez-Montón, C., Roura, S., Llucià-Valldeperas, A., et al., 2015. In vitro comparative study of two decellularization protocols in search of an optimal myocardial scaffold for recellularization. Am. J. Transl. Res. 7, 558–573.

Phelps, C.J., Koike, C., Vaught, T.D., Boone, J., Wells, K.D., Chen, S.H., et al., 2003. Production of α1,3-galactosyltransferase-deficient pigs. Science 299, 411–414.

Radisic, M., Christman, K.L., 2013. Materials science and tissue engineering: repairing the heart. Mayo Clin. Proc. 88, 884–898.

Raeder, R.H., Badylak, S.F., Sheehan, C., Kallakury, B., Metzger, D.W., 2002. Natural anti-galactose α1,3 galactose antibodies delay, but do not prevent the acceptance of extracellular matrix xenografts. Transpl. Immunol. 10, 15–24.

Ravetch, J.V., Bolland, S., 2001. IgG Fc receptors. Annu. Rev. Immunol. 19, 275–290.

Reing, J.E., Brown, B.N., Daly, K.A., Freund, J.M., Gilbert, T.W., Hsiong, S.X., et al., 2010. The effects of processing methods upon mechanical and biologic properties of porcine dermal extracellular matrix scaffolds. Biomaterials 31, 8626–8633.

Sabari, J., Lax, D., Connors, D., Brotman, I., Mindrebo, E., Butler, C., et al., 2011. Fibronectin matrix assembly suppresses dispersal of glioblastoma cells. PLoS One 6, e24810.

Shakibaei, M., 1998. Inhibition of chondrogenesis by integrin antibody in vitro. Exp. Cell Res. 240, 95–106.

Sicari, B.M., Dziki, J.L., Siu, B.F., Medberry, C.J., Dearth, C.L., Badylak, S.F., 2014. The promotion of a constructive macrophage phenotype by solubilized extracellular matrix. Biomaterials 35, 8605–8612.

Song, L., Murphy, S.V., Yang, B., Xu, Y., Zhang, Y., Atala, A., 2014. Bladder acellular matrix and its application in bladder augmentation. Tissue Eng. B 20, 163–172.

Stone, K.R., Ayala, G., Goldstein, J., Hurst, R., Walgenbach, A., Galili, U., 1998. Porcine cartilage transplants in the cynomolgus monkey. III. Transplantation of α-galactosidase-treated porcine cartilage. Transplantation 65, 1577–1583.

Stone, K.R., Abdel-Motal, U.M., Walgenbach, A.W., Turek, T.J., Galili, U., 2007a. Replacement of human anterior cruciate ligaments with pig ligaments: a model for anti-non-gal antibody response in long-term xenotransplantation. Transplantation 83, 211–219.

Stone, K.R., Walgenbach, A.W., Turek, T.J., Somers, D.L., Wicomb, W., Galili, U., 2007b. Anterior cruciate ligament reconstruction with a porcine xenograft: a serologic, histologic, and biomechanical study in primates. Arthroscopy 23, 411–419.

Stone, K.R., Walgenbach, A., Galili, U., 2017. Induced remodeling of porcine tendons to human anterior cruciate ligaments by α-gal epitope removal and partial crosslinking. Tissue Eng. B 23, 412–419.

Takahagi, Y., Fujimura, T., Miyagawa, S., Natashia, H., Shigehisa, T., Shirakura, R., et al., 2005. Production of α1,3-galactosyltransferase gene knockout pigs expressing both human decay-accelerating factor and *N*-acetylglucosaminyltransferase III. Mol. Reprod. Dev. 71, 331–338.

Tanemura, M., Maruyama, S., Galili, U., 2000. Differential expression of α-gal epitopes (Galα1-3Galβ1-4GlcNAc-R) on pig and mouse organs. Transplantation 69, 187–190.

Tearle, R.G., Tange, M.J., Zannettino, Z.L., Katerelos, M., Shinkel, T.A., Van Denderen, B.J., et al., 1996. The α-1,3-galactosyltransferase knockout mouse. Implications for xenotransplantation. Transplantation 61, 13–19.

Thall, A.D., Maly, P., Lowe, J.B., 1995. Oocyte Galα1,3Gal epitopes implicated in sperm adhesion to the zona pellucida glycoprotein ZP3 are not required for fertilization in the mouse. J. Biol. Chem. 270, 21437–21440.

Tseng, Y.L., Kuwaki, K., Dor, F.J., Shimizu, A., Houser, S., Hisashi, Y., et al., 2005. α1,3-galactosyltransferase gene-knockout pig heart transplantation in baboons with survival approaching 6 months. Transplantation 80, 493–500.

van Lookeren Campagne, M., Wiesmann, C., Brown, E.J., 2007. Macrophage complement receptors and pathogen clearance. Cell. Microbiol. 9, 2095–2102.

Vashi, A.V., White, J.F., McLean, K.M., Neethling, W.M., Rhodes, D.I., Ramshaw, J.A., et al., 2015. Evaluation of an established pericardium patch for delivery of mesenchymal stem cells to cardiac tissue. J. Biomed. Mater. Res. A 103, 1999–2005.

Wang, B., Borazjani, A., Tahai, M., Curry, A.L., Simionescu, D.T., Guan, J., et al., 2010. Fabrication of cardiac patch with decellularized porcine myocardial scaffold and bone marrow mononuclear cells. J. Biomed. Mater. Res. A 94, 1100–1110.

Wigglesworth, K.M., Racki, W.J., Mishra, R., Szomolanyi-Tsuda, E., Greiner, D.L., Galili, U., 2011. Rapid recruitment and activation of macrophages by anti-gal/α-gal liposome interaction accelerates wound healing. J. Immunol. 186, 4422–4432.

Williams, D.A., Rios, M., Stephens, C., Patel, V.P., 1991. Fibronectin and VLA-4 in haematopoietic stem cell-microenvironment interactions. Nature 352, 438–441.

Wilson, A.C., 1985. The molecular basis of evolution. Sci. Am. 253, 164–173.

Wong, S.P., Rowley, J.E., Redpath, A.N., Tilman, J.D., Fellous, T.G., Johnson, J.R., 2015. Pericytes, mesenchymal stem cells and their contributions to tissue repair. Pharmacol. Ther. 151, 107–120.

Wydra, F.B., York, P.J., Johnson, C.R., Silvestri, L., 2016. Allografts for ligament reconstruction: where are we now? Am. J. Orthop. 45, 446–452.

Wynn, T.A., Ramalingam, T.R., 2012. Mechanisms of fibrosis: therapeutic translation for fibrotic disease. Nat. Med. 18, 1028–1040.

Xu, H., Wan, H., Zuo, W., Sun, W., Owens, R.T., Harper, J.R., et al., 2009. A porcine-derived acellular dermal scaffold that supports soft tissue regeneration: removal of terminal galactose-α-(1,3)-galactose and retention of matrix structure. Tissue Eng. A 15, 1807–1819.

Yamada, K., Yazawa, K., Shimizu, A., Iwanaga, T., Hisashi, Y., Nuhn, M., et al., 2005. Marked prolongation of porcine renal xenograft survival in baboons through the use of α1,3-galactosyltransferase gene-knockout donors and the co-transplantation of vascularized thymic tissue. Nat. Med. 11, 32–34.

Yeh, P., Ezzelarab, M., Bovin, N., Hara, H., Long, C., Tomiyama, K., et al., 2010. Investigation of potential carbohydrate antigen targets for human and baboon antibodies. Xenotransplantation 17, 197–206.

Yu, P.B., Parker, W., Everett, M.L., Fox, I.J., Platt, J.L., 1999. Immunochemical properties of anti-Galα1-3Gal antibodies after sensitization with xenogeneic tissues. J. Clin. Immunol. 19, 116–126.

Zhang, Y., He, Y., Bharadwaj, S., Hammam, N., Carnagey, K., Myers, R., et al., 2009. Tissue-specific extracellular matrix coatings for the promotion of cell proliferation and maintenance of cell phenotype. Biomaterials 30, 4021–4028.

Zhu, A., Monahan, C., Zhang, Z., Hurst, R., Leng, L., Goldstein, J., 1995. High-level expression and purification of coffee bean α-galactosidase produced in the yeast *Pichia pastoris*. Arch. Biochem. Biophys. 324, 65–70.

Zhu, A., Leng, L., Monahan, C., Zhang, Z., Hurst, R., Lenny, L., Goldstein, J., 1996. Characterization of recombinant α-galactosidase for use in seroconversion from blood group B to O of human erythrocytes. Arch. Biochem. Biophys. 327, 324–329.

CHAPTER

14

Post Infarction Regeneration of Ischemic Myocardium by Intramyocardial Injection of α-Gal Nanoparticles

OUTLINE

INTRODUCTION

Myocardial infarction (MI) is the most frequent cause of death worldwide. Coronary arterial occlusion results in injury to the myocardium of a size dependent on the location of the occlusion. The ischemia caused downstream of the occlusion results in excessive cardiomyocytes death and the development of an inflammatory process that leads to clearing the cell debris and invasion of proliferating fibroblasts. These fibroblasts ultimately form collagenous scar tissue, which prevents rupture of the injured ventricular wall. The fibrotic scar tissue formed at the infarcted area is permanent, and it further promotes reactive fibrosis in the uninjured myocardium adjacent to the ischemic tissue, which is

The Natural Anti-Gal Antibody as Foe Turned Friend in Medicine
http://dx.doi.org/10.1016/B978-0-12-813362-0.00014-2

detrimental to the contractility of the ventricle (Talman and Ruskoaho, 2016; Prabhu and Frangogiannis, 2016). The injured myocardium has a very limited ability of regeneration either by differentiation of stem cells and progenitor cells into cardiomyocytes, or by cell division of uninjured cardiomyocytes (Malliaras et al., 2016; Cai and Molkentin, 2017). Because of this limited regenerative capacity of the cardiac tissue, extensive studies have been performed with the aim of inducing post-MI myocardial regeneration by a variety of regenerative therapeutic approaches. These have included application of cell suspensions, implantation of *in vitro* engineered tissue constructs (e.g., myocardial patch bio-implants discussed in Chapter 13) to the injured area of the myocardium (Parsa et al., 2016; Feric and Radisic, 2016), and injection of bio-materials developed to enhance cardiac regeneration and remodeling post-MI (Radisic and Christman, 2013). This chapter suggests a hypothetical use of α-gal nanoparticles for inducing rapid recruitment of macrophages into the ischemic myocardium and their activation into anti-inflammatory, pro-healing macrophages. These macrophages may further induce stem cells recruitment and differentiation into cardiomyocytes that repopulate the ischemic myocardium and restore the contractility of the ventricle.

Documented spontaneous myocardial regeneration in small acute infarcts suggests that the dead cardiomyocytes can be partly replaced by cardiomyocytes derived from cells recruited from uninjured heart tissue (progenitor cells), from dividing noninjured cardiomyocytes, or from the bone marrow (i.e., mesenchymal stem cells). Macrophages are pivotal cells in healing of myocardium, as in healing of other injuries, such as wounds and burns (Chapter 12). Macrophage significance in myocardium regeneration in mice was demonstrated in studies on depletion of macrophages in mice with injured left ventricle (van Amerongen et al., 2007; Ben-Mordechai et al., 2013). Depletion of macrophages was found to impair healing of the myocardium and increase mortality. In addition to debriding of dead cells, macrophages were shown to be activated and induce vascularization in the ischemic tissue (Herold et al., 2004), promote ischemic myocardium healing (Minatoguchi et al., 2004; Yano et al., 2006; Dewald et al., 2005), mediate cardiomyocyte protection (Chazaud et al., 2003; Trial et al., 2004), and secrete cytokines that recruit stem cells, thereby initiating myocardial regeneration (Eisenberg et al., 2003; Ren et al., 2003; Leor et al., 2006; Nahrendorf et al., 2007; Lambert et al., 2008). Stem cells reaching the injured myocardium are believed to receive cues from adjacent healthy cells, microenvironment, and extracellular matrix (ECM) of the injured myocardium. These cues direct the differentiation of recruited stem cells into cardiomyocytes (Radisic and Christman, 2013). If the extent of the infarct exceeds the regenerative response of recruited stem cells, the default repair mechanism which ensues causes fibrosis of the ischemic myocardium and irreversible scar formation. Such fibrosis occurs in many post-MI patients, leading to deleterious cardiac remodeling, ischemic cardiomyopathy, and ultimately to heart failure.

Extensive preclinical research has been conducted with cell-based therapy for regeneration of ischemic myocardium by local (intramyocardial, intracoronary) or systemic (intravenous) delivery of stem cells (Orlic et al., 2001; Anversa and Nadal-Ginard, 2002; Murry et al., 2002; Kumar et al., 2005; Leor et al., 2007). The clinical experience with stem cells treatment in patients with ischemic myocardium has demonstrated only modest short-term improvements (Mozid et al., 2014; Behbahan et al., 2015; Choudry et al., 2016). This limited success may be partly the result of necrotic or fibrotic post-MI myocardium that is not conducive to stem cells proliferation and differentiation into cardiomyocytes. As indicated above, this chapter proposes an alternative method for the recruitment of endogenous stem cells into the ischemic myocardium, using α-gal nanoparticles. The recruited stem cells are assumed to differentiate into cardiomyocytes that repopulate the injured tissue and restore structure and function of the injured myocardium.

α-Gal nanoparticles injected into ischemic myocardium are expected to induce several sequential processes, similar to those illustrated in Fig. 7B in Chapter 13. Chapter 13 describes a suggested treatment using a myocardial ECM patch bio-implant containing biodegradable α-gal nanoparticles. Such a bio-implant may be studied for regeneration of ischemic myocardium, which underwent fibrosis, resulting in formation of a scar. The hypothetical therapy with α-gal nanoparticles proposed in present chapter, for inducing regeneration of ischemic myocardium, is based on the same principle as that in Chapter 13. The treatment described in Chapter 13 suggests that anti-Gal interaction with α-gal nanoparticles within myocardial ECM patch bio-implant activates the complement system. The generated complement cleavage chemotactic factors induce extensive migration of macrophages into the ECM bio-implant. These macrophages are activated by Fc/Fcγ receptors interaction with anti-Gal-coated α-gal nanoparticles and secrete cytokines/growth factors that may recruit stem cells and progenitor cells into the ECM. The myocardial ECM bio-implant is expected to induce differentiation of these stem cells into cardiomyocytes. In this chapter, it is suggested that a similar mechanism may induce regeneration of ischemic myocardium if the myocardium is injected with α-gal nanoparticles, shortly after the MI and before the fibrosis process initiates in the injured tissue, i.e., immediately after or within few days post-MI.

PRELIMINARY OBSERVATIONS ON INTRAMYOCARDIAL INJECTION OF α-GAL NANOPARTICLES

Chapter 12 describes the rapid recruitment and activation of macrophages by α-gal nanoparticles that are applied to wounds and the resulting acceleration of wound healing by cytokines/growth factors secreted from the activated macrophages. This accelerated healing of wounds further results in accelerated restoration of normal structure of the injured skin, thus avoiding fibrosis and scar formation (Fig. 7 in Chapter 12). These observations raised the possibility that an analogous accelerated regeneration of post-MI ischemic myocardium may be achieved by injection of α-gal nanoparticles into the injured tissue (Fig. 1). Such injection may restore the normal structure and function of the ischemic myocardium prior to the initiation of fibrosis, thereby preventing scar formation.

This chapter is included in the "Future Studies" section of the book because the experimental studies assessing efficacy of this proposed therapy have not been performed, as yet. However, preliminary studies on injection of α-gal nanoparticles into mouse heart implants have provided initial information on macrophage recruitment into ischemic myocardium. These studies were performed in anti-Gal producing α1,3galactosyltransferase knockout mice (GT-KO mice). Hearts obtained from these mice were injected intramyocardial *in vitro* with 1 mg α-gal nanoparticles or with saline. The injected hearts were implanted subcutaneously in recipient GT-KO mice and were not connected to the circulation of the recipients.

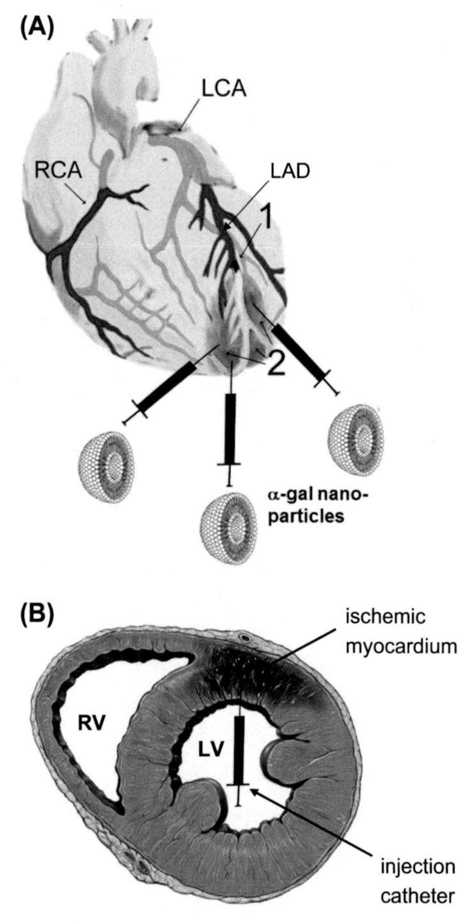

FIGURE 1 Suggested post myocardial infarction (MI) therapy with α-gal nanoparticles. (A) Illustration of the suggested therapy for post-MI regeneration of ischemic myocardium by intramyocardial injection of α-gal nanoparticles. The blockage in the left anterior descending artery (LAD), which stops the flow of blood, is labeled as (1). The affected myocardium undergoing ischemia is shown as the gray area, labeled (2). Following reperfusion, the ischemic myocardium is injected with α-gal nanoparticles. These nanoparticles bind the natural anti-Gal antibody and activate the complement system to generate complement cleavage chemotactic peptides that recruit macrophages into the injected sites. The macrophages are activated following Fc/Fcγ receptor interaction with anti-Gal coating the α-gal nanoparticles. These macrophages secrete anti-inflammatory, pro-healing cytokines/growth factors. The secreted cytokines recruit stem cells, which are induced by the myocardium extracellular matrix and the microenvironment to differentiate into cardiomyocytes that repopulate the ischemic myocardium and restore its function. *LCA*, left coronary artery; *RCA*, right coronary artery. (B) Schematic illustration of transendocardial injection of α-gal nanoparticles into ischemic myocardium by a catheter. The guiding catheter is navigated via the aorta into the left ventricle, and α-gal nanoparticles are delivered by an injection catheter into several sites, under fluoroscopy. LV, left ventricle; RV, right ventricle. *Adapted from Wikipedia, with permission.*

The hearts were explanted after 2 or 4 weeks and subjected to histological evaluation. As shown in Fig. 2A, hearts injected with α-gal nanoparticles and examined after 2 weeks displayed extensive recruitment of macrophages, despite the absence of any connection to the recipient's vascular system. This recruitment is likely to be the result of complement activation by anti-Gal binding to α-gal nanoparticles, and generation of chemotactic complement

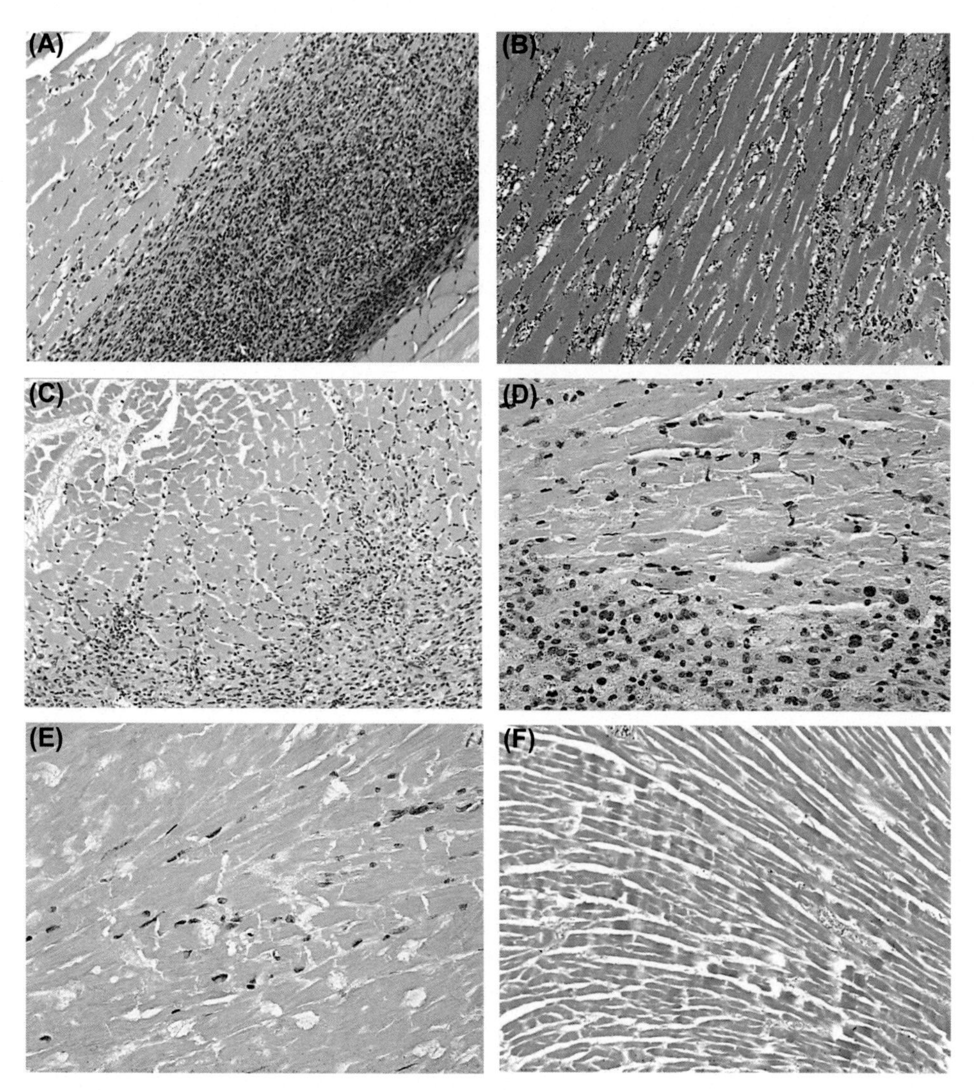

FIGURE 2 Macrophages recruited by α-gal nanoparticles into α1,3galactosyltransferase knockout (GT-KO) mouse hearts injected *in vitro* with saline (B) or with α-gal nanoparticles (A, C–F), prior to subcutaneous implantation of the hearts into anti-Gal producing GT-KO mice. Hearts were explanted 2 weeks (A, B) and 4 weeks post implantation (C–F), sectioned, and stained with hematoxylin and eosin (H&E). (A) Heart injected with 1 mg α-gal nanoparticles displays extensive recruitment of macrophages, 2 weeks post implantation (×100). (B) Heart injected with saline, 2 weeks post implantation. The infiltrating cells are neutrophils, and the myocardium is necrotic (×100). (C) Heart implant injected with α-gal nanoparticles and viewed 4 weeks post implantation, displays macrophages migrating into areas of myocardium that were not injected (×100). (D) As in (C), demonstrating migration of macrophages among cardiomyocytes (×200). (E) A region far from the injected area demonstrating migrating macrophages among conserved cardiomyocytes that lack nuclei (×200). Some of the cardiomyocytes display striation. Intercalated discs are observed, as well. (F) A region far from the injection site of α-gal nanoparticles displaying conserved structure of cardiomyocytes lacking nuclei, despite the complete absence of macrophages (×200). In (E) and (F), no necrosis of cardiomyocytes is observed 4 weeks post implantation of the ischemic mouse hearts despite the lack of nuclei in the cells. *Adapted from Galili, U., 2013b. Discovery of the natural anti-Gal antibody and its past and future relevance to medicine. Xenotransplantation 20, 138–147, with permission.*

cleavage peptides similar to that observed in skin injected with these nanoparticles (Fig. 3 in Chapter 12). In hearts injected only with saline, neutrophils, but no macrophages, were found in the myocardium (Fig. 2B). In addition, the cardiomyocytes displayed an acidophilic necrotic morphology.

No heart implant tissue was found after 4 weeks in mice implanted subcutaneously with saline-injected hearts. In contrast, heart implants injected with α-gal nanoparticles retained 20%–50% of their original size, 4 weeks post implantation. Histological analysis indicated that the macrophages recruited into these hearts began migrating into the noninjected areas (Fig. 2C and D). The most striking feature in these hearts was the conservation of the cardiomyocytes structure, including distinct intercalated discs (Fig. 2D and E) and striation in the cytoplasm. Interestingly, the cardiomyocytes structure was conserved despite the absence of nuclei (Figs. 2E and 2F), which were not visible even at the 2-week time point. These observations suggest that the activated macrophages secrete cytokines that suppress neutrophil-mediated inflammation, conserve the structure of cardiomyocytes for several weeks, and prevent their necrosis, even in the absence of nuclei. The nuclei are likely to be destroyed shortly after the ischemia of the myocardium by lysosomal nucleases causing fragmentation of chromosomal DNA. Conservation of cardiomyocytes structure without nuclei was observed even in areas that were completely devoid of infiltrating macrophages, suggesting diffusion of the conserving cytokines from areas with high concentration of activated macrophages (Fig. 2E). This cardiomyocytes structure conservation for several weeks may be of significance in the suggested process of myocardium regeneration mediated by α-gal nanoparticles. The lack of necrosis is likely to reflect the intactness of the ECM, and a microenvironment that may be conducive to differentiation of recruited stem cells into functional cardiomyocytes.

As discussed below, GT-KO pigs may serve as a suitable animal model for studying post-MI regeneration of ischemic myocardium by α-gal nanoparticles treatment because these pigs naturally produce anti-Gal with characteristics similar to those of human anti-Gal (Dor et al., 2004; Fang et al., 2012; Galili, 2013a). Furthermore, pigs are large enough to enable delivering the nanoparticles into the ischemic myocardium by a catheter navigated into the ventricle. Such a catheter may be used for transendocardial injections into the ischemic myocardium (Fig. 1B). In attempt to determine whether α-gal nanoparticles induce recruitment of macrophages into the myocardium of GT-KO pigs, a guide catheter was navigated through the aorta and the left atrium into the left ventricle of these pigs. Approximately 10 mg of α-gal nanoparticles were injected into the healthy myocardium, by a microcatheter, under continuous fluoroscopy. The pigs were euthanized after 5 and 7 days. The injection area was identified, sectioned, and processed for histological analysis. The injection site (seen as empty space in Fig. 3A) in a heart studied 5 days postinjection was surrounded by infiltrating mononuclear cells, which display the morphology of activated macrophages, similar to that observed in the skin of GT-KO mice injected with α-gal nanoparticles (Fig. 3 in Chapter 12). In the heart studied 7 days post injection of the nanoparticles, the recruited macrophages seemed to migrate between cardiomyocytes possibly away from the injected site (Fig. 3B). These preliminary studies suggest that, similar to the observations in GT-KO mice (Fig. 2), macrophages recruited to the injection site of α-gal nanoparticles, begin after several days to migrate into noninjected areas in the myocardium.

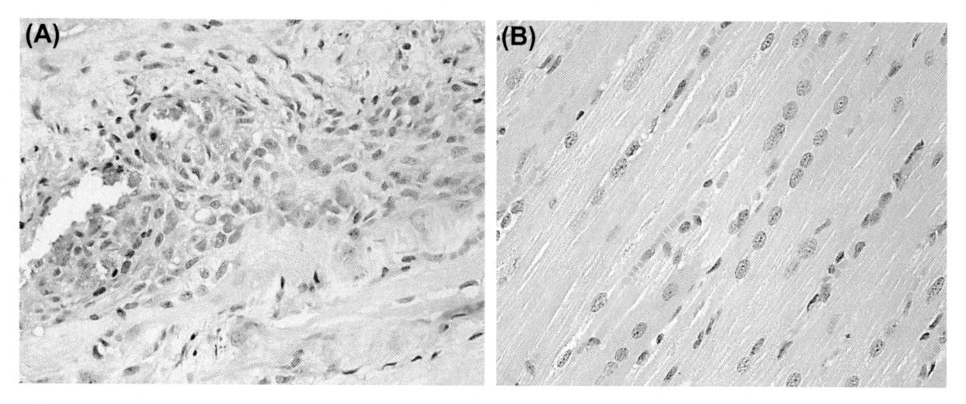

FIGURE 3 Recruitment of macrophages into α1,3galactosyltransferase knockout pig healthy heart by transendocardial catheter injection of ~10mg α-gal nanoparticles. (A) Injection area 5days postadministration of the nanoparticles. Multiple macrophages migrating to the injection site, identified as the empty area. (B) Myocardium at a distance from the injection site, 7days post administration of α-gal nanoparticles. The nuclei arranged in rows among the cardiomyocytes seem to be of macrophages migrating away from the α-gal nanoparticles injection site (H&E, ×200).

SUGGESTED INDUCTION OF POST MYOCARDIAL INFARCTION REGENERATION BY INTRAMYOCARDIAL INJECTION OF α-GAL NANOPARTICLES

The observations in Figs. 2 and 3 suggest that post reperfusion injection of α-gal nanoparticles into ischemic myocardium may initiate a sequence of processes as those illustrated in Fig. 7B in Chapter 13. Anti-Gal binding to these nanoparticles activates the complement system to generate complement cleavage chemotactic peptides that induce rapid recruitment of macrophages. These macrophages are likely to be activated by Fc/Fcγ receptor interaction with anti-Gal coating the nanoparticles (visualized in Fig. 5 in Chapter 12) and are induced to secrete a variety of cytokines/growth factors, including stem cell recruiting cytokines. Secretions by the activated macrophages also seem to include uncharacterized substances that delay necrosis of the ischemic tissue and conserve the structure of dead cardiomyocytes. Such conservation may maintain the intactness of the ECM and an "authentic" myocardial microenvironment, both conducive to differentiation of stem cells into functional cardiomyocytes.

As shown in Figs. 4B and 9 in Chapter 12, the macrophages activated by anti-Gal/α-gal nanoparticles interaction secrete VEGF, which induces local angiogenesis, as well as other anti-inflammatory, pro-healing cytokines (Wigglesworth et al., 2011; Hurwitz et al., 2012). Studies on the macrophages recruited into polyvinyl alcohol (PVA) sponge discs indicated that most of the recruited macrophages are M2 polarized that are positive for interleukin-10 and arginase, and negative for interleukin-12 (in preparation). These recruited macrophages were further found to induce recruitment of stem cells which differentiated following interaction with fragmented ECM within PVA sponge discs (Fig. 6 in Chapter 13) (Galili, 2015). M2 macrophages were reported to mediate the reconstruction of post-MI ischemic myocardium (Ma et al., 2013; Ben-Mordechai et al., 2013; Frangogiannis, 2014). Taken together, the studies in Chapters 12–14 raise the possibility that α-gal nanoparticles injected into post-MI ischemic myocardium may modulate the inflammatory response of physiologically infiltrating pro-inflammatory

M1 macrophages into extensive recruitment and activation of anti-inflammatory pro-healing M2 macrophages. These macrophages secrete cytokines that induce vascularization, conserve the dead cardiomyocytes structure, delay necrosis, and recruit stem cells, thereby shifting the inflammatory processes within the ischemic myocardium from fibrosis and scar formation, into regenerative processes.

The inflammatory changes occurring in ischemic myocardium are highly dynamic, so that changes leading to fibrosis may occur within several days post-MI (Frangogiannis, 2014). Thus, one could assume that injection of α-gal nanoparticles into the ischemic myocardium, by a catheter navigated into the ventricle (Fig. 1B), should be performed within a short period post-MI (i.e., immediately after reperfusion or within few days). An early injection may increase the probability of having intact cardiomyocytes, ECM, and microenvironment that can induce effective stimulation of recruited stem cells to differentiate into cardiomyocytes. The observed rapid recruitment of macrophages, within 24h following intradermal injection of α-gal nanoparticles (Fig. 3A in Chapter 12), suggests that injection of the nanoparticles into ischemic myocardium is likely to result in recruitment of the macrophages within a similar period. If studies in experimental animal models indicate that injection of α-gal nanoparticles is followed by partial or complete myocardium regeneration, it would be important to determine the optimal time point for administering the α-gal nanoparticles into the injured myocardium for obtaining maximum regenerative effects. Administration of α-gal nanoparticles into ischemic myocardium following blockage of the left anterior descending (LAD) artery (Fig. 1) may be most feasible by the use of a catheter navigated through the aorta and left atrium into the left ventricle, followed by transendocardial injection as schematically illustrated in Fig. 1B and performed in Fig. 3. The injection may be performed in multiple points of the ischemic area, similar to studies in which stem cells were injected in 10–20 sites within ischemic myocardium of the left ventricle (Hatzistergos et al., 2010; Smits et al., 2003; Mushtaq et al., 2014). The experimental models, which may enable addressing the various issues of myocardial regeneration by α-gal nanoparticles, are described below.

EXPERIMENTAL MODELS FOR STUDYING MYOCARDIUM REGENERATION POST MYOCARDIAL INFARCTION

Similar to the other three chapters in this section, the therapy suggested in this chapter for the regeneration of ischemic myocardium by α-gal nanoparticles is hypothetical at present and requires experimental demonstration of its efficacy and safety. As mentioned above, the two nonprimate mammalian models that lack α-gal epitopes and thus are suitable for performing such studies are GT-KO mice and GT-KO pigs. GT-KO mice produce the anti-Gal antibody following immunization with xenogeneic cell membranes presenting α-gal epitopes, such as rabbit red blood cell membranes (LaTemple and Galili, 1998) or porcine kidney membranes homogenate (Tanemura et al., 2000). MI may be simulated in GT-KO mice by surgical opening of the chest cavity under anesthesia and blocking for 30min the blood flow in the LAD artery by ligation. Subsequent reperfusion of the ischemic myocardium is achieved by removing the ligation knot (De Celle et al., 2004; Xu et al., 2014). Injection of α-gal nanoparticles into the ischemic area of the left ventricle is performed right after reperfusion, and then the chest is closed. In untreated hearts, the affected left ventricular wall myocardium is converted into a thin fibrotic wall within a period of 3–4 weeks. In mice treated by intramyocardial injection of α-gal nanoparticles, the initial damage may be confirmed by impaired contractility or decreased cardiac output, followed by regeneration identified as restoration of the normal cardiac contractility and output, as determined by echocardiography. Such regeneration is further correlated with the histology of the myocardium at different time points. Ligation followed by intramyocardial saline injection should result in irreversible impairment of cardiac functions and confirmed histologically as conversion of the left ventricular wall into fibrotic tissue (stained blue for collagen in Trichrome staining).

Studies simulating the suggested therapy by transendocardial catheter injection of α-gal nanoparticles may be performed in GT-KO pigs. Ischemia of the left ventricular myocardium can be accomplished with a balloon catheter that blocks the LAD artery for 90–120 min. Subsequently, the balloon catheter is replaced with an injection catheter that is navigated into the left ventricular lumen. The catheter may be further used for transendocardial injection of α-gal nanoparticles into multiple sites of the ischemic myocardium under continuous fluoroscopy, to distribute the nanoparticles throughout the infracted myocardium (Hatzistergos et al., 2010). If the α-gal nanoparticles induce regeneration of the ischemic myocardium, the initial ischemia injury and the subsequent regeneration may be monitored for changes in cardiac functions by echocardiography. The extent of fibrosis at different time points may be further assessed by cardiac magnetic resonance imaging(MRI). As in the suggested mouse studies above, increase contractility and cardiac output may ultimately be correlated with histology of the regenerating myocardium. Control GT-KO pigs in which the ischemia is treated by injection of saline, or of nanoparticles lacking α-gal epitopes, are expected to display irreversible fibrosis of the ischemic myocardium, by echocardiography, MRI and by histology of the ischemic myocardium, within several weeks post ischemia.

POTENTIAL RISKS

It may be premature to consider risks associated with post-MI α-gal nanoparticles therapy. If successful in inducing regeneration of the ischemic myocardium, the effects of the extensive recruitment of macrophages by α-gal nanoparticles on heart physiology should be determined, in particular in view of a recent report on macrophages that facilitate electrical conduction in the heart (Hulsmans et al., 2017). However, one potential risk of this theoretical therapy, which should be kept in mind, is associated with the possibility of delivering the α-gal nanoparticles by transendocardial injection within the left ventricle. The use of this injection method has been demonstrated in studies on injecting cell populations such as myoblasts or stem cells (e.g., bone marrow cells) into multiple sites in the ischemic myocardium (Smits et al., 2003; Mushtaq et al., 2014). Accidental release of such cells into the left ventricular lumen is likely to be of no risk. However, accidental release of 10–100 mg of α-gal nanoparticles into the circulation may have deleterious effects because the effective binding of anti-Gal to these nanoparticles may induce systemic activation of the complement system. Such complement activation may result in systemic activation of mast cells and basophils (Erdei et al., 1997; Finkelman et al., 2016), which could further induce systemic vasodilation and a range of allergy-like reactions. To study this hypothetical effect,

50–100 mg α-gal nanoparticles were released into the lumen of the left ventricle in GT-KO pigs. This release resulted in extensive vasodilation in the pig as indicated by the reddening of the skin all over the body, within 20 s. In addition, the pulse of the pigs increased from 90 to 160–230 bpm and was associated with arrhythmia that resulted in death of one of the three GT-KO pigs studied. Injection of lidocaine decreased the pulse to normal levels and restored the normal color of the skin in the remaining two pigs within 10–15 min (unpublished observations). The risk of accidental release of α-gal nanoparticles into the lumen of the ventricle may be avoided by using a catheter that anchors well within the myocardium, and a controlled injection mechanism that allows for release of small volumes of the α-gal nanoparticles suspension.

No Conclusions section is included since the proposed therapy in this chapter is hypothetical.

References

Anversa, P., Nadal-Ginard, B., 2002. Myocyte renewal and ventricular remodeling. Nature 415, 240–243.

Behbahan, I.S., Keating, A., Gale, R.P., 2015. Bone marrow therapies for chronic heart disease. Stem Cells 33, 3212–3227.

Ben-Mordechai, T., Holbova, R., Landa-Rouben, N., Harel-Adar, T., Feinberg, M.S., Abd Elrahman, I., et al., 2013. Macrophage subpopulations are essential for infarct repair with and without stem cell therapy. J. Am. Coll. Cardiol. 62, 1890–1901.

Cai, C.L., Molkentin, J.D., 2017. The elusive progenitor cell in cardiac regeneration: slip slidin' away. Circ. Res. 120, 400–406.

Chazaud, B., Sonnet, C., Lafuste, P., Bassez, G., Rimaniol, A.C., Poron, F., et al., 2003. Satellite cells attract monocytes and use macrophages as a support to escape apoptosis and enhance muscle growth. J. Cell Biol. 163, 1133–1143.

Choudry, F., Hamshere, S., Saunders, N., Veerapen, J., Bavnbek, K., Knight, C., 2016. A randomized double-blind control study of early intra-coronary autologous bone marrow cell infusion in acute myocardial infarction: the REGENERATE-AMI clinical trial. Eur. Heart J. 37, 256–263.

De Celle, T., Cleutjens, J.P., Blankesteijn, W.M., Debets, J.J., Smits, J.F., Janssen, B.J., 2004. Long-term structural and functional consequences of cardiac ischaemia-reperfusion injury in vivo in mice. Exp. Physiol. 89, 605–615.

Dewald, O., Zymek, P., Winkelmann, K., Koerting, A., Ren, G., Abou-Khamis, T., et al., 2005. CCL2/monocyte chemoattractant protein-1 regulates inflammatory responses critical to healing myocardial infarcts. Circ. Res. 96, 881–889.

Dor, F.J., Tseng, Y.L., Cheng, J., Moran, K., Sanderson, T.M., Lancos, C.J., et al., 2004. α1,3-Galactosyltransferase gene-knockout miniature swine produce natural cytotoxic anti-Gal antibodies. Transplantation 78, 15–20.

Eisenberg, L.M., Burns, L., Eisenberg, C.A., 2003. Hematopoietic cells from bone marrow have the potential to differentiate into cardiomyocytes in vitro. Anat. Rec. A Discov. Mol. Cell Evol. Biol. 274, 870–882.

Erdei, A., Kerekes, K., Pecht, I., 1997. Role of C3a and C5a in the activation of mast cells. Exp. Clin. Immunogenet. 14, 16–18.

Fang, J., Walters, A., Hara, H., Long, C., Yeh, P., Ayares, D., et al., 2012. Anti-gal antibodies in α1,3-galactosyltransferase gene-knockout pigs. Xenotransplantation 19, 305–310.

Feric, N.T., Radisic, M., 2016. Strategies and challenges to myocardial replacement therapy. Stem Cells Transl. Med. 5, 410–416.

Finkelman, F.D., Khodoun, M.V., Strait, R., 2016. Human IgE-independent systemic anaphylaxis. J. Allergy Clin. Immunol. 137, 1674–1680.

Frangogiannis, N.G., 2014. The inflammatory response in myocardial injury, repair, and remodeling. Nat. Rev. Cardiol. 11, 255–265.

Galili, U., 2013a. α1,3Galactosyltransferase knockout pigs produce the natural anti-Gal antibody and simulate the evolutionary appearance of this antibody in primates. Xenotransplantation 20, 267–276.

Galili, U., 2013b. Discovery of the natural anti-Gal antibody and its past and future relevance to medicine. Xenotransplantation 20, 138–147.

Galili, U., 2015. Avoiding detrimental human immune response against mammalian extracellular matrix implants. Tissue Eng. Part B 21, 231–241.

Hatzistergos, K.E., Quevedo, H., Oskouei, B.N., Hu, Q., Feigenbaum, G.S., Margitich, I.S., et al., 2010. Bone marrow mesenchymal stem cells stimulate cardiac stem cell proliferation and differentiation. Circ. Res. 107, 913–922.

Herold, J., Pipp, F., Fernandez, B., Xing, Z., Heil, M., Tillmanns, H., Braun-Dullaeus, R.C., 2004. Transplantation of monocytes: a novel strategy for in vivo augmentation of collateral vessel growth. Hum. Gene Ther. 15, 1–12.

Hulsmans, M., Clauss, S., Xiao, L., Aguirre, A.D., King, K.R., Hanley, A., et al., 2017. Macrophages facilitate electrical conduction in the heart. Cell 169, 510–522.

Hurwitz, Z., Ignotz, R., Lalikos, J., Galili, U., 2012. Accelerated porcine wound healing with α-gal nanoparticles. Plastic Reconstr. Surg. 129, 242–251.

Kumar, D., Kamp, T.J., LeWinter, M.M., 2005. Embryonic stem cells: differentiation into cardiomyocytes and potential for heart repair and regeneration. Coron. Artery Dis. 16, 111–116.

Lambert, J.M., Lopez, E.F., Lindsey, M.L., 2008. Macrophage roles following myocardial infarction. Int. J. Cardiol. 130, 147–158.

LaTemple, D.C., Galili, U., 1998. Adult and neonatal anti-Gal response in knock-out mice for α1,3galactosyltransferase. Xenotransplantation 5, 191–196.

Leor, J., Rozen, L., Zuloff-Shani, A., Feinberg, M.S., Amsalem, Y., Barbash, I.M., et al., 2006. Ex vivo activated human macrophages improve healing, remodeling, and function of the infarcted heart. Circulation 114 (1 Suppl.), I94–I100.

Leor, J., Gerecht, S., Cohen, S., Miller, L., Holbova, R., Ziskind, A., et al., 2007. Human embryonic stem cell transplantation to repair the infarcted myocardium. Heart 93, 1278–1284.

Ma, Y., Halade, G.V., Zhang, J., Ramirez, T.A., Levin, D., Voorhees, A., et al., 2013. Matrix metalloproteinase-28 deletion exacerbates cardiac dysfunction and rupture after myocardial infarction in mice by inhibiting M2 macrophage activation. Circ. Res. 112, 675–688.

Malliaras, K., Vakrou, S., Kapelios, C.J., Nanas, J.N., 2016. Innate heart regeneration: endogenous cellular sources and exogenous therapeutic amplification. Expert. Opin. Biol. Ther. 16, 1341–1352.

Minatoguchi, S., Takemura, G., Chen, X.H., Wang, N., Uno, Y., Koda, M., et al., 2004. Acceleration of the healing process and myocardial regeneration may be important as a mechanism of improvement of cardiac function and remodeling by post-infarction granulocyte colony-stimulating factor treatment. Circulation 109, 2572–2580.

Mozid, A.M., Holstensson, M., Choudhury, T., Ben-Haim, S., Allie, R., Martin, J., et al., 2014. Clinical feasibility study to detect angiogenesis following bone marrow stem cell transplantation in chronic ischaemic heart failure. Nucl. Med. Commun. 35, 839–848.

Murry, C.E., Whitney, M.L., Reinecke, H., 2002. Muscle cell grafting for the treatment and prevention of heart failure. J. Card. Fail. 8 (6 Suppl.), S532–S541.

Mushtaq, M., DiFede, D.L., Golpanian, S., Khan, A., Gomes, S.A., Mendizabal, A., et al., 2014. Rationale and design of the percutaneous stem cell injection delivery effects on neomyogenesis in dilated cardiomyopathy (the POSEIDON-DCM study): a phase I/II, randomized pilot study of the comparative safety and efficacy of transendocardial injection of autologous mesenchymal stem cell vs. allogeneic mesenchymal stem cells in patients with non-ischemic dilated cardiomyopathy. J. Cardiovasc. Transl. Res. 7, 769–780.

Nahrendorf, M., Swirski, F.K., Aikawa, E., Stangenberg, L., Wurdinger, T., Figueiredo, J.L., et al., 2007. The healing myocardium sequentially mobilizes two monocyte subsets with divergent and complementary functions. J. Exp. Med. 204, 3037–3047.

Orlic, D., Kajstura, J., Chimenti, S., Bodine, D.M., Leri, A., Anversa, P., 2001. Transplanted adult bone marrow cells repair myocardial infarcts in mice. Ann. N.Y. Acad. Sci. 938, 221–229.

Parsa, H., Ronaldson, K., Vunjak-Novakovic, G., 2016. Bioengineering methods for myocardial regeneration. Adv. Drug Deliv. Rev. 96, 195–202.

Prabhu, S.D., Frangogiannis, N.G., 2016. The biological basis for cardiac repair after myocardial infarction: from inflammation to fibrosis. Circ. Res. 119, 91–112.

Radisic, M., Christman, K.L., 2013. Materials science and tissue engineering: repairing the heart. Mayo Clin. Proc. 88, 884–898.

Ren, G., Dewald, O., Frangogiannis, N.G., 2003. Inflammatory mechanisms in myocardial infarction. Curr. Drug Targets Inflamm. Allergy 2, 242–256.

Smits, P.C., van Geuns, R.J., Poldermans, D., Bountioukos, M., Onderwater, E.E., Lee, C.H., et al., 2003. Catheter-based intramyocardial injection of autologous skeletal myoblasts as a primary treatment of ischemic heart failure: clinical experience with six-month follow-up. J. Am. Coll. Cardiol. 42, 2063–2069.

Talman, V., Ruskoaho, H., 2016. Cardiac fibrosis in myocardial infarction-from repair and remodeling to regeneration. Cell Tissue Res. 365, 563–581.

Tanemura, M., Yin, D., Chong, A.S., Galili, U., 2000. Differential immune response to α-gal epitopes on xenografts and allografts: implications for accommodation in xenotransplantation. J. Clin. Investig. 105, 301–310.

Trial, J., Rossen, R.D., Rubio, J., Knowlton, A.A., 2004. Inflammation and ischemia: macrophages activated by fibronectin fragments enhance the survival of injured cardiac myocytes. Exp. Biol. Med. 229, 538–545.

van Amerongen, M.J., Harmsen, M.C., van Rooijen, N., Petersen, A.H., van Luyn, M.J., 2007. Macrophage depletion impairs wound healing and increases left ventricular remodeling after myocardial injury in mice. Am. J. Pathol. 170, 818–829.

Wigglesworth, K.M., Racki, W.J., Mishra, R., Szomolanyi-Tsuda, E., Greiner, D.L., Galili, U., 2011. Rapid recruitment and activation of macrophages by anti-Gal/α-gal liposome interaction accelerates wound healing. J. Immunol. 186, 4422–4432.

Xu, Z., Alloush, J., Beck, E., Weisleder, N., April 10, 2014. A murine model of myocardial ischemia-reperfusion injury through ligation of the left anterior descending artery. J. Vis. Exp. (86).

Yano, T., Miura, T., Whittaker, P., Miki, T., Sakamoto, J., Nakamura, Y., et al., 2006. Macrophage colony-stimulating factor treatment after myocardial infarction attenuates left ventricular dysfunction by accelerating infarct repair. J. Am. Coll. Cardiol. 47, 626–634.

CHAPTER

15

Regeneration of Injured Spinal Cord and Peripheral Nerves by α-Gal Nanoparticles

OUTLINE

INTRODUCTION

The ability of injured spinal cord or of peripheral nerves to regenerate is very limited. This can be inferred from the poor prognosis following spinal cord and peripheral nerve injuries involving severed axons. Such injuries often result in irreversible permanent loss of neurologic function. Pathophysiological events associated with spinal cord injuries, such as loss of blood supply, inflammation, and demyelination, all decrease ability of the microenvironment at the injury site to induce regeneration of the injured nerve tissue (Mautes et al., 2000; Fassbender et al., 2011). Regeneration of the injured nerve is achieved, if sprouts developing at the severed end of the axon grow across the lesion gap and "find" endoneurial tubes of the distal axonal segment (referred to, here as "distal segment") in which they further grow to restore the full length and function of the injured neurons. The probability for the growth of a sprout into a distal segment is low because this growth does not occur at a specific orientation toward the distal part of the served nerve

http://dx.doi.org/10.1016/B978-0-12-813362-0.00015-4

(Silver and Miller, 2004; Dray et al., 2009). Moreover, the sprout growth and reconnection with a distal segment are processes limited by time, because of a simultaneous process in which pro-inflammatory M1 macrophages reaching the lesion, mediate inflammation that results in fibrosis and scar formation (Kigerl et al., 2009; Gensel and Zhang, 2015). The fibrotic tissue prevents further growth of sprouts across the lesion into the distal portion of the nerve and does not enable the axonal extension required for nerve regeneration (Silver and Miller, 2004; David et al., 2012). These constraints imply that the probability for regeneration of severed nerves may be increased by creating conditions in the injured spinal cord or injured peripheral nerve that increase the number of axonal sprouts growing into the lesion. This chapter describes a suggested hypothetical method for increasing the number of axonal sprouts with α-gal nanoparticles.

ASSOCIATION BETWEEN AXONAL SPROUTS GROWTH AND ANGIOGENESIS

Studies monitoring the events occurring following induced small spinal cord injuries demonstrated multiple architectural changes, such as axons dieback hundreds of micrometers within 30 min after trauma (Kerschensteiner et al., 2005). Many of the axons "attempt" regeneration within 6–24 h after lesion formation. This growth response, although robust, seems to fail as a result of the inability of axons to navigate in the proper direction. In later stages, the severed axons are capable only of low level of sprouting that provides poor recovery and regeneration because the lesion is filled with scar tissue (Silver and Miller, 2004). This fibrosis and scar formation is analogous to scar formation in healing of untreated wounds (Chapter 12; Wigglesworth et al., 2011; Galili, 2017) or in ischemic myocardium post myocardial infarction (Chapter 14).

The study of Dray et al. (2009) using time-lapse dual-color high-resolution imaging, allowed simultaneous early visualization of post traumatic vascular responses, sprouting of axons and neurovascular interactions in the spinal cord of mice. In that study, the minimal injury to dorsal root ganglion axons in the spinal cord was caused by piercing with the end of a 26G needle. This minimal injury enabled the regrowth of axonal sprouts across the lesion gap and reconnection with distal segments, resulting in complete regeneration and restoration of the nerve structure within 60 days. Dray et al. (2009) found that many of the axonal sprouts grew in close vicinity to newly formed small blood vessels. This growth pattern suggested that newly formed small blood vessels provide oxygen, nutrients, and possibly scaffold of laminin and collagen to the growing sprouts. Because there was no specific orientation to the blood vessels generated by the neo-vascularization and the growing axonal sprouts along them, only a small proportion of the sprouts "succeeded" in reconnecting with distal segments.

The observations on the dependency of axonal sprouts growth on neo-vascularization of the lesion site are in agreement with studies reporting on angiogenesis, which precedes recovery following spinal cord injury, and with studies indicating that axon regeneration is related to angiogenesis (Zhang and Guth, 1997; Woerly et al., 2004; Rauch et al., 2009). Furthermore, induction of angiogenesis by administration of vascular endothelial growth factor (VEGF) into spinal cord injuries was reported to enhance regeneration, which was associated with the induced revascularization of the lesion site (Facchiano et al., 2002; Widenfalk et al., 2003; Ferrara and Kerbel, 2005; De Laporte et al., 2011; Yu et al., 2016). Because reconnection of regenerating axons with distal segments of the injured neurons are random encounter events, it was suggested that high density of active sprouts should increase the probability of productive reconnections, and that such high density may be achieved by induction of angiogenesis in the neural lesion (Ferrara and Kerbel, 2005; Dray et al., 2009). As suggested in this chapter, these objectives may be achieved by physiologic mechanisms set in motion following administration of α-gal nanoparticles to spinal cord or peripheral nerve lesion sites.

MACROPHAGES CONTRIBUTION TO NERVE REGENERATION

The accumulation of macrophages in nerve injuries has suggested that macrophages facilitate the inflammatory processes, which ultimately result in fibrosis, scar formation, and irreversible prevention of neuron regeneration, analogous to fibrosis and scar formation in healing skin wounds. This notion is based on studies, in which macrophages have been depleted or inactivated after spinal cord injury. In such experimental models, injured nerves were found to display increased regeneration and improvement in various neural functions in the absence of macrophages (Oudega et al., 1999; Popovich et al., 1999). However, macrophages were also found to have beneficial effects on nerve tissues following injury (Prewitt et al., 1997; Rapalino et al., 1998; Schwartz, 2010; Kwon et al., 2013). It has been proposed that macrophages reaching the nerve lesion are pro-inflammatory M1 polarized macrophages that promote the inflammatory reaction and scar formation. The M1 polarized macrophages usually reside within the injured spinal cord for at least several weeks and mediate the irreversible damage to the nerves (Schwartz, 2010; Peng et al., 2016). If, however, these macrophages are replaced or converted early enough into anti-inflammatory, pro-healing M2 macrophages, the local inflammatory response may be suppressed, and the lesion may become well revascularized. Under such circumstances, the probability that the neurons may regenerate increases because of growth of axonal sprouts across the lesion gap into distal segments (Kigerl et al., 2009; Shechter et al., 2013; Gensel and Zhang, 2015). As suggested below, α-gal nanoparticles applied to the lesion site may induce recruitment of M2 macrophages, which secrete VEGF and thus, induce localized angiogenesis that may ultimately increase the density of axonal sprouts.

SUGGESTED REGENERATION OF INJURED NERVES BY α-GAL NANOPARTICLES

The suggested hypothetical therapy by α-gal nanoparticles for inducing regeneration of injured nerves is based on the following observations:

1. As discussed above, regeneration of injured nerves is enhanced by local elevation in VEGF concentration, which induces revascularization of the lesion site by multiple capillaries. Such regeneration is further enhanced by M2 polarized macrophages that are recruited by various methods into the lesion site.
2. α-Gal nanoparticles administered into various injuries rapidly bind the natural anti-Gal antibody and activate the complement system (Fig. 1B in Chapter 12). The generated chemotactic complement cleavage peptides C5a and C3a recruit macrophages to the area of administered nanoparticles (Fig. 3 in Chapter 12). Analysis of the macrophages recruited *in vivo* by α-gal nanoparticles into polyvinyl alcohol sponge disc implants indicated that these are M2 polarized macrophages that produce IL10 and arginase, but not IL12 (in preparation).

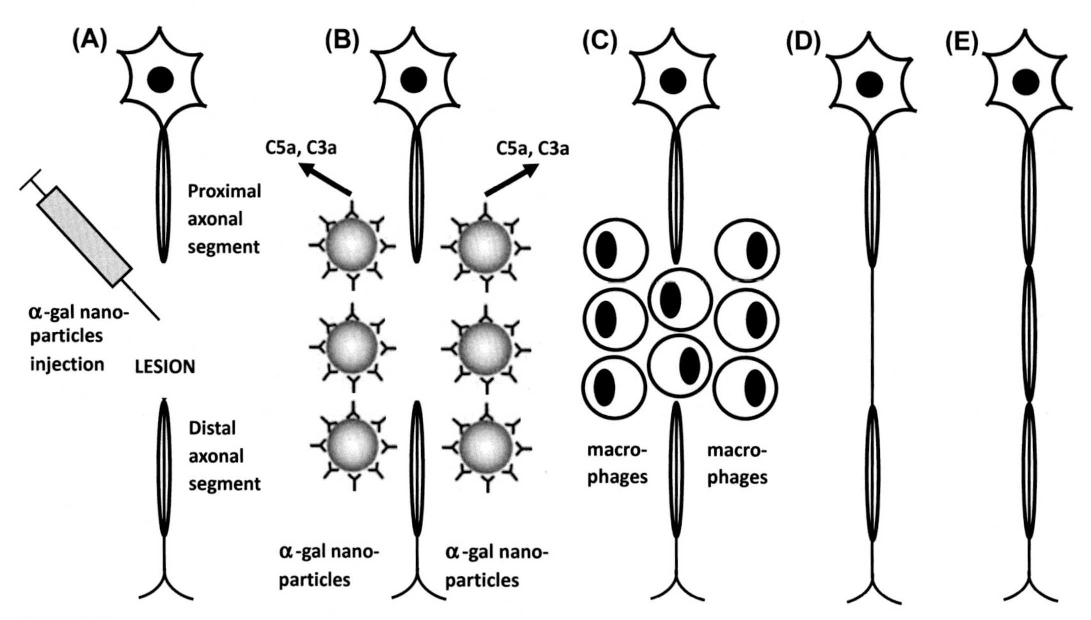

FIGURE 1 Illustration of the suggested hypothetical therapy for regeneration of injured nerves by administration of α-gal nanoparticles. (A) α-Gal nanoparticles are administered to a spinal cord injury or a peripheral nerve injury by injection or within a semisolid medium. (B) Binding of anti-Gal (illustrated as IgG molecules) to the multiple α-gal epitopes on the nanoparticles activates the complement system to generate complement cleavage chemotactic peptides C5a and C3a. (C) The chemotactic complement peptides recruit macrophages to the lesion. The macrophages binding anti-Gal coated α-gal nanoparticles via Fc/Fcγ receptor interaction are polarized into M2 macrophages and are activated to secrete vascular endothelial growth factor (VEGF) and anti-inflammatory, pro-healing cytokines/growth factors for prolonged periods. (D) VEGF secreted by the activated macrophages induces neovascularization within the lesion. Few of the multiple axonal sprouts growing along the many newly formed capillaries, reconnect with distal segments, resulting in growth of the severed axons. (E) The newly grown axons are myelinated, thereby completing the regeneration of the injured neurons.

3. α-Gal nanoparticles bind to the recruited macrophages via interaction between the Fc portion of anti-Gal coating these nanoparticles and Fcγ receptors on macrophages (Fig. 5 in Chapter 12). This Fc/Fcγ receptor interaction activates the macrophages to be M2 polarized macrophages that secrete anti-inflammatory, pro-healing cytokines/growth factors, including VEGF (Fig. 4B in Chapter 12; Wigglesworth et al., 2011). The angiogenic effects of VEGF released into wounds treated with α-gal nanoparticles are presented in Fig. 9 in Chapter 12 (Hurwitz et al., 2012). In addition, these macrophages secrete cytokines that suppress neutrophil mediated inflammation and conserve the structure of dead cells in injured tissues (Fig. 2 in Chapter 14).

In view of these observations, it is hypothesized that application of α-gal nanoparticles to spinal cord injuries may result in the activation of several physiologic processes that are schematically illustrated in Fig. 1 and which may lead to the regeneration of the injured neurons. The processes are labeled as in Fig. 1 and are thought to be as follows: (A) α-Gal nanoparticles are applied to the injury in the spinal cord. (B) α-Gal nanoparticles in the lesion bind the natural anti-Gal antibody (presented as IgG molecules coating the nanoparticles). This antigen/

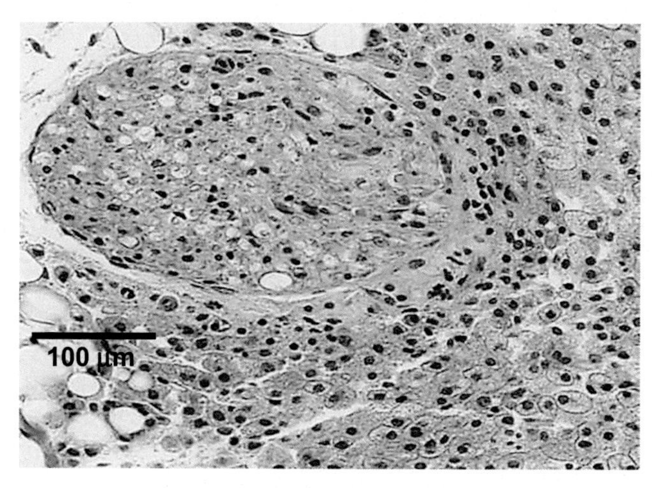

FIGURE 2 Recruitment of macrophages around a branch of an uninjured sciatic nerve in the leg of anti-Gal producing α1,3galactosyltransferase knockout mouse, 4 days after injection of α-gal nanoparticles near the nerve. Most of the recruited macrophages surround the nerve and few of the macrophages migrate into the nerve (Hematoxylin and eosin, ×200).

antibody interaction activates the complement system and generates C5a and C3a chemotactic complement cleavage peptides. (C) The complement chemotactic peptides recruit macrophages into the lesion. The macrophages are activated to secrete for prolonged periods VEGF and other cytokines/growth factors that induce angiogenesis. (D) The multiple capillaries formed under VEGF induction further induce growth of many axonal sprouts, some of which grow across the lesion gap and reconnect with distal segments (endoneurial tubes of the distal section of the axon), thereby enabling full regeneration of neurons. (E) The reconnected axons growing into distal segments undergo myelination, thus completing the regeneration process.

Administration of the α-gal nanoparticles into the spinal cord injury site may be performed by injection or applied in a conduit containing semisolid medium (e.g., hydrogel or plasma clot) to prevent dispersion of the nanoparticles. An example for α-gal nanoparticles applied within a plasma clot to wounds, and the resulting recruitment of macrophages is presented in Fig. 11 in Chapter 12.

The same hypothetical therapy with α-gal nanoparticles, as that in spinal cord injuries, may apply to induction of regeneration in injured peripheral nerves. In various nerves, injection into the injury site may not be feasible because the nerve is too narrow or is severed. In such injured nerves, the α-gal nanoparticles may be applied in a semisolid medium (e.g., hydrogel) placed within a conduit that surrounds the lesion, thus preventing dispersion of the nanoparticles. An example of α-gal nanoparticles recruiting macrophages to an area surrounding a nerve is presented in Fig. 2. The nanoparticles were injected into the hind leg of an anti-Gal producing α1,3galactosyltransferase knockout (GT-KO) mouse, near a branch of the sciatic nerve. The nerve and the surrounding tissue were examined 4 days post injection. Because the injection of α-gal nanoparticles was performed near the nerve (cross section of the nerve has an elliptical shape), most of the macrophages surround the nerve and relatively few penetrated into the nerve. However, if the α-gal nanoparticles also are administered into the injured nerve, it is probable that many of the recruited macrophages will migrate into the nerve. The observation in Fig. 2 suggests that macrophages may migrate effectively into injured spinal cord or peripheral nerves treated with α-gal nanoparticles. It remains to be determined whether such migration enhances the regeneration of injured nerves, as proposed in the hypothesis described above and illustrated in Fig. 1. As discussed in Chapter 13, GT-KO mice and GT-KO pigs are likely to serve as suitable experimental animal models for studying regeneration of spinal cord and nerve injuries by α-gal nanoparticles.

No Conclusions section is included since the proposed therapy in this chapter is hypothetical.

References

David, S., Zarruk, J.G., Ghasemlou, N., 2012. Inflammatory pathways in spinal cord injury. Int. Rev. Neurobiol. 106, 27–52.

De Laporte, L., des Rieux, A., Tuinstra, H.M., Zelivyanskaya, M.L., De Clerck, N.M., Postnov, A.A., et al., 2011. Vascular endothelial growth factor and fibroblast growth factor 2 delivery from spinal cord bridges to enhance angiogenesis following injury. J. Biomed. Mater. Res. A 98, 372–382.

Dray, C., Rougon, G., Debarbieux, F., 2009. Quantitative analysis by in vivo imaging of the dynamics of vascular and axonal networks in injured mouse spinal cord. Proc. Natl. Acad. Sci. U.S.A. 106, 9459–9464.

Facchiano, F., Fernandez, E., Mancarella, S., Maira, G., Miscusi, M., D'Arcangelo, D., et al., 2002. Promotion of regeneration of corticospinal tract axons in rats with recombinant vascular endothelial growth factor alone and combined with adenovirus coding for this factor. J. Neurosurg. 97, 161–168.

Fassbender, J.M., Whittemore, S.R., Hagg, T., 2011. Targeting microvasculature for neuroprotection after SCI. Neurotherapeutics 8, 240–251.

Ferrara, N., Kerbel, R.S., 2005. Angiogenesis as a therapeutic target. Nature 438, 967–974.

Galili, U., 2017. α-Gal nanoparticles in wound and burn healing acceleration. Adv. Wound Care 6, 81–91.

Gensel, J.C., Zhang, B., 2015. Macrophage activation and its role in repair and pathology after spinal cord injury. Brain Res. 1619, 1–11.

Hurwitz, Z., Ignotz, R., Lalikos, J., Galili, U., 2012. Accelerated porcine wound healing with α-gal nanoparticles. Plast. Reconstr. Surg. 129, 242–251.

Kerschensteiner, M., Schwab, M.E., Lichtman, J.W., Misgeld, T., 2005. In vivo imaging of axonal degeneration and regeneration in the injured spinal cord. Nat. Med. 11, 572–577.

Kigerl, K.A., Gensel, J.C., Ankeny, D.P., Alexander, J.K., Donnelly, D.J., Popovich, P.G., 2009. Identification of two distinct macrophage subsets with divergent effects causing either neurotoxicity or regeneration in the injured mouse spinal cord. J. Neurosci. 29, 13435–13444.

Kwon, M.J., Kim, J., Shin, H., Jeong, S.R., Kang, Y.M., Choi, J.Y., et al., 2013. Contribution of macrophages to enhanced regenerative capacity of dorsal root ganglia sensory neurons by conditioning injury. J. Neurosci. 33, 15095–15108.

Mautes, A.E., Weinzierl, M.R., Donovan, F., Noble, L., 2000. Vascular events after spinal cord injury: contribution to secondary pathogenesis. J. Phys. Ther. 80, 673–687.

Oudega, M., Vargas, C.G., Weber, A.B., Kleitman, N., Bundge, M.B., 1999. Long-term effects of methylprednisolone following transection of adult rat spinal cord. Eur. J. Neurosci. 11, 2453–2464.

Peng, Z., Gao, W., Yue, B., Jiang, J., Gu, Y., et al., November 15, 2016. Promotion of neurological recovery in rat spinal cord injury by mesenchymal stem cells loaded on nerve-guided collagen scaffold through increasing alternatively activated macrophage polarization. J. Tissue Eng. Regen. Med. http://dx.doi.org/10.1002/term.2358. [Epub ahead of print].

Popovich, P.G., Guan, Z., Wei, P., Huitinga, I., van Rooijen, N., Stokes, B.T., 1999. Depletion of hematogenous macrophages promotes partial hindlimb recovery and neuroanatomical repair after experimental spinal cord injury. Exp. Neurol. 158, 351–365.

Prewitt, C.M., Niesman, I.R., Kane, C.J., Houlé, J.D., 1997. Activated macrophage/microglial cells can promote the regeneration of sensory axons into the injured spinal cord. Exp. Neurol. 148, 433–443.

Rapalino, O., Lazarov-Spiegler, O., Agranov, E., Velan, G.J., Yoles, E., Fraidakis, M., et al., 1998. Implantation of stimulated homologous macrophages results in partial recovery of paraplegic rats. Nat. Med. 4, 814–821.

Rauch, M.F., Hynes, S.R., Bertram, J., Redmond, A., Robinson, R., Williams, C., 2009. Engineering angiogenesis following spinal cord injury: a coculture of neural progenitor and endothelial cells in a degradable polymer implant leads to an increase in vessel density and formation of the blood-spinal cord barrier. Eur. J. Neurosci. 29, 132–145.

Schwartz, M., 2010. "Tissue-repairing" blood-derived macrophages are essential for healing of the injured spinal cord: from skin-activated macrophages to infiltrating blood-derived cells? Brain Behav. Immun. 24, 1054–1057.

Shechter, R., Miller, O., Yovel, G., Rosenzweig, N., London, A., Ruckh, J., et al., 2013. Recruitment of beneficial M2 macrophages to injured spinal cord is orchestrated by remote brain choroid plexus. Immunity 38, 555–569.

Silver, J., Miller, J.H., 2004. Regeneration beyond the glial scar. Nat. Rev. Neurosci. 5, 146–156.

Widenfalk, J., Lipson, A., Jubran, M., Hofstetter, C., Ebendal, T., Cao, Y., et al., 2003. Vascular endothelial growth factor improves functional outcome and decreases secondary degeneration in experimental spinal cord contusion injury. Neuroscience 120, 951–960.

Wigglesworth, K.M., Racki, W.J., Mishra, R., Szomolanyi-Tsuda, E., Greiner, D.L., Galili, U., 2011. Rapid recruitment and activation of macrophages by anti-Gal/α-gal liposome interaction accelerates wound healing. J. Immunol. 186, 4422–4432.

Woerly, S., Doan, V.D., Sosa, N., de Vellis, J., Espinosa-Jeffrey, A., 2004. Prevention of gliotic scar formation by NeuroGel allows partial endogenous repair of transected cat spinal cord. J. Neurosci. Res. 75, 262–272.

Yu, S., Yao, S., Wen, Y., Wang, Y., Wang, H., Xu, Q., 2016. Angiogenic microspheres promote neural regeneration and motor function recovery after spinal cord injury in rats. Sci. Rep. 6, 33428.

Zhang, Z., Guth, L., 1997. Experimental spinal cord injury: Wallerian degeneration in the dorsal column is followed by revascularization, glial proliferation, and nerve regeneration. Exp. Neurol. 147, 159–171.

CHAPTER

16

Inhalation of α-Gal/Sialic Acid Liposomes for Decreasing Influenza Virus Infection

OUTLINE

INTRODUCTION

The previous chapters discuss how harnessing of the natural anti-Gal antibody/α-gal epitope interaction may be directed toward amplification of the immune response to tumor-associated antigens and to microbial antigens in various vaccines and for repair and regeneration of various tissues. This chapter describes a hypothetical therapy that harnesses the natural anti-Gal antibody for attenuating an ongoing viral infection by influenza (flu) virus and accelerating the generation of a protective immune response against the infecting virus. This therapy uses a novel type of liposomes called α-gal/sialic acid (SA) liposomes, which may be inhaled as an aerosol at early stages of the infection and which present two different carbohydrate epitopes: The α-gal epitope for binding the natural anti-Gal antibody to the liposomes and SA epitopes that bind flu virus to the liposomes (Fig. 1). As suggested in this chapter, binding of both anti-Gal and flu virus with the α-gal/SA liposomes may slow the progression of the viral infection and convert the infecting virus into an effective vaccine that prevents progression of the viral infection to stages, which cause morbidity and even mortality of infected patients.

The Natural Anti-Gal Antibody as Foe Turned Friend in Medicine
http://dx.doi.org/10.1016/B978-0-12-813362-0.00016-6

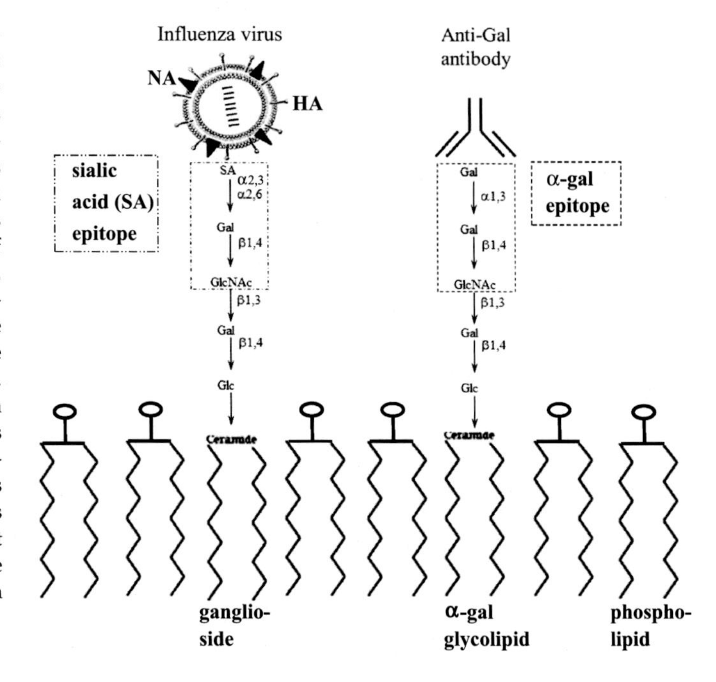

FIGURE 1 Illustration of the carbohydrate epitopes on α-gal/sialic acid (SA) liposomes. The α-gal epitope (Galα1-3Galβ1-4GlcNAc-R) is at the nonreducing end of the carbohydrate chain of a glycolipid. This epitope readily binds the natural anti-Gal antibody. The SA epitope is linked to the carbohydrate chain of SA glycolipids (gangliosides) as α2-3, α2-6, or other α-linkages. The carbohydrate chains of both glycolipids are linked to ceramide molecules that are anchored in the phospholipids bilayer. Hemagglutinin (HA) on influenza virus mediates binding of the virus to SA epitopes on the α-gal/SA liposomes and to such epitopes on various cells. *NA*, neuraminidase. The virus is schematically described with eight pieces of segmented negative sense RNA illustrated as small bars within the virus.

Flu is a contagious respiratory disease causing annual outbreaks. Flu virus causes each season 3–5 million cases of severe illness around the world, resulting in 250,000–500,000 deaths (WHO Fact sheet, November 2016). The virus is spread by microdroplets (aerosol) and is distributed from patients to healthy individuals, by sneezing, coughing, or talking. The virus penetrating the upper and lower airways binds to SA epitopes functioning as receptors on ciliated respiratory epithelium cells via the hemagglutinin (HA) envelope protein on the virus in mammals (Unverzagt et al., 1994) and birds (Thompson et al., 2006). Following its binding to SA epitopes, the virus penetrates the cells and releases its RNA-8 genetic pieces. After multiplication within infected cells, the core structure is covered by the cell membrane containing HA and neuraminidase (NA). The full virus detaches from the cell following the activity of viral NA that releases the virus HA from binding to cell surface SA epitopes. Inhalation of NA inhibitors aerosol (e.g., Tamiflu) is used for preventing the latter step and thus, slowing the progression of virus infection. The efficacy of these NA inhibitor drugs in inducing an effective slowing of the flu virus infection is still controversial because some clinical studies reported no beneficial effects, whereas others reported some clinical effects.

Currently used inactivated flu virus vaccines are the product of the 2+6 reassortment containing HA and NA genes from the vaccine target strain and the remaining genes from A/Puerto Rico/8/34-H1N1 (PR8) flu virus strain described in Chapter 9. These vaccines display suboptimal efficacy, as determined by the finding that ~25%–50% of immunized individuals (particularly in elderly populations) contract the disease during the flu season (Webster, 2000).

From the time of infection by flu virus, there is a "race" between the virus produced in increasing numbers in cells of the respiratory tract epithelium and the immune system that is activated to generate protective humoral and cellular immune responses against the infecting virus. Slowing the infection (i.e., inhibition of virus growth) at early stages of the infection is critical for enabling the immune system to mount a timely combination of humoral and cellular protective immune responses that prevent further increase in the virus burden. The humoral immune response is comprised of anti-flu virus IgA antibodies and to lesser extent IgG antibodies that neutralize the virus and prevent further infection of healthy cells. The cellular immune response consists of flu virus specific T cells that kill virus infected cells, thereby contributing to prevention of further infection of healthy cells by the virus. It is suggested that α-gal/SA liposomes may cause effective destruction of the virus by macrophages and enhance the immunogenicity of viral antigens, thereby decrease the time for eliciting a protective anti-virus immune response in treated individuals, as well as in infected chickens.

STRUCTURE AND FUNCTIONS OF α-GAL/SIALIC ACID LIPOSOMES

Structure and production of α-gal/sialic acid liposomes

α-Gal/SA liposomes are biodegradable liposomes presenting multiple carbohydrate epitopes of two types: α-gal epitopes (Galα1-3Galβ1-4GlcNAc-R) on the carbohydrate chain of α-gal glycolipids and SA epitopes on SA glycolipids (also called gangliosides) (Fig. 1). As further shown in Fig. 1, the suggested therapeutic activity of α-gal/SA liposomes is based on the specific binding of flu virus via its HA to SA epitopes on the membranes of various cells, including those of the respiratory tract epithelium. This interaction is the first step in the process of infecting cells and subsequently proliferating within them (Johnson et al., 1964; Skehel and Wiley, 2000). In addition, as discussed in the various chapters of this book, the natural anti-Gal antibody, which is produced in all humans as IgG, IgA, and IgM antibody molecules, binds specifically to α-gal epitopes on α-gal/SA liposomes as well as it binds to α-gal liposomes and nanoparticles, described in Chapter 12. Combination of these two characteristics of the liposomes (i.e., binding both flu virus and anti-Gal antibody) is the basis for the suggested hypothetical use of the α-gal/SA liposomes in treatment of patients with flu.

Production of liposomes presenting both α-gal epitopes and SA epitopes is similar to production of α-gal liposomes described in Chapter 12. The washed rabbit red blood cell (RBC) membranes obtained by hypoosmotic lysis of 1 L packed rabbit RBC are subjected to extraction for 2 h in chloroform:methanol 1:1 solution followed by continued overnight extraction in 1:2 solution of chloroform:methanol (Galili et al., 2007, 2010; Wigglesworth et al., 2011). Washed human blood group O RBC membranes, obtained from 100 mL packed human RBC also undergo similar extraction in chloroform:methanol to provide SA glycolipids (gangliosides). The extracted mixtures of phospholipids, glycolipids, and cholesterol from the two RBC membrane preparations are mixed and dried in a round flask by a rotary evaporator. Subsequently, saline is added to the dried mixture, which is subjected to sonication for the formation of α-gal/SA liposomes. It should be stressed that rabbit RBC also were found to have few gangliosides (Takemae et al., 2010) in their cell membrane. Thus, experimental analysis of flu virus binding to liposomes produced from extracts of rabbit RBC can determine if such liposomes have the same binding activity as that of liposomes produced from mixed extracts of rabbit and human RBC membranes. In addition, α-gal/SA liposomes may be produced from mixtures of synthetic α-gal glycolipids, synthetic gangliosides (both available for purchasing), and phospholipids, which are dissolved in an organic solvent, dried, and sonicated into liposomes in saline or PBS. The SA on SA epitopes may be linked to the penultimate carbohydrate in α2-3 or α2-6 bond. Because some flu virus strains were reported to bind preferably to terminal SAα2-3 residues and other to SAα2-6 residues of the SA epitope (Rogers and Paulson, 1983; Unverzagt et al., 1994; Matrosovich and Klenk, 2003; Mochalova et al., 2003), it is suggested that α-gal/SA liposomes may be optimized for binding multiple strains by having gangliosides that carry SAα2-3 terminal residues, as well as those with SAα2-6 terminal residues.

Proposed activities of α-gal/sialic acid liposomes inhaled following influenza virus infection

The hypothetical therapy described in this chapter may be studied in experimental animal models producing the anti-Gal antibody, such as α1,3galactosyltransferase knockout (GT-KO) mice and GT-KO pigs, prior to preclinical evaluation in primates. The suggested therapy consists of inhalation of α-gal/SA liposomes by a nebulizer containing a suspension of such biodegradable liposomes. It is hypothesized that inhalation of the α-gal/SA liposomes into an experimental animal model that is infected with flu virus or in humans, at early stages of the infection, may result in a process of anti-viral activity comprised of several sequential steps that are illustrated in Fig. 2: Step 1. The inhaled α-gal/SA liposomes land in the mucus and surfactant linings of the respiratory tract and bind flu virus in the airways. This binding, mediated by the interaction between the HA of the virus and the multiple SA epitopes on the α-gal/SA liposomes, prevents further infection of epithelial cells of the airways by the bound virus. Step 2. The natural anti-Gal antibody secreted into the mucus and surfactant binds to the multiple α-gal epitopes on the α-gal/SA liposomes and activates the complement system. As with the application of α-gal nanoparticles to wounds (described in Chapter 12), complement activation by anti-Gal interaction with α-gal/SA liposomes results in production of chemotactic complement cleavage peptides such as C5a and C3a. Step 3. As in wound healing, the chemotactic complement cleavage peptides induce recruitment of macrophages to the inhaled α-gal/SA liposomes. Step 4. The interaction between the Fc portion of anti-Gal opsonizing α-gal/SA liposomes and Fc receptors (both Fcα receptors and Fcγ receptors [FcR]) on macrophages induces phagocytosis of the liposomes and the virus bound to them by these macrophages. Step 5. The macrophages, functioning as antigen presenting cells (APC), destroy the virus in the lysosomal compartment, process the immunogenic flu virus antigens and transport them to the regional lymph nodes. In the lymph nodes, the macrophages present the processed flu virus immunogenic peptides to the flu virus specific T cells. The CD4+ activated T helper cells provide help to B cells for producing protective antibodies that neutralize and destroy infecting flu virus in the respiratory tract and in other infected tissues. The activated CD8+ T cells proliferate and differentiate into cytotoxic T cells that kill cells infected by flu virus, thereby preventing further proliferation of the infecting virus in patients.

Overall, anti-Gal-mediated uptake of flu virus adhering to α-gal/SA liposomes, by macrophages, may decrease the amount of virus in the airways and reduce the damage to the cells lining the respiratory tract. Moreover, the rapid and effective processing and transport of the flu virus immunogenic peptides to lymph nodes of the respiratory tract may convert the virus

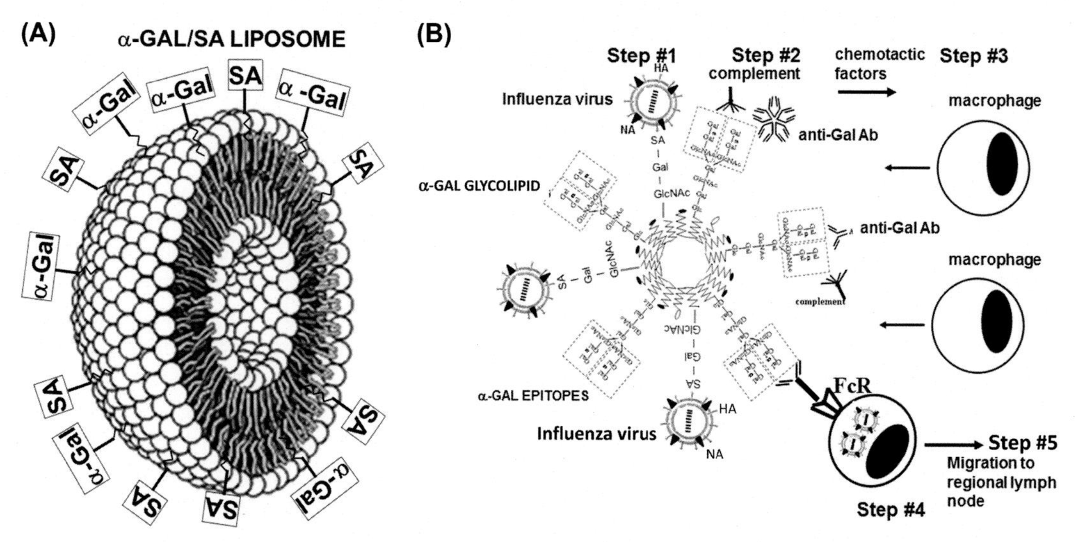

FIGURE 2 Schematic illustration of an α-gal/sialic acid (SA) liposome (A) and the hypothesized sequential steps of the process that such liposomes mediate in the respiratory tract of influenza virus infected individuals (B). (A) The α-gal/SA liposome wall is a phospholipid bilayer in which α-gal glycolipids and gangliosides (SA glycolipids) are anchored via their ceramide tail. (B) Following landing of inhaled α-gal/SA liposomes into the mucus and surfactant of the respiratory tract, they bind the influenza virus via HA/SA interaction (Step 1) and further bind the anti-Gal antibody to their α-gal epitopes. Anti-Gal binding activates the complement system, which generates complement cleavage chemotactic peptides (Step 2). These chemotactic peptides recruit macrophages to the α-gal/SA liposomes (Step 3). The macrophages internalize α-gal/SA liposomes, and the influenza virus bound to them, via Fc/Fc receptor interaction with anti-Gal opsonizing the α-gal/SA liposomes (Step 4). These macrophages transport the internalized virus to the regional lymph nodes (Step 5), destroy the virus, process, and present the virus immunogenic peptides for the activation of influenza virus specific CD4+ and CD8+ T cells. *HA*, hemagglutinin; *NA*, influenza virus neuraminidase. *(A) Partially adapted from Wikipedia with permission. (B) Adapted from Galili, U., 2016. Inhalation of α-gal/sialic acid liposomes: a novel approach for inhibition of influenza virus infection. Future Virol. 11, 95–99, with permission.*

into an effective vaccine, like vaccines presenting α-gal epitopes described in Chapter 9. If the α-gal/SA liposomes are inhaled at early stages of the flu virus infection, the rapid transport, processing, and presentation of flu virus antigens by the macrophages may activate the immune system to mounts a protective immune response that stops the progression of the infection within a shorter time than the physiologic immune response to the viral infection. Such an accelerated immune response may reduce morbidity and prevent development of conditions that lead to lethal outcomes of such infections (e.g., bacterial pneumonia).

Observations supporting the hypothesis on α-gal/sialic acid liposomes

As emphasized above, the proposed α-gal/SA liposomes therapy is hypothetical, i.e., there are no direct experimental data at present that support the hypothesis illustrated in Fig. 2. However, there are several observations, which may indirectly support this hypothesis. The predicted binding of flu virus to SA epitopes on α-gal/SA liposomes (Step 1) is supported by well-established function of SA epitopes as "docking" receptors for the virus on cell membranes (Johnson et al., 1964; Skehel and Wiley, 2000). Moreover, natural and synthetic SA epitopes were reported to inhibit binding of flu virus to SA epitopes on cell membranes (Unverzagt et al., 1994; Mochalova et al., 2003; Matrosovich and Klenk, 2003; Olofsson and Bergström, 2005), further supporting the assumption that this virus can bind to SA presented on the liposomes.

Binding of anti-Gal to inhaled α-gal/SA liposomes results in activation of the complement system and generation of chemotactic complement cleavage peptides in Step 2, which further induce recruitment of macrophages to these liposomes, as predicted in Step 3. The possible occurrence of these two steps is supported by observations presented in Chapter 12 on the interaction of anti-Gal with α-gal liposomes (Galili et al., 2010; Wigglesworth et al., 2011). As shown in Chapter 12, cutaneous administration of α-gal nanoparticles (submicroscopic α-gal liposomes) into anti-Gal producing GT-KO mice results in binding of this antibody to the multiple α-gal epitopes on the α-gal nanoparticles and activation of the complement system because of this antigen/antibody interaction. This complement activation resulted in the formation of complement chemotactic cleavage peptides that induce rapid recruitment of macrophages toward the area in which α-gal liposomes and α-gal nanoparticles were injected (Fig. 3 in Chapter 12) (Galili et al., 2010; Wigglesworth et al., 2011).

Step 4 predicts that α-gal/SA liposomes with the flu virus attached to their SA epitopes will be effectively internalized into macrophages as a result of Fc/FcR interaction between anti-Gal opsonizing these liposomes and FcR of the macrophages. Effective anti-Gal-mediated binding of liposomes to macrophages could be demonstrated with α-gal nanoparticles, as shown in Fig. 5 in Chapter 12. The phagocytosis of cells presenting α-gal epitopes and opsonized by anti-Gal is further demonstrated in Fig. 1B in Chapter 10 with anti-Gal-coated lymphoma cells that are internalized by macrophages.

Step 5 in Fig. 1 predicts that macrophages internalizing the α-gal/SA liposomes and the flu virus attached to them will be effectively transported to the regional lymph nodes, and that the viral antigens will be processed and presented for the activation of virus specific T cells. This assumption is supported by the observations in Chapter 9, indicating that ovalbumin (OVA) liposomes opsonized by anti-Gal are internalized *in vivo* by APC, which transport them to the regional lymph nodes and present the immunodominant OVA SIINFEKL peptide to CD8+ T cells (Fig. 7 in Chapter 9). No such effective transport, processing, and presentation of immunogenic SIINFEKL was observed in mice lacking the anti-Gal antibody (Abdel-Motal et al., 2009). Furthermore, immunization with inactivated flu virus presenting α-gal epitopes, and forming immune complexes with anti-Gal, elicits anti-virus antibody production at titers that are ~100-fold higher than immunization with flu virus lacking α-gal epitopes (Fig. 3 in Chapter 9) (Abdel-Motal et al., 2007). All these observations suggest that inhalation of α-gal/SA liposomes in individuals that are at early stages of flu infection will result in binding of the virus in the airways and binding of anti-Gal to these liposomes, recruitment of macrophages to the liposomes and effective Fc/FcR uptake of the liposomes and the virus attached to them by the recruited macrophages. This will be followed by transport of the internalized and processed viral antigens, by macrophages to regional lymph nodes. The induction of an accelerated protective anti-virus immune response may prevent the progression of the flu virus infection, decrease the length of the disease period, lower morbidity, and decrease the probability of secondary complications, which may have lethal consequences.

POTENTIAL TOXICITIES OF α-GAL/SIALIC ACID LIPOSOMES

Like many drugs, the use of α-gal/SA liposomes may theoretically be toxic if the dose administered is too high. Extensive activation of complement may result in vasodilation and secretion of fluids and mucus in amounts that may obstruct some of the airways and affect respiration. Thus, if the scenario described in Fig. 2 is found to be effective in preventing progression of flu virus infection, it would further require identification of the optimal dose range of α-gal/SA liposomes that can be inhaled without causing adverse effects because of excessive activation of the complement system in the respiratory tract.

Another risk factor that should be addressed is in individuals who produce anti-Gal IgE and thus, are allergic to α-gal epitopes in meat, as discussed in Chapter 7 (Chung et al., 2008; Commins et al., 2012; Platts-Mills et al., 2015). The proportion of individuals producing anti-Gal IgE is manyfold higher in the Southern regions of the United States than in the Northern regions (Chung et al., 2008). These observations raise the concern that anti-Gal IgE producers may have allergic reactions in the respiratory tract, following inhalation of α-gal/SA liposomes. It is of interest to note that no respiratory allergies (e.g., asthma) were found to be associated with anti-Gal IgE in individuals allergic to α-gal epitopes (Commins et al., 2012). Nevertheless, if the treatment with α-gal/SA liposomes is found effective in preventing progression of flu virus infection, the use of this therapy in humans may require performing a skin test with these liposomes prior to their inhalation. Such test may exclude individuals who could be allergic to α-gal/SA liposomes.

TREATMENT OF INFECTED CHICKENS WITH α-GAL/SIALIC ACID LIPOSOMES

Among mammals, the natural anti-Gal antibody is produced only humans, apes, and Old World monkeys, whereas other mammals synthesize the α-gal epitope and thus cannot produce this antibody (Chapters 1 and 2) (Galili et al., 1987, 1988). Therefore, the therapeutic effect of α-gal/SA liposomes may be found in anti-Gal producing primates and not in other mammals (unless they are engineered to lack α-gal epitopes as in GT-KO mice and GT-KO pigs). However, the α-gal epitope is absent in birds and in other nonmammalian vertebrates because the gene encoding α1,3galactosyltransferase—the enzyme that synthesizes α-gal epitopes, appeared only in mammals (prior to divergence of marsupials from placental mammals, as discussed in Chapter 2) and is absent in nonmammalian vertebrates. Chickens and other birds lack the α-gal epitope (Galili et al., 1988) and produce the natural anti-Gal antibody (McKenzie et al., 1999; Cotter et al., 2005; Cotter and Van Eerden, 2006; Minozzi et al., 2008).

Flu virus infecting birds was found to bind to SA epitopes on bird respiratory epithelium like the adhesion of this virus to airways epithelial cells in mammals (Thompson et al., 2006). Because anti-Gal is produced in chickens, it may be possible that inhalation of α-gal/SA liposomes by chickens infected with flu virus will have therapeutic anti-flu virus effects, similar to those suggested in Fig. 2 for humans infected with this virus. Presently, flu virus infections of chicken flocks require the immediate destruction of the infected flock. Thus, it would be of interest to study the effects of aerosolized α-gal/SA liposomes on progression of flu virus infection in chickens in parallel to the study of these liposomes in GT-KO mice. If successful, administration of aerosolized α-gal/SA liposomes to chicken flocks may harness the natural anti-Gal antibody in these birds for eliciting an accelerated protective immune response against flu virus infection. If the flu virus is completely eliminated within treated chicken (including immune sanctuaries), success of the α-gal/liposomes treatment in infected flocks may raise the possibility that they may not have to be destroyed.

No Conclusions section is included since the proposed therapy in this chapter is hypothetical.

References

Abdel-Motal, U.M., Guay, H.M., Wigglesworth, K., Welsh, R.M., Galili, U., 2007. Immunogenicity of influenza virus vaccine is increased by anti-Gal-mediated targeting to antigen-presenting cells. J. Virol. 81, 9131–9141.

Abdel-Motal, U.M., Wigglesworth, K., Galili, U., 2009. Mechanism for increased immunogenicity of vaccines that form in vivo immune complexes with the natural anti-Gal antibody. Vaccine 27, 3072–3082.

Chung, C.H., Mirakhur, B., Chan, E., Le, Q.T., Berlin, J., Morse, M., et al., 2008. Cetuximab-induced anaphylaxis and IgE specific for galactose-α-1,3-galactose. N. Engl. J. Med. 358, 1109–1117.

Commins, S.P., Kelly, L.A., Rönmark, E., James, H.R., Pochan, S.L., Peters, E.J., et al., 2012. Galactose-α-1,3-galactose-specific IgE is associated with anaphylaxis but not asthma. Am. J. Respir. Crit. Care. Med. 185, 723–730.

Cotter, P.F., Ayoub, J., Parmentier, H.K., 2005. Directional selection for specific sheep cell antibody responses affects natural rabbit agglutinins of chickens. Poult. Sci. 84, 220–225.

Cotter, P.F., Van Eerden, E., 2006. Natural anti-Gal and *Salmonella*-specific antibodies in bile and plasma of hens differing in diet efficiency. Poult. Sci. 85, 435–440.

Galili, U., 2016. Inhalation of α-gal/sialic acid liposomes: a novel approach for inhibition of influenza virus infection. Future Virol. 11, 95–99.

Galili, U., Clark, M.R., Shohet, S.B., Buehler, J., Macher, B.A., 1987. Evolutionary relationship between the anti-Gal antibody and the Galα1-3Gal epitope in primates. Proc. Natl. Acad. Sci. U.S.A. 84, 1369–1373.

Galili, U., Shohet, S.B., Kobrin, E., Stults, C.L.M., Macher, B.A., 1988. Man, apes, and Old World monkeys differ from other mammals in the expression of α-galactosyl epitopes on nucleated cells. J. Biol. Chem. 263, 17755–17762.

Galili, U., Wigglesworth, K., Abdel-Motal, U.M., 2007. Intratumoral injection of α-gal glycolipids induces xenograft-like destruction and conversion of lesions into endogenous vaccines. J. Immunol. 178, 4676–4687.

Galili, U., Wigglesworth, K., Abdel-Motal, U.M., 2010. Accelerated healing of skin burns by anti-Gal/α-gal liposomes interaction. Burns 36, 239–251.

Johnson, C.A., Pekas, D.J., Winzler, R.J., 1964. Neuraminidase and influenza virus infection in embryonated eggs. Science 143, 1051–1052.

Matrosovich, M., Klenk, H.D., 2003. Natural and synthetic sialic acid-containing inhibitors of influenza virus receptor binding. Rev. Med. Virol. 13, 85–97.

McKenzie, I.F., Patton, K., Smit, J.A., Mouhtouris, E., Xing, P., Myburgh, et al., 1999. Definition and characterization of chicken Gal α(1,3)Gal antibodies. Transplantation 67, 864–870.

Minozzi, G., Parmentier, H.K., Mignon-Grasteau, S., Nieuwland, M.G., Bed'hom, B., Gourichon, D., et al., 2008. Correlated effects of selection for immunity in White Leghorn chicken lines on natural antibodies and specific antibody responses to KLH and *M. butyricum*. BMC Genet. 9, 5.

Mochalova, L., Gambaryan, A., Romanova, J., Tuzikov, A., Chinarev, A., Katinger, D., et al., 2003. Receptor-binding properties of modern human influenza viruses primarily isolated in Vero and MDCK cells and chicken embryonated eggs. Virology 313, 473–480.

Olofsson, S., Bergström, T., 2005. Glycoconjugate glycans as viral receptors. Ann. Med. 37, 154–172.

Platts-Mills, A.E., Schuyler, A.J., Hoyt, A.E., Commins, S.P., 2015. Delayed anaphylaxis involving IgE to galactose-α-1,3galactose. Curr. Allergy Asthma Rep. 15, 512.

Rogers, G.N., Paulson, J.C., 1983. Receptor determinants of human and animal influenza virus isolates: differences in receptor specificity of the H3 hemagglutinin based on species of origin. Virology 127, 361–373.

Skehel, J.J., Wiley, D.C., 2000. Receptor binding and membrane fusion in virus entry: the influenza hemagglutinin. Annu. Rev. Biochem. 69, 531–569.

Takemae, N., Ruttanapumma, R., Parchariyanon, S., Yoneyama, S., Hayashi, T., Hiramatsu, H., 2010. Alterations in receptor- binding properties of swine influenza viruses of the H1 subtype after isolation in embryonated chicken eggs. J. Gen. Virol. 91, 938–948.

Thompson, C.I., Barclay, W.S., Zambon, M.C., Pickles, R.J., 2006. Infection of human airway epithelium by human and avian strains of influenza A virus. J. Virol. 80, 8060–8068.

Unverzagt, C., Kelm, S., Paulson, J.C., 1994. Chemical and enzymatic synthesis of multivalent sialoglycopeptides. Carbohydr. Res. 251, 285–301.

Webster, R.G., 2000. Immunity to influenza in the elderly. Vaccine 18, 1686–1689.

WHO Media Center Fact sheet "Influenza (Seasonal)" November 2016 (online).

Wigglesworth, K.M., Racki, W.J., Mishra, R., Szomolanyi-Tsuda, E., Greiner, D.L., Galili, U., 2011. Rapid recruitment and activation of macrophages by anti-Gal/α-gal liposome interaction accelerates wound healing. J. Immunol. 186, 4422–4432.

Index

Note: 'Page numbers followed by "f" indicate figures, "t" indicate tables.'

D

E

F

G

H

Printed in the United States
By Bookmasters